Applied Mathematics Advanced Level

W. E. Williams
Professor of Mathematics, University of Surrey

A. Waltham
Head of Mathematics, Hampton School

HODDER AND STOUGHTON
LONDON SYDNEY AUCKLAND TORONTO

In the same series

Pure Mathematics for Advanced Level, J. A. H. Shepperd and C. J. Shepperd (This book is available either as a complete volume or in two parts.)
Further Pure Mathematics for Advanced Level, J. A. H. Shepperd and C. J. Shepperd
Statistics for Advanced Level Mathematics, I. Gwyn Evans

British Library Cataloguing in Publication Data

Williams, W. E. (William Elwyn)
Applied mathematics for advanced level.
1. Mathematics
I. Title II. Waltham, A.
510 QA39.2

ISBN 0 340 35243 4

First printed 1985

Typeset in Times New Roman (Monophoto) by Macmillan India Ltd., Bangalore

Printed in Great Britain for
Hodder and Stoughton Educational,
a division of Hodder and Stoughton Ltd.,
Mill Road, Dunton Green, Sevenoaks, Kent TN13 2YD,
by Page Bros (Norwich) Ltd.

Contents

Preface

Applied Mathematics for Advanced Level includes the mechanics content of current pure and applied mathematics Advanced Level G.C.E. syllabuses and the probability content of syllabuses, such as the present London Syllabus B, which include elementary probability as well as mechanics. In view of the variation in the mechanics content of present 'single' and 'double' subject syllabuses it will be found that the material covered will, in some cases, be more than that required for the 'single subject'.

The approach adopted is to set out as clearly, and as rigorously as possible at this level, the basic principles governing elementary mechanics. A large number of worked examples have been included in order to illustrate the application of the principles and, where appropriate, methods of approaching general types of examination questions are included. There are graded exercises at the end of each section and the miscellaneous exercises at the end of the chapters are taken almost entirely from the past papers of the Associated Examining Board (AEB), the University of Cambridge Local Examinations Syndicate (C), the Joint Matriculation Board (JMB), the University of London University Entrance and Schools Examinations Council (L), the Delegacy of Local Examinations of the University of Oxford (O) and the Oxford and Cambridge Schools Examination Board (O and C). We gratefully acknowledge these Boards' permission to reproduce their questions.

W. E. Williams
A. Waltham

1 Probability

We are all familiar with situations in which, as a result of some action or experiment which we are about to carry out (such as rolling a die or tossing a coin) there are a number of different outcomes. In this chapter methods of calculating probabilities of specific outcomes of such experiments are explained in §1.1–1.4. This is followed by the explanation, in §1.5, of how the probabilities of simple events can be combined to give probabilities of compound events. The chapter concludes with a study of the probability tree in §1.6 and an introduction to the binomial distribution in §1.8.

1.1 Possibility space

An interviewer conducting a Gallup poll asks an elector whether she will vote for the Conservative party, for the S.D.P-Liberal Alliance, for Labour, whether she will abstain, or whether she does not know how she will vote. The set of five categories, {votes Conservative, votes Alliance, votes Labour, abstains, does not know}, constitutes what the investigator hopes is the complete list of possible responses. The set is an example of a *possibility space.*

If a die is rolled three times and the number of sixes that show up is recorded, then the associated possibility space is $\{0, 1, 2, 3\}$. If a die is rolled once and the number showing is recorded, the possibility space is $\{1, 2, 3, 4, 5, 6\}$.

A possibility space is related to an experiment or trial in which the experimenter decides to record particular outcomes of the trial. The possibility space is the set of all the possible outcomes of the trial that the experimenter wishes to record. The set must be exhaustive – that is, it must include all the possible outcomes. The possibility space is sometimes referred to as the *sample space* but, in this text, the term possibility space will be adhered to. Note that a trial may have different possibility spaces depending on what the experimenter wishes to record. Thus, if a coin is tossed twice and the number of heads showing is recorded, the required space is $\{0, 1, 2\}$. If, on the other hand, the different ordered pairs of heads (H) and tails (T) in which the two coins turn up is required, the space is $\{HH, HT, TH, TT\}$.

1.2 Events

Any subset of the possibility space is an *event*. For example, in the Gallup poll the set {votes Conservative, votes Alliance, votes Labour} – that is, the voting categories – is one event. For this possibility space there are 32 (2^5) possible events – a set of 5 elements having 2^5 possible subsets.

If two dice are rolled and the scores on each die are recorded, the possibility space can be expressed by the following rectangular array.

		Second die					
		1	2	3	4	5	6
	1	(1, 1)	(1, 2)	(1, 3)	(1, 4)	(1, 5)	(1, 6)
First	2	(2, 1)	(2, 2)	(2, 3)	(2, 4)	(2, 5)	(2, 6)
die	3	(3, 1)	(3, 2)	(3, 3)	(3, 4)	(3, 5)	(3, 6)
	4	(4, 1)	(4, 2)	(4, 3)	(4, 4)	(4, 5)	(4, 6)
	5	(5, 1)	(5, 2)	(5, 3)	(5, 4)	(5, 5)	(5, 6)
	6	(6, 1)	(6, 2)	(6, 3)	(6, 4)	(6, 5)	(6, 6)

We may be interested in a total score of 7. This results from the set {(6, 1), (5, 2), (4, 3), (3, 4), (2, 5), (1, 6)} and this set constitutes the event, a total score of 7.

EXERCISE 1.2

1 Detail the set of ordered pairs from the rectangular array for the event, a total score of 5. Detail also the set of ordered pairs for the event, the difference between the scores on the two dice is 2.

2 State the possibility spaces in each of the following trials:
(a) cutting two cards from a pack of 52 cards and recording (i) the number of aces, (ii) the combination of suits, (iii) the combination of face values.
(b) choosing from an urn containing three black, four white and five red balls (i) one ball and recording the colour, (ii) two balls from the same urn and recording the combination of colours, (iii) two balls and recording the colours as an ordered pair.
(c) rolling three dice and recording the total score.
(d) choosing one domino from a complete set of dominoes and recording (i) whether it is a double, (ii) the total score.

1.3 Quantifying probability

Suppose that a needle is thrown down on to a horizontal sheet of paper, which is covered by a set of equidistant parallel lines, the needle being shorter than the distance between adjacent parallel lines. Thus, the needle can only lie across at most one line at a time. We wish to record whether the needle falls so that it lies across a line or not. The possibility space is {lies across, does not lie across}, and we will call the event {needle lies across a line}, event A.

Fig. 1.1

When the needle is thrown repeatedly, the proportion of trials in which the outcome is A appears to approach a limiting value. A typical sequence of values would be: after 100 trials there were 39 'successes', after 1 000, 343 'successes', after 10 000, 3 452 'successes' and after 100 000, 34 486 'successes'.

However, we cannot measure the proportion in this manner. It will be assumed, for the time being, that the limiting value is defined. Although this value is a property of a long sequence of trials, it does seem to have some application in attempting to predict the outcome of one trial.

Definition The limiting value will be defined to be the probability of event A and will be denoted by $P(A)$. It follows at once that the possibility space is $\{A, A'\}$, where A' means not A, and hence $P(A') = 1 - P(A)$.

In practice, most values of probability are estimated from experience. Thus, if it is stated that a certain course of medical treatment has an 80 per cent probability of success, this would imply that out of a large number of patients who have been treated with this course, 80 per cent were cured. This makes assumptions about the homogeneity of the population, and the probability would not necessarily apply to the treatment of an individual patient, who might be more or less resistant to the particular infection than the average.

1.4 Argument from symmetry

In a number of problems we are given the probability of particular events. However, there is one set of problems in which the probability of events can be estimated. In certain experiments a measure of the probability involved can be obtained by consideration of the symmetry of the possible results. The argument refers to idealisations of real experiments.

If a perfectly symmetric cube is thrown in a random manner, then it is equally likely that any one of the six faces will fall uppermost. We, therefore, argue that $P(6) = 1/6$. Similarly, in cutting a card at random from a pack of 52 cards, $P(\text{ace}) = 1/13$. The phrase 'at random' will be taken as justifying the use of the argument from symmetry. A random choice is one in which each of the cards is equally likely to be chosen.

Care is necessary when assessing symmetry. If two dice are rolled and the total score is recorded, the probability space is $\{2, 3, 4, \ldots, 12\}$. But we can see from the array in §1.2 that P (total score $= 6$) is not $1/11$. The subset $\{(1, 5), (2, 4), (3, 3), (4, 2), (5, 1)\}$ is the required event and, thus, $P(6) = 5/36$. There is symmetry between the 36 individual ordered pairs (r, s) $1 \leqslant r \leqslant 6$, $1 \leqslant s \leqslant 6$, but not between the 11 possible scores 2, 3, $\ldots$, 12. Before tackling examples based on the argument from symmetry, we need a brief introduction to permutations and combinations.

Permutations and combinations

Suppose that we wish to find the number of different arrangements of two digits taken from the set of ten digits, $\{0, 1, 2, \ldots, 9\}$, without repetition. We have ten choices for the first digit, and nine choices for the second digit after the first has been chosen. Thus, we have 10×9 ordered pairs of digits. If the number of arrangements of three digits without repetition is required, then there will be $10 \times 9 \times 8$ different ordered triplets. If we use factorial notation, where $5! = 5 \times 4 \times 3 \times 2$ and $n! = n(n-1)(n-2) \ldots \times 3 \times 2$, then $10 \times 9 \times 8$ may be written as $10!/7!$.

The number of arrangements of r items chosen from a set of n different items is $n!/(n-r)!$. Note that this is the number of different *arrangements* – a different order gives a different arrangement.

We now return to the problem of choosing three digits from the ten. Suppose that we wish to choose different *sets* of digits in which the order is irrelevant. Such sets are called *combinations*. In this case, amongst the $10 \times 9 \times 8$ different ordered triplets, 987, 978, 897, 879, 798 and 789 are all the same set of digits. Thus, each different combination will be repeated six times in the $10 \times 9 \times 8$ ordered triplets.

$$\text{Number of different combinations} = \frac{10 \times 9 \times 8}{6} = \frac{10!}{7!3!}.$$

If we select r items from a list of n different items, at first we have $n(n-1)(n-2) \ldots (n-r+1)$ different arrangements of these r items. But each set will be repeated $r!$ times corresponding to the $r!$ ways in which a set of r items can be ordered. Thus,

$$\text{number of different sets} = \frac{n!}{(n-r)!r!}.$$

The number of ways of selecting r items from a set of n different items is

$$\frac{n!}{(n-r)!r!}$$

and this is written as $\binom{n}{r}$ or ${}^nC_r$.

EXAMPLE 1 *How many choices of two digits can be made from the set* $\{1, 2, 3, 4, 5\}$ *without repetition?*

$$\text{Number of choices} = \binom{5}{2} = \frac{5!}{3!2!} = \frac{5 \times 4}{2} = \mathbf{10}.$$

Note that these are: (1, 2), (1, 3), (1, 4), (1, 5), (2, 3), (2, 4), (2, 5), (3, 4), (3, 5), (4, 5).

EXAMPLE 2 *How many different hands of* 13 *cards can be dealt to one person, containing exactly eight hearts? Leave the answer in factorials.*

The hand must consist of eight hearts and five cards chosen from the remaining three suits of diamonds, spades and clubs. The eight hearts can be chosen in $\binom{13}{8}$ ways. The remaining 5 can be chosen in $\binom{39}{5}$ ways.

As each selection of hearts can be associated with each selection of the remaining five,

$$\text{the total number of different hands} = \binom{13}{8} \times \binom{39}{5} = \mathbf{\frac{13!}{8!\,5!} \times \frac{39!}{34!\,5!}}.$$

EXERCISE 1.4A

1 Find the number of different hands of three cards that can be dealt from a pack of 52 cards.

2 How many different hands of 13 cards can be dealt to one person containing exactly six spades. Leave the answer in factorials.

3 A diagonal of a convex n-sided polygon is a line that joins two vertices which are not adjacent. How many diagonals has a nine-sided convex polygon?

4 A fruit machine has four windows, in each of which one of eight different pictures can appear. How many different *arrangements* of the pictures are possible?

5 In how many different ways can four dice show a 1, a 2, a 3 and a 4?

6 A committee decides to meet twice each week. In how many different ways can the two days for meeting be chosen, assuming they meet on different days?

EXAMPLE 3 *Three cards are to be drawn from a pack of* 52 *cards without replacement. Find the probability of obtaining one king, one queen and one jack in any order.*

The number of possible choices of a king (K) is 4, of a queen (Q) is 4 and of a jack (J) is 4. Hence, the number of different choices of a $K, Q, J = 4^3$.

$$\text{The number of possible choices of three cards from } 52 = \binom{52}{3} = \frac{52 \times 51 \times 50}{1 \times 2 \times 3}.$$

Hence,

$$P(K, Q, J) = \frac{4^3 \times 6}{52 \times 51 \times 50} \approx \mathbf{0{\cdot}0029}.$$

EXAMPLE 4 *A die is rolled four times. Find the probability of obtaining exactly three 6s.*

The number of different results with exactly three 6s $= 5 \times 4$. (There are five choices for the score on the die which is not a 6 and there are four choices for this particular die to be the one not showing a 6.) There are 6^4 different results in the possibility space. Thus,

$$P\,(\text{exactly three 6s}) = \frac{5 \times 4}{6^4} \approx \mathbf{0{\cdot}0154}.$$

EXAMPLE 5 *A die is to be rolled four times. Find the probability of getting a total score of (a) 6, (b) 19.*

(a) A total of 6 can be obtained in the following unordered ways 1, 1, 1, 3 or 1, 1, 2, 2. There are 4 ways of arranging 1, 1, 1, 3 and $\binom{4}{2}$ ways of arranging 1, 1, 2, 2. Thus,

$$P\,(6) = \frac{\mathbf{10}}{\mathbf{6^4}}.$$

(b) To obtain a total of 19, first consider 6, 6, 6, 1 and list the other possibilities as

6, 6, 5, 2,
6, 6, 4, 3,
6, 5, 5, 3,
6, 5, 4, 4,
5, 5, 5, 4.

The number of possible arrangements are as follows.
For 6, 6, 6, 1 there are 4 choices for the die showing 1.
For 6, 6, 5, 2 there are $\binom{4}{2}$ choices for the dice showing 6, and 2 ways of placing the 5 and the 2 between the remaining dice. Thus the number of arrangements is $\binom{4}{2} \times 2 = 12$.
Similarly, there are 12 ways for 6, 6, 4, 3, for 6, 5, 5, 3, and for 6, 5, 4, 4, and 4 ways for 5, 5, 5, 4, making a total of 56 possibilities. Thus,

$$P\,(19) = \frac{\mathbf{56}}{\mathbf{6^4}}.$$

EXERCISE 1.4B

1 A die is to be rolled twice. When the total of the scores showing on the two rolls is an odd number, twice this number is recorded. When this total score is even, then this number is recorded. Draw up the possibility space showing the recorded score for each ordered pair of possible scores. Hence obtain the probabilities that (a) the recorded score will be 6, (b) the recorded score will be 10.

2 Two coins are to be tossed. Draw up the possibility space showing the possible ordered pairs of heads and tails. Hence obtain the probabilities that (a) 2 heads will turn up; (b) exactly one head will turn up.

3 Two cards are to be cut from a pack of 52 cards without replacement. Find the probability of obtaining two honours exactly – that is, any two cards from the set of {aces, kings, queens, jacks and tens}.

4 Three cards are to be cut from a pack of 52 without replacement. Find the probability that there will be exactly two aces.

5 A die is to be rolled three times. Find the probability of obtaining a total score of (a) 6, (b) 12.

6 In a certain multichoice examination there are five choices for the answer to each question. If two questions are considered, how many different choices of answers can be made? When both questions are answered in a random manner, what is the probability that (a) both questions will be correct, (b) one and only one answer will be correct?

7 If three questions in the multichoice examination of the last example are to be answered at random, find the probability that (a) all three will be correct, (b) exactly two will be correct, (c) all will be wrong.

8 Two digits are to be selected at random with replacement from the set $\{0, 1, 2, \ldots, 9\}$. Find the probability that (a) the sum will be greater than or equal to 6, (b) the difference will be less than or equal to 2.

9 Three cards are to be cut from a pack of 52, the card being replaced each time. Calculate the probability that (a) two kings and an ace will be drawn, (b) no aces will be drawn.

What are the corresponding probabilities if the cards to be cut are not replaced after each cut?

10 A poker hand is a set of five cards dealt from a pack of 52. Find the probability of obtaining the following hands at poker:

(a) a royal flush (ace, king, queen, jack and a ten of the same suit),
(b) four of a kind (four cards of the same value),
(c) a full house (one set of three cards of the same value and one pair of the same value),
(d) a flush (five cards of the same suit but not in sequence),
(e) a straight flush (five cards in sequence of the same suit with the ace, if used, counting as high or low).

11 A trial consists of rolling a red die and a blue die. The score T resulting from the trial is defined as the sum of the numbers showing when the numbers on the red and blue dice are the same, but as the product of these numbers when they are different. Find (a) $P(T = 6)$, (b) $P(T = 8)$, (c) $P(T = 12)$.

If the scores from two trials are to be added together, find the probability of getting more than 45. *(L)*

12 If two dice are to be thrown simultaneously, find the probability that
(a) each die shows an even number,
(b) the total of the numbers thrown is either 7 or 10.

13 In the game of snooker, the yellow, green, brown, blue and pink balls are identical in size and worth 2, 3, 4, 5, and 6 points respectively. The five balls are placed in a bag and from this bag two balls are to be selected simultaneously at random. The points value of the two selected balls are added together to give a score X. Copy and complete the table showing the possible scores of X and the respective probabilities $P(X)$ of getting these scores.

X	5	6	7	8	9	10	11
$P(X)$	0·1			0·2			

Using your completed table, find the probability that (a) X is odd, (b) X is a multiple of 3.

A game is devised between two players in which the first player selects two balls at random from the bag containing the five balls and obtains a total score, X. The second player obtains his total score, Y, by adding the scores of the three remaining balls. Calculate the probability that Y is less than X. (*L*)

14 Two unbiassed dice are to be thrown simultaneously. Let X be the random variable denoting the numerical difference between the score on the two dice. Copy and complete the following table showing the probability $P(X)$ of obtaining each possible value of X.

X	0	1	2	3	4	5	6
$P(X)$		5/18			1/9		

15 Two numbers are to be chosen at random with replacement from each of the sets $\{1, 2, 3, \ldots, 9\}$, $\{1, 3, 5, 7, 9, 12, 13, 14, 16\}$. Calculate the probability that (a) all four numbers are even, (b) their sum will be even. (*L*)

16 Two cards are to be selected without replacement from a set of six cards numbered 1, 2, 2, 2, 3, 3. Find the probability that the sum of the numbers of the two cards selected in this way is 4.

17 An urn contains six balls – one red, one yellow, one green, one blue and two white. An experiment consists of drawing the balls one after another from the urn and placing them in a row on a table. How many different colour sequences is it possible to obtain in this way?

The experiment is repeated a large number of times and a result is to be selected at random. Find the probability

(a) that the red and green balls will not be adjacent,
(b) that the two white balls will occur at the two ends of the row,
(c) that the blue ball will be adjacent to both white balls, that is, that it will be in between them. (*L*)

18 A die is to be rolled three times. Find the probabilities of obtaining a total score of (a) 4, (b) 6, (c) 8.

19 A box A contains four black and two white beads of identical shape and size, and a box B contains one black and five white beads.

(i) A random selection of one bead from each box is to be made and the colours of the two beads are to be recorded. Draw a diagram to illustrate the possibility space for this experiment, and mark on it the set of points associated with each of the following events:

(a) two black beads will be obtained,
(b) at least one black bead will be obtained.

Find the probability of event (b) occurring.

20 The numbers 1, 2, 3, 4, are written one on each face of a regular tetrahedron. Two more identical tetrahedra are labelled in the same way. The three tetrahedra are to be allowed to fall on a table in such a way that any face of each has the same chance of resting on the table. The score for each tetrahedron is the number written on that face which is in contact with the table. Copy and complete the following table showing the probabilities of all possible total scores after a single throw of the three tetrahedra.

total score	3	4	5	6	7	8	9	10	11	12
probability	1/64			10/64		12/64				

Deduce the probability that, for a single throw of the three tetrahedra,
(a) the total score will be 3 or 4,
(b) the total score will be less than 9,
(c) the score for each tetrahedra will be the same,
(d) the total score for two tetrahedra will be the same as the score for the third. (*L*)

Agreement between theory and practice

In the seventeenth century, it was fashionable to gamble with dice. There was in France a famous gambler, named the Chevalier de Méré, who observed that it paid him in the long run to bet on a 6 turning at least once in four throws but that it paid him to bet against a double-six turning up once in 24 throws of two dice. This seemed illogical to him, so he wrote to Blaise Pascal, the mathematician and philosopher. Pascal's reflection on this problem is said to have been the beginning of the mathematical theory of probability. Let us solve the Chevalier's problem.

$$P\text{ (no. sixes in one throw)} = 5/6.$$
$$P\text{ (no. sixes in four throws)} = (5/6)^4 \approx 0{\cdot}482.$$
$$P\text{ (at least one six in four throws)} \approx 1 - 0{\cdot}482 = 0{\cdot}518.$$
$$P\text{ (one double six in one throw of two dice)} = 1/36.$$
$$P\text{ (no. double sixes in 24 throws of two dice)} = (35/36)^{24} \approx 0{\cdot}509.$$

It is interesting that such small differences from 0·5 should have been observed by the gambler in practice. He must have used a large number of trials in his experimentation.

1.5 Compound events

The intersection $A \cap B$ (both A and B)

Suppose we consider again the rolling of a die with the possibility space as the score, that is, the set $S = \{1, 2, 3, 4, 5, 6\}$. If event A is the set of odd numbers

$$A = \{1, 3, 5\},$$

and if event B is the set of prime numbers, then

$$B = \{2, 3, 5\}$$

so that

$$A \cap B = \{3, 5\}.$$

$A \cap B$ is the set of outcomes in which both A and B occur.

If $A \cap B = \phi$, then events A and B are said to be *mutually exclusive*. If A occurs, then B does not and vice versa.

In the case of rolling a die, if C is the event of obtaining an even number, and D is obtaining a 3 or a 5, then $C \cap D = \phi$ and C and D are mutually exclusive.

The union $A \cup B$ (either A or B or both)

If we want the probability, in the previous trial, of obtaining either A or B or both, we require the event $A \cup B$. If any one of the outcomes 1, 2, 3, or 5 occurs, then we have a member of $A \cup B$.

Set diagrams (Venn)

It is convenient to represent possibility spaces and events by sets of points, where each point represents just one possible outcome of a trial. Suppose the possibility space is represented by a rectangular array of points. Assuming that each point (outcome) in the possibility space is equally likely, then if $n(A)$ is the number of points in A and $n(B)$ the number of points in B, $P(A) = n(A)/N$, and $P(B) = n(B)/N$, where N is the total number of points in the possibility space.

Thus in Fig. 1.2(a) in which events A and B are mutually exclusive,

$$P(A \cup B) = \frac{n(A)+n(B)}{N} = \frac{n(A)}{N} + \frac{n(B)}{N} = P(A)+P(B).$$

The event A or B occurs if any of the $n(A)+n(B)$ events occur. Hence, when A and B are mutually exclusive,

$$P(A \cup B) = P(A)+P(B).$$

A more general demonstration of this result follows from the definition of probability as the limit of the ratio of the number of successes to the

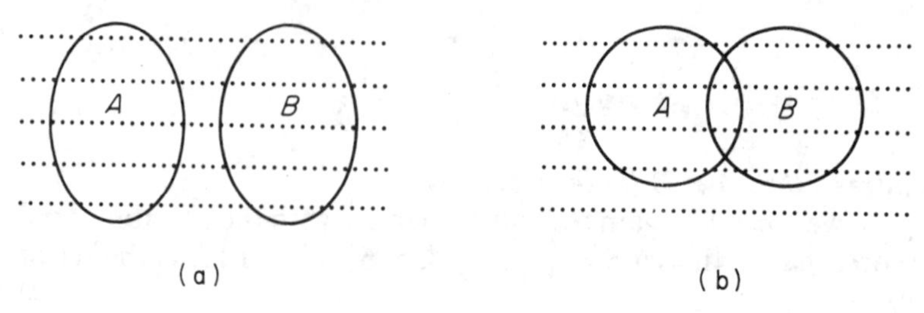

Fig. 1.2

number of trials in an extended number of trials. If in a number of repeated trials the number of times A occurs is n_A and the number of times B occurs is n_B, then $n_A/N \to P(A)$ and $n_B/N \to P(B)$.

$$\frac{n_A + n_B}{N} \to P(A \cup B),$$

N being the number of trials. Thus, if A and B are mutually exclusive,

$$P(A \cup B) = P(A) + P(B). \qquad 1.1$$

If A and B are not mutually exclusive, then $n_A + n_B$ includes those outcomes twice in which both A and B occur. Hence,

$$n(A \cup B) = n(A) + n(B) - n(A \cap B)$$

which leads to

$$\mathrm{P}(\mathrm{A} \cup B) = P(A) + P(B) - P(A \cap B). \qquad 1.2$$

This result also holds when A and B are mutually exclusive as then $P(A \cap B) = 0$, and equation 1.2 is equivalent to equation 1.1 in this case.

Conditional probability

If A and B are two events from one possibility space, the probability that B will occur when it is known that A has occurred is sometimes required. The probability that B will occur given that A has occurred is written as $P(B|A)$. An example of this is the probability of getting an ace in cutting one card when it is known that an honour (Ace, King, Queen, Jack or ten) has been drawn. Here $P(B|A) = 1/5$.

Referring to Fig. 1.2(b), in the case when A has occurred, the possibility space is limited to event A. If B occurs, then both A and B occur. Using the argument from the number of points in the sets:

$$P(B|A) = n(A \cap B)/n(A)$$

which leads to

$$P(A \cap B)/P(A),$$

$$P(B|A) = P(A \cap B)/P(A) \text{ and}$$

$$P(A \cap B) = P(A) \, . \, P(B|A)$$

Independence

If the events A and B are such that the probability of B occurring is the same whether A occurs or not, that is, if $P(B|A) = P(B)$, then event B is independent of A. In this case,

$$P(A \cap B) = P(A) \, . \, P(B).$$

As $A \cap B = B \cap A$,

$$P(B \cap A) = P(B) \, . \, P(A).$$

But $$P(B \cap A) = P(B).P(A|B).$$
Hence $$P(A|B) = P(A)$$
and thus A is independent of B.

Rule

When A and B are independent, $P(A \cap B) = P(A).P(B)$ and conversely. This rule is used to test the independence of two events.

EXAMPLE 1 *Two dice are to be rolled. Event A is the scoring of a 2 or a 3 on the first die. Event B is the scoring of a total of 9. Event C is the scoring of a total of 7. Investigate whether*
(a) events A and B are independent,
(b) events A and C are independent.

(a) $P(A) = 1/3$, $P(B) = 4/36 = 1/9$, using the array in §1.2.

$$P(A \cap B) = P((3, 6)) = 1/36.$$
$$P(A \cap B) \neq P(A).P(B).$$

Hence, A and B are **not independent**.
(b) $P(C) = 6/36 = 1/6$.

$$P(A \cap C) = P[(2, 5), (3, 4)] = 1/18 = P(A).P(C).$$

Hence, A and C are **independent**.

EXAMPLE 2 *A and B are events in a possibility space.* $P(A \cup B) = 2/3$. $P(A) = \frac{1}{2}$. *Calculate* $P(B)$ *in the two cases*
(a) when A and B are independent,
(b) when A and B are mutually exclusive.

(a) $P(A \cup B) = P(A) + P(B) - P(A \cap B)$. Let $P(B) = x$. As A and B are independent

$$P(A \cap B) = x.P(A)$$

and so

$$2/3 = \tfrac{1}{2} + x - \tfrac{1}{2}x$$
$$\Leftrightarrow \tfrac{1}{2}x = 1/6 \Leftrightarrow x = 1/3.$$
$$P(B) = \mathbf{1/3}.$$

(b) When A and B are mutually exclusive

$$P(A \cup B) = P(A) + P(B).$$

Hence,

$$P(B) = 2/3 - 1/2 = \mathbf{1/6}.$$

EXAMPLE 3 *Four cards are to be cut from a pack of 52 cards. Find the probability that exactly three aces will be cut when the cut cards are not replaced.*

First method Using combinations, the number of different choices of four cards with exactly three aces = 4 × 48. [There are four choices for the ace omitted and 12 choices for the card which is not an ace.]

The number of possible choices of 4 cards $= \binom{52}{4}$.

Hence $$P(3 \text{ aces}) = 192\Big/\binom{52}{4} = \frac{192 \times 2 \times 3 \times 4}{52 \times 51 \times 50 \times 49} = 7.09 \times 10^{-4}.$$

Second method We regard each cut separately and we have the following possible ways of getting exactly three aces (A): A, A, A, N; N, A, A, A; A, N, A, A; A, A, N, A; where A represents an ace and N represents a card which is not an ace.

$$P(\text{A, A, A, N}) = \frac{1}{13} \times \frac{3}{51} \times \frac{2}{50} \times \frac{48}{49}.$$

$$P(\text{N, A, A, A}) = \frac{48}{52} \times \frac{4}{51} \times \frac{3}{50} \times \frac{2}{49}.$$

$$P(\text{A, N, A, A}) = \frac{4}{52} \times \frac{48}{51} \times \frac{3}{50} \times \frac{2}{49}.$$

$$P(\text{A, A, N, A}) = \frac{4}{52} \times \frac{3}{51} \times \frac{48}{50} \times \frac{2}{49}.$$

Each of these different orders has the same probability. Thus,

$$P(3 \text{ aces}) = 4 \times \frac{1}{13} \times \frac{3}{51} \times \frac{2}{50} \times \frac{48}{49} = 7.09 \times 10^{-4}.$$

EXERCISE 1.5

1 A and B are events in a possibility space. State the connection between $P(A \cup B)$, $P(A)$, $P(B)$ and $P(A \cap B)$.

If $P(A) = \frac{1}{2}$ and $P(B) = \frac{1}{3}$, calculate $P(A \cap B)$ and $P(A \cup B)$ when (a) A and B are independent, (b) A and B are mutually exclusive.

2 In an experiment there are two bags, H and K, which contain marbles. In H there are three red and seven black marbles and in K there are six red and four black marbles. An unbiased die is to be rolled and, if a score of at least 5 is obtained, a marble is to be selected at random from H; otherwise a marble is selected at random from K. If event A is 'a red marble is selected' and event B is 'a score of 5 or 6 is obtained on the die', calculate (a) $P(B)$, (b) $P(A)$, (c) $P(A \cap B)$, (d) $P(A|B)$, (e) $P(A' \cap B)$, where A' is the event 'not A'. (*L*)

3 Two cards are to be cut at random from a pack of 52 cards, without replacement. Find the probability of obtaining (a) two aces, (b) at least one ace, (c) two hearts.

4 A set of 26 cards, each containing one letter of the alphabet, is to be divided among gentlemen R, S and T so that

R has letters A to F,
S has letters G to N,
T has the 12 letters O to Z.

Each draws one card (quite independently) from his set of cards, in the hope that the word DIP will be formed. Find the probability (i) that D is drawn, but not I or P, (ii) that D and I are drawn but not P, (iii) that at least one of D, I, P is drawn, (iv) that precisely two of D, I, P are drawn. (*C*)

5 Cards are to be drawn, with replacement, from a pack of 52 cards. Find the probability that
(a) the first two cards drawn are both clubs,
(b) the third card is the first club to be drawn,
(c) exactly four cards out of the first five drawn are clubs,
(d) only two clubs are drawn in the first six cards and they are different cards. (*L*)

6 Two cards are to be drawn from a pack of playing cards, without replacement. Find the probability that the first is a 2 and just one of them is a heart. (*L*)

7 The event that Mary goes to a dance is M and the event that Linda goes to the dance is L. The events L and M are independent and $P(L' \cap M') = \frac{1}{4}$, $P(L) + P(M) = 23/24$. Find the probability that both Linda and Mary will go to the dance. (*L*)

8 The random events A, B and C are defined in a finite sample space D. The events A and B are mutually exclusive and the events A and C are independent. $P(A) = 1/5$, $P(B) = 1/10$, $P(A \cup C) = 7/15$ and $P(B \cup C) = 23/60$. Evaluate $P(A \cap B)$, $P(A \cup B)$, $P(A \cap C)$ and $P(B \cap C)$ and state whether B and C are independent. (*L*)

9 The following information is given about events A, B and C: $P(A) = \frac{1}{2}$, $P(B) = \frac{1}{3}$, $P(C) = \frac{1}{3}$, $P(A \cup B \cup C) = 1$, $P(B \cap C) = 0$ and $P(C \cap A) = 0$. Show that (i) $P(A \cup B) = P(B \cup C)$, (ii) $P(A \cap B) = P(A) \times P(B)$, (iii) $P(A \cup C) = P(A) + P(C)$. Find $P(A \cap B' \cap C')$. (*L*)

10 It is given that the probabilities that a man and his wife will live for a further 20 years are 1/5 and 1/4 respectively. Assuming that these events are independent, determine the probability that (a) both will be alive in 20 years time, (b) at least one will be alive in 20 years time. (*L*)

11 Explain what is meant by the statements that two events A and B are independent. Let A be the event that a family has children of both sexes. Let B be the event that a family has at most one boy. Assuming that the probability of the birth of a boy and of a girl are equal, determine whether the events A and B are independent if the family has (a) three children, (b) two children. (*L*)

12 In a firm's car park at leaving time, there are 8 Ford, 6 Vauxhall and 10 Leyland cars. If no more cars enter the car park and cars leave one at a time at random, find the probabilities that
(a) the first car to leave is a Ford,
(b) the last car to leave is a Leyland,
(c) all Vauxhall cars leave one after another,
(d) when there are only three cars left in the car park, there is just one of each make.

13 Given that events A and B are independent, events B and C are mutually exclusive, and $P(A) = 0{\cdot}4$, $P(A \cup B) = 0{\cdot}58$, $P(C) = 2P(B)$, find
(a) the value of $P(B)$,
(b) the greatest possible of $P(A \cap C)$,
(c) the greatest possible value of $P(A \cup C)$. (*L*)

14 The events A, B and C are exhaustive and mutually exclusive. Given that A occurs with probability $\frac{1}{3}$ and B occurs with probability $\frac{1}{4}$, obtain the probability that (a) A occurs and B does not, (b) C only occurs, (c) both A and C occur. (*L*)

15 Calculate the probability that both a head and a tail can be seen when four coins are placed at random on a table. (*L*)

1.6 The probability tree

The product law of probability, $P(A \cap B) = P(A) . P(B|A)$, leads to an interesting method of assessing all the possible outcomes of a set of trials. As an example, suppose that it is desired to find the probability that a randomly chosen family containing three children will have two boys and one girl, assuming that the probability of a randomly chosen birth being that of a boy is 0·52 (taken from the *Annual Abstract of Statistics*).

Fig. 1.3 is an example of a probability tree. At each birth there are two possibilities, a boy (B) or a girl (G). Thus, there is a branching path. The possibility space is

$$\{\text{BBB, BBG, BGB, BGG, GBB, GBG, GGB, GGG}\}.$$

The probability that one of these paths will be followed, for example, GBG, is the product of the probabilities in that path, that is, $0{\cdot}48 \times 0{\cdot}52 \times 0{\cdot}48$.

The event set giving two boys and a girl is {BBG, BGB and GBB}.

$$P(\text{BBG}) = 0{\cdot}52 \times 0{\cdot}52 \times 0{\cdot}48 = P(\text{BGB}) = P(\text{GBB}).$$

Hence,

$$P(\text{two boys and a girl}) = 3 \times (0{\cdot}52)^2 \times 0{\cdot}48 \approx 0{\cdot}389.$$

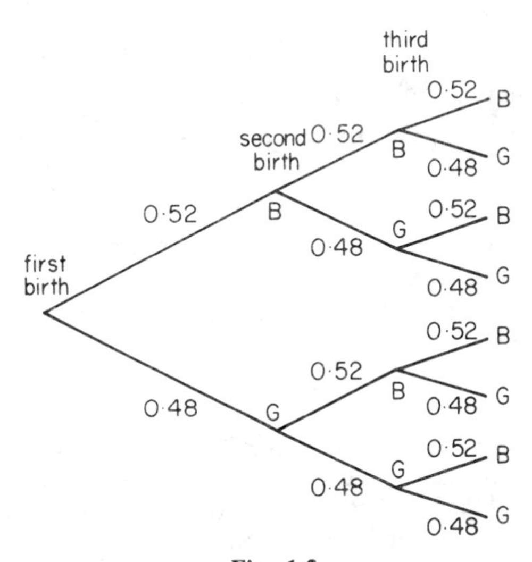

Fig. 1.3

EXAMPLE 1 *Assume that in a test series between the West Indies and England the probabilities that in each game the West Indies win, England win and the match is drawn are respectively* $\frac{1}{2}$, $\frac{1}{4}$ *and* $\frac{1}{4}$. *Find the probability that in a series of three matches the West Indies will win two or more matches.*

We first draw the probability tree (Fig. 1.4), in which W = a win for the West Indies, D = a draw and E = a win for England. The branches that result in two or more wins for the West Indies are WWW, WWD, WWE, WDW, WEW, DWW and EWW.

$$P(\text{WWW}) = \tfrac{1}{2} \times \tfrac{1}{2} \times \tfrac{1}{2},$$

$$P(\text{WWD}) = \tfrac{1}{2} \times \tfrac{1}{2} \times \tfrac{1}{4}$$

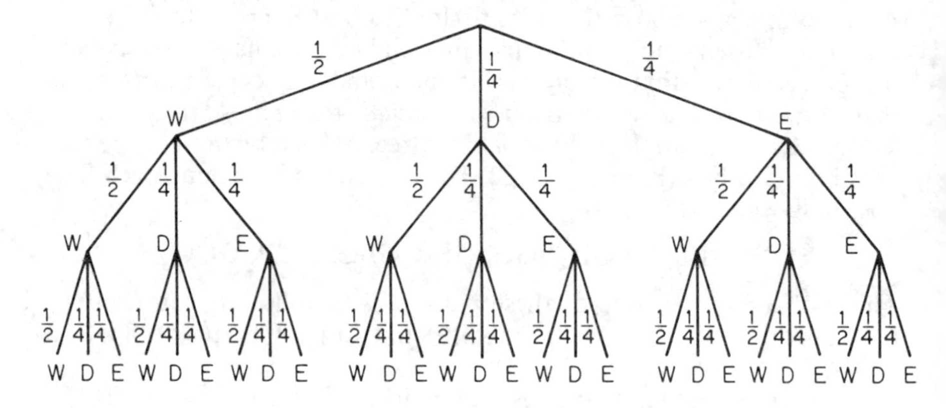

Fig. 1.4

and each of $P(\text{WWE})$, $P(\text{WDW})$, $P(\text{WEW})$, $P(\text{DWW})$, $P(\text{EWW})$ is equal to $\frac{1}{2} \times \frac{1}{2} \times \frac{1}{4}$. Hence,

$$P\,(\text{two or more wins for the West Indies}) = 1/8 + 6 \times 1/16 = \tfrac{1}{2}.$$

The corresponding probability for two or more wins for England is

$$P\,(\text{EEE, EED, EEW, EDE, EWE, DEE, WEE}) = 1/64 + 3 \times \tfrac{1}{4} \times \tfrac{1}{4} \times \tfrac{1}{2} + 3 \times \tfrac{1}{4} \times \tfrac{1}{4} \times \tfrac{1}{4} = 5/32.$$

It is not always necessary in the solution of a problem to construct the complete probability tree, but the method does provide a clear way of listing all the possible outcomes of a set of trials.

EXERCISE 1.6

1 A bag contains seven black and three white marbles. Three marbles are to be chosen at random and in succession, each marble being replaced after it has been taken out of the bag. Draw a tree diagram to show all possible selections.

From your diagram, or otherwise, calculate to two significant figures the probability of choosing

(a) three black marbles,
(b) a white marble, a black marble and a white marble, in that order,
(c) two white marbles and a black marble in any order,
(d) at least one black marble.

State an event from this experiment which, together with the event described in (d), would be both exhaustive and mutually exclusive. (*L*)

2 The probability of the player serving at tennis winning a particular point is $\frac{2}{3}$. By using a tree diagram, or otherwise, calculate the probability that
(a) the server wins a game by 4 points to 1 point,
(b) the server loses a game by 1 point to 4 points,
(c) the game reaches 3 points all.
(Note that in (a) the server wins the last point and in (b) the server loses the last point. Also, to win a game at least 4 points must be won).

3 A bag initially contains one red ball and two blue balls. A trial consists of selecting a ball at random, noting its colour and replacing it together with an additional ball of the same colour. Given that three trials are made, draw a tree diagram illustrating the various possibilities. Hence, or otherwise, find the probability that (a) at least one blue ball will be drawn, (b) exactly one blue ball will be drawn. (*L*)

4 Three machines A, B and C account respectively for 55%, 30% and 15% of the output of a certain product. Of the items produced by machine A, 1% are defective and the corresponding percentages for B and C are 1.5% and 3% respectively. What is the overall percentage of defective items? (*L*)

5 Three balls are to be drawn in succession from a bag containing five red balls and three blue balls, each ball being replaced in the bag before the next draw is made.
(a) What is the probability that all three balls will be of the same colour?
(b) What is the probability that at least one ball of each colour will be drawn?

The experiment is now repeated but this time the balls are not replaced after drawing. Answer questions (a) and (b) for this second experiment. (*L*)

6 In each round of a certain game, a player can score 1, 2 or 3 only. Copy and complete the table below, which shows the scores and two of the respective probabilities of these being scored in a single round

score	1	2	3
probability	4/7		1/7

Draw a tree diagram to show all the possible total scores and their respective probabilities after a player has completed two rounds.

Find the probability that a player has (a) a score of 4 after two rounds, (b) an odd-numbered score after two rounds. (*L*)

7 In a game of 'triads' a player can score 2, 3 or 4, with probabilities $\frac{1}{2}$, $\frac{1}{3}$ or 1/6 respectively, in a single turn. Draw a tree diagram to show all possible score combinations for a player having two turns, and calculate the associated probabilities. Hence, or otherwise, calculate
(a) the probability that a player has a total of 6 after two turns,
(b) the probability that a player has a total score which is odd after two turns,
(c) the probability that a player has a total score in excess of 9 after two turns,
(d) the probability that a player has a total score after three turns of 6.

8 Nine discs numbered from 1 to 9 are placed in a bag and three discs are then to

be drawn at random without replacement. The number on the first disc drawn is denoted by n, and the sum of the numbers on the three discs is denoted by S.
(a) Find the probability that $S = 10$.
(b) Find the probability that both $n = 2$ and $S = 10$.
(c) Given that $n = 2$, find the probability that $S = 10$.
(d) Given that $n \neq 2$, find the probability that $S = 10$.

1.7 Calculating conditional probability

The rule $P(A \cap B) = P(A) \times P(B|A)$ can be used to determine $P(B|A)$ – the probability of B given that A has occurred. From this rule we deduce

$$P(B|A) = \frac{P(A \cap B)}{P(A)}.$$

The probability tree is useful when considering conditional probability, as shown in the following example.

EXAMPLE 1 *There are two bags S and T. Bag S contains three red and four black counters. Bag T contains four red and three black counters. A counter is drawn at random from bag S and placed in bag T. Then a counter is drawn at random from bag T and placed in bag S. Given that S now contains three red and four black counters, find the probability that the first counter drawn was a red one.*

The probability tree is shown in Fig. 1.5 where R represents a red counter and B represents a black counter.

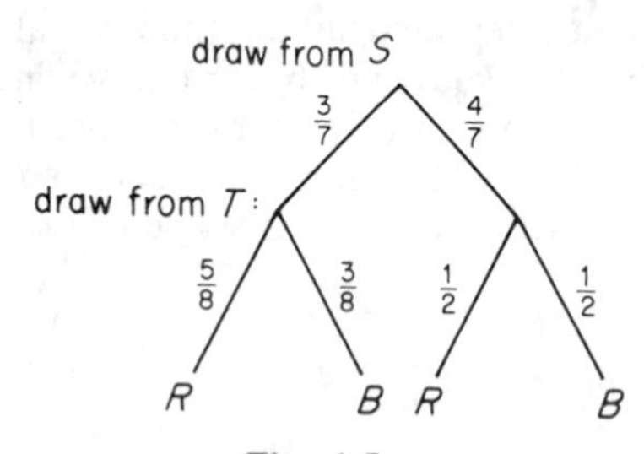

Fig. 1.5

Let X be the event that S has three red and four black counters.

$$P(X) = \frac{3}{7} \times \frac{5}{8} + \frac{4}{7} \times \frac{1}{2}.$$

Let Y be the event that the first counter is red.

$$P(Y \cap X) = \frac{3}{7} \times \frac{5}{8}.$$

$$P(Y|X) = \frac{3/7 \times 5/8}{3/7 \times 5/8 + 4/7 \times 1/2} = \mathbf{\frac{15}{31}}.$$

A probability tree can be compared with a pin-ball machine, the probability associated with each branch representing the proportions in which the balls pass along that branch. We can interpret the problem as follows.

We are only concerned with the RR branch and the BB branch, each yielding three red and four black counters in bag S. The proportion in which the RR branch is followed is $3/7 \times 5/8$. The BB branch is followed in the proportion $4/7 \times 1/2$. Thus, of the branches yielding 3R and 4B in bag A, $\dfrac{15/56}{15/56 + 2/7}$ is the fraction which followed RR, and hence this fraction is equal to the probability that the first draw was an R.

EXERCISE 1.7

1 Referring to question 4 of Exercise 1.5, if an item selected at random is found to be defective, what is the probability that it came from machine A? (*L*)

2 There are two bags, A and B. Bag A contains four red counters and three white counters. Bag B contains three red and four white counters. A counter is drawn at random from bag A and placed in bag B, then a counter is drawn at random from bag B and found to be red. What is the probability that the first counter drawn was red?

3 There are two dice, a biased and an unbiased one. The respective probabilities of rolling 1, 2, 3, 4, 5, 6, with the biased die are $1/9, 1/6, \frac{1}{4}, 1/9, 1/12, \frac{1}{4}$. One of the dice is chosen at random and rolled twice, giving a total score of 6. Find the probability that the die used was the biased one.

4 A game is to be played according to the following rules. A card is cut from a complete pack of 52. If the card is an honour (A, K, Q, J, 10), a score of 5 is recorded. If the card cut is one of the set of $\{9, 8, 7, 6,\}$ a score of 3 is recorded. If a 5, 4 or a 3 is obtained a score of 1 is recorded. Otherwise the recorded score is 0. The card is returned to the pack after each cut. After two cuts the total recorded score is 6. What is the probability that the first card cut was an honour?

5 The respective probabilities of a win by the West Indies, of a draw and of a win by England, in each game of a test series is assumed to be $\frac{1}{2}, 1/6, \frac{1}{3}$. If after three games, England are leading by one game, find the probability that England won the first game when the result of the first game is not known.

6 Each of three identical boxes A, B and C has two drawers. Box A contains a prize in each drawer. Box B contains a prize in one drawer only. Box C does not contain any prize.

A box is to be chosen at random, and a drawer is opened and found to be empty. Find the probability that a prize will be found

(a) if the other drawer in the same box is opened,

(b) if one of the other two boxes is chosen at random and a drawer is opened. (*L*)

7 A company makes beach balls which are coloured either blue or yellow. The balls are produced at two factories P and Q, and P produces five-eighths of the total output. One-tenth of the balls produced at P are blue and half of the balls produced at Q are blue. Balls from both factories are randomly mixed and packaged before being distributed to retailers. Show that when a man enters a shop and selects a ball at random, the probability that it is blue is $\frac{1}{4}$.

Two shops S_1 and S_2 sell the balls, and both shops have two balls, taken from a random sample of all balls produced, in stock. Find the probabilities that (a) neither shop has a yellow ball, (b) there is at least one yellow ball in each shop.

Given that the two balls in shop S_1 come from P and that the two balls in shop S_2 come from Q, find the probability that there is exactly one blue ball in each shop. *(L)*

1.8 The Binomial probability distribution

Suppose that a probability tree is drawn for the rolling of an unbiased die three times and recording the number of sixes obtained. The tree is shown in Fig. 1.6, where S represents the number of successes and F represents the number of failures.

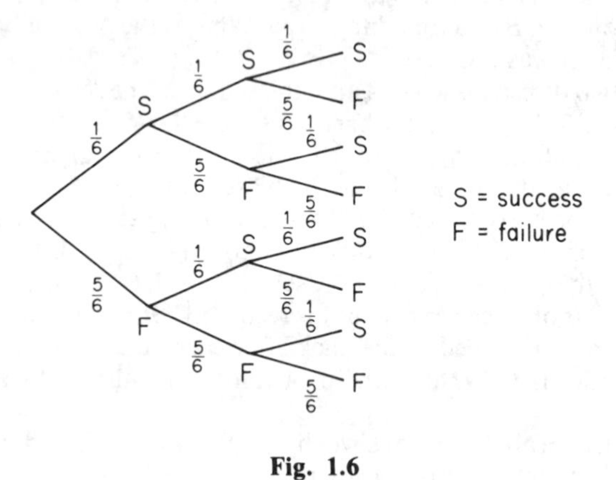

Fig. 1.6

We draw up a table of the possibilities with their associated probabilities.

No. of successes (X)	$P(X)$
0	$(5/6)^3$
1	$3\,(5/6)^2\,(1/6)$
2	$3\,(5/6)\,(1/6)^2$
3	$(1/6)^3$

The table which lists all the possible outcomes with the associated probabilities is an example of a *probability distribution.*

Note that the probabilities are the terms of the binomial expansion $(5/6+1/6)^3$. It is a particular example of the binomial distribution.

If four telephone numbers were to be chosen at random and the last digit was examined, the probability of obtaining 0, 1, 2, 3, 4 nines would be as shown in the following table.

X = no. of nines	$P(X)$
0	$(0{\cdot}9)^4$
1	$4\,(0{\cdot}9)^3\,.\,(0{\cdot}1)$
2	$6\,(0{\cdot}9)^2\,.\,(0{\cdot}1)^2$
3	$4\,(0{\cdot}9)\,.\,(0{\cdot}1)^3$
4	$(0{\cdot}1)^4$

These probabilities are the terms of $(0{\cdot}9+0{\cdot}1)^4$. The results may be obtained by means of another probability tree.

An argument that can be more easily generalized is as follows. In one trial,

$$P\ (\text{not a nine}) = 0{\cdot}9.$$

Hence,

$$P\ (\text{no nines in four trials}) = (0{\cdot}9)^4.$$

The probability of obtaining: not a nine, a nine, not a nine, not a nine, in that order, is

$$0{\cdot}9 \times 0{\cdot}1 \times 0{\cdot}9 \times 0{\cdot}9 = (0{\cdot}9)^3\,.\,(0{\cdot}1).$$

But it is possible to get one nine exactly in four different ways, for the nine can occur in any one of the four trials. Hence

$$P\ (\text{one nine in four trials}) = 4\,(0{\cdot}9)^3\ (0{\cdot}1).$$

Similarly, to get two nines there are six different orders – the $\binom{4}{2}$ ways in which the two positions for the nine can be chosen.

Definition

A general statement of the binomial distribution is as follows. If there are to be n repeated trials which are independent and in each of which the probability of success is constant and equal to p, and q is $1-p$ (the probability of failure in one trial), then the probability of obtaining r successes is $P(r)$ where $P(r) = \binom{n}{r} p^r q^{n-r}$.

Proof

The probability of obtaining r successes in any assigned order, for example on the first, third, sixth, . . . , trials is

$$p \times q \times p \times q \times q \times p \times \ \ldots = p^r q^{n-r}.$$

But there are $\binom{n}{r}$ different orders of obtaining r successes. Hence,

$$P\ (r \text{ successes in any order}) = \binom{n}{r} p^r q^{n-r}.$$

EXAMPLE 1 *Assume that each time a marksman shoots at a target, the probability that he scores a bull is 1/5, and each shot is independent. Find the probabilities that if such a marksman fires at the target, he will score (a) a bull, not a bull, a bull, not a bull, a bull, in that order, (b) he will score three bulls in his five shots, (c) his fifth shot will be his third bull.*

(a) Let B represent a bull and N, not a bull.

$$P(BNBNB) = 1/5 \times 4/5 \times 1/5 \times 4/5 \times 1/5 = \mathbf{0{\cdot}00512}.$$

(b) P (3 bulls in any order) $= \binom{5}{3}(1/5)^3(4/5)^2 = \mathbf{0{\cdot}0512}.$

(c) If his fifth shot is to be his third bull, then he will have to score two bulls in four shots and then get a bull on his fifth.

$$P\ (2 \text{ bulls in 4 shots}) = \binom{4}{2}(1/5)^2(4/5)^2.$$

Therefore,

$$P\ (\text{fifth shot is third bull}) = \binom{4}{2}(1/5)^2(4/5)^2(1/5) = \mathbf{0{\cdot}03072}.$$

EXAMPLE 2 *An interviewer stops six people. For each person that she stops, the probability that the person will reply is $\frac{2}{3}$. Find the probabilities that*
(a) she will receive exactly four replies,
(b) she will receive at least four replies.
Given that when she receives a reply the probability that the person stopped will complete a questionnaire is 2/5, estimate, to two significant figures, the probability that she will have received five completed questionnaires. (*L*)

There are six trials with a probability of $\frac{2}{3}$ for success in each trial.

(a) P (4 successes in 6 trials) $= \binom{6}{4}\left(\frac{2}{3}\right)^4\left(\frac{1}{3}\right)^2 = \frac{6 \times 5}{2} \times 2^4 \times \left(\frac{1}{3}\right)^6 = \mathbf{\frac{80}{243}}.$

(b) P (at least 4 successes) $= P(4) + P(5) + P(6) = 80/243 + 6 \times 2^5/3^6 + (\frac{2}{3})^6$

$$= \mathbf{496/729}.$$

To receive exactly five completed questionnaires, she must either receive five replies, all of whom complete, or receive six replies, five of whom complete the questionnaire.

P (5 replies, followed by 5 completions) $= \binom{6}{5} \times \left(\frac{2}{3}\right)^5\left(\frac{1}{3}\right) \times (2/5)^5 \approx 0{\cdot}00270.$

P (6 replies, followed by exactly 5 completions) $= \left(\frac{2}{3}\right)^6 \times 6 \times (2/5)^5 \times (3/5)$

$$\approx 0{\cdot}00324.$$

So, the total probability is $0{\cdot}00270 + 0{\cdot}00324 = \mathbf{0{\cdot}00594}$.

EXERCISE 1.8

1 Three unbiased dice are to be thrown. Find the probabilities of obtaining the following numbers of sixes (a) 0, (b) 1, (c) 2, (d) 3.

2 A trial consists of throwing an unbiased die, and a success occurs when a six is obtained. If the die is to be thrown five times, calculate the probability of (a) five successes, (b) at least three failures. (*L*)

3 A marksman is to fire five rounds at a small target. If the probability of his hitting the target with any one shot is 0·4, calculate the probability that he hits the target (a) exactly once, (b) at least twice. (*L*)

4 In a game of darts the probability that a certain player gets a treble 20 with any one throw is $\frac{1}{4}$. Find the probabilities of this player obtaining the following numbers of treble 20s in three throws (a) 0, (b) 1, (c) 2, (d) 3.

5 Suppose that wet and fine days occur independently with a probability of a fine day being 2/7. Find the probability that in a random week there will be exactly three fine days.

6 A polyanthus seed mixture containing 20 per cent blue strain, the remainder being mixed colours not containing blue, is sown by a nursery. The resulting plants are transplanted into boxes containing eight plants each. Obtain the probability that a randomly chosen box will contain (a) exactly six blue plants, (b) less than three blue plants.

7 An archer shoots six arrows at a target. Given that the probability that each arrow hits the target is constant and equal to $\frac{1}{3}$, find the probability that
(a) he scores a hit, miss, hit, miss, hit, miss, in that order,
(b) he scores three hits exactly in his six shots,
(c) his sixth shot is the third hit.

8 It is known that 30 per cent of an apple crop has been attacked by a pest. Assuming that the choice of the apple to be attacked by the pest is random, find the probability that out of four apples picked at random (a) exactly two will have been attacked, (b) at least one will have been attacked.

9 A sampling procedure to test whether the manufacture of a particular product is satisfactory is as follows. A sample of ten is tested. If all the items in the sample are sound, the product is passed. If two or more are defective, the product is rejected. If only one of the sample is found to be defective, then a further sample of ten is taken. If all of this second sample are sound the product is passed, otherwise it is rejected. If the probability of a random item being defective is 0·1, obtain an expression for the probability that the product fails to pass the inspection test.

MISCELLANEOUS EXERCISE 1

1 A committee, consisting of nine persons, is to be selected from seven Englishmen and six Scotsmen. If *at least* four Scotsmen must be included, show that the committee can be formed in 560 different ways. If these 560 ways are equally likely, calculate the probability that the Scotsmen will be in a majority on the committee.

2 Six numbers are to be selected at random without replacement, from a set of five positive numbers and four negative numbers, and multiplied together. Calculate the probability that the product will be negative. (*L*)

3 Sixteen association football players, from a number of different clubs, form a squad from which 11 players in the national team are to be selected.
(a) Find the number of different teams which could be selected from the squad, irrespective of the positions in which the men play.
(b) Given that three of the squad are goalkeepers and a team contains just one goalkeeper, find the number of different ways in which the team can now be selected, irrespective of the positions in which the remaining men play.
(c) Just four players in the squad, none of whom is a goalkeeper, belong to the Liverpool club. A team is to be selected at random from the squad of three goalkeepers and 13 other players. Calculate the probability that all four of the players from the Liverpool club will be included in the team. (*L*)

4 Using only the digits 8, 7, 5, 1, without repetition, calculate
(a) the number of different four digit numbers which can be formed,
(b) the number of odd numbers exceeding 700 which can be formed,
(c) the probability that if a number is chosen at random from all the one, two, three and four digit numbers which can be formed from the given digits, without repetition, it will exceed 800. (*L*)

5 (i) If four letters are chosen from a group of ten different letters, find the number of different arrangements which are possible.
(ii) Find the number of ways in which all the letters of the word STATISTICS can be arranged.
(iii) If one of the arrangements mentioned in (ii) is chosen at random, show that the probability that it begins and ends with an I is 1/45. (*L*)

6 Nine trees are to be planted, five on the north side and four on the south side of a road.
(a) Find the number of different ways in which this can be done if the trees are all of different species.
(b) If the trees in (a) are to be planted at random, find the probability that two particular trees will be next to each other on the same side of the road.
(c) If there are three cypresses, four plum trees and two magnolias, find the number of different ways in which these could be planted, assuming that trees of the same species are indistinguishable.
(d) If the trees in (c) are to be planted at random, find the probability that the two magnolias will be on opposite sides of the road. (*L*)

7 Four cards are to be drawn at random from a pack of 52, one at a time with replacement. Find the probability that
(a) no heart will be drawn,
(b) four hearts will be drawn,
(c) two hearts and two diamonds will be drawn (in any order),
(d) one card from each suit will be drawn. (*L*)

8 A child, using his printing set, arranges individual letters to make up the three words SOOTY AND SWEEP.
(a) If the five letters of the word SOOTY will fall out to form a five-letter word (not necessarily meaningful), how many different words are possible? How many of these end in the letter 'O'?
(b) If the two 'O's, the 'A' and the two 'E's fall out and are to be put back in random order into the positions previously filled by these five letters, calculate the probability of the word SWEEP being correct. In addition, calculate the probability of the word SWEEP being correct, given that at least one of the two 'O's has been correctly replaced. (*L*)

9 Four out of a batch of 40 manufactured articles are known to be defective. If a sample of four is to be drawn at random from the batch, find the probability that it will contain
(i) exactly two defective articles,
(ii) not more than one defective article. (*O&C*)

10 A hand of five cards is to be drawn from a pack of 52 playing cards. Find the probability of drawing
(a) five cards of the same suit,
(b) a pair plus a triple (for example, two 7s and three aces).
If a hand consists of two aces, a king, a jack and a 4, and a player keeps the pair of aces but discards the other three cards and draws three more cards from the remaining 47, find the probability of obtaining a pair plus a triple. (*L*)

11 (i) Two cards are to be taken, without replacement, from a well-shuffled pack of 52 playing cards. Calculate the probability that
(a) both will be aces,
(b) at least one king or queen will be obtained.
(ii) In a game, a turn consists of drawing a card from a pack of 52 cards and rolling one of two dice. If the card is a heart, the die rolled is a red one numbered 2, 4, 6, 8, 10, 12; if not, a blue die numbered 1, 2, 3, 4, 5, 6 is rolled. Find the probability of obtaining a *total score* of 6 on the dice in two turns. (*L*)

12 In a particular community, the probability of a man chosen at random having brown eyes is $2/5$. If a random sample of six men is to be taken, find the probability that
(a) exactly three men,
(b) at least three men, will have brown eyes.
A second random sample of eight men is to be chosen at random. Find the probability that less than two of the sample will have brown eyes. (*L*)

13 Four people are to be taken at random from all the people born in June 1959. Find the probability that
(a) two at least will have the same birthday,
(b) at least three will have the same birthday. (*L*)

14 Cards are to be drawn at random and with replacement from an ordinary pack of playing cards until three spades have been drawn. Find the probability that the number of draws required will be
(a) exactly six,
(b) at least six.
Find an expression for $P(n)$, the probability that exactly n cards will have to be drawn to get three spades ($n \geqslant 3$). Hence, or otherwise, find the most likely value (or values) of n. (*L*)

15 Two normal dice, one coloured red and the other coloured blue, are to be rolled and the sum of their scores recorded. Find the probability that
(a) the sum will be divisible by 3; (b) the sum of 15 will occur, given that the red die shows a number less than 5; (c) the sum will be less than 7, given that the blue die shows 3; (d) the red die will show a number greater than the blue die. (*L*)

16 Given that $P(A) = \frac{1}{2}$, $P(B) = \frac{1}{3}$, $P(C) = \frac{1}{4}$, and that events A and B are independent, events A and C are independent, and events B and C are mutually exclusive, find

(a) $P(A \cup C)$,
(b) $P(A \cap B)$,
(c) $P(A' \cap B' \cap C')$,
giving in each case sufficient explanation to show how your results are obtained.

17 If 5 per cent of the articles produced by a machine are defective, find expressions for the probability that out of 25 articles to be selected at random
(a) exactly three will be defective, (b) none will be defective, (c) two or more will be defective.

What is the most likely number of defectives to be found? What would be the most likely number of defectives to be found if a random sample of 100 articles were selected? *(L)*

18 A die is thrown until both a six and a one have been obtained. Show that the probability that n throws will be required, where n takes the value 2, 3, 4, . . ., is $\frac{1}{3}[(5/6)^{n-1} - (2/3)^{n-1}]$. *(L)*

19 If A, B, C and D are independent random events and $P(A) = 0{\cdot}1$, $P(B) = 0{\cdot}2$, $P(C) = 0{\cdot}3$ and $P(D) = 0{\cdot}4$, calculate $P(A \cup B \cup C)$ and $P(A \cup B \cup C \cup D)$, giving sufficient explanation to show how your results are obtained.

20 Four coins and a die are to be tossed together and S is defined as the sum of the number of heads together with the number showing on the die. Find (a) $P(S = 6)$, (b) $P(S > 7)$, (c) $P(2 \leqslant S \leqslant 5)$, (d) $P(S = 8$ and two heads show). *(L)*

21 It is known that 1 per cent of the very large number of cheques presented to a bank contains errors of some sort. Assuming the errors are randomly distributed, use a binomial distribution to find an expression for the probability that amongst a batch of 800 cheques, exactly ten will contain errors. Calculate, to two significant figures, the probability that there will be no errors in the next 200 cheques examined. After 1 000 cheques have been examined, what is the probability that the next cheque to contain an error will be the tenth one of those subsequently examined? *(L)*

22 A pack of playing cards is shuffled and dealt into four hands of 13 cards each. Calculate the probability that
(a) one of the hands will contain all four aces,
(b) no hand will contain more than one ace. *(L)*

23 In a television panel game, each player is asked a series of questions until either he answers one incorrectly or he has answered three correctly; his score for the turn is the number of correct answers he has given, with a bonus mark if he gives three correct answers. Assume that the questions are independent and that the chance of giving a correct answer to any question is p. Obtain the mean score for a turn.

Show that for $p = \frac{1}{2}$ the chance that a team of three players will obtain a total score over 3 after one turn each is 25/64. *(C)*

24 Playing a certain 'one-armed bandit', which is advertised to 'increase your money tenfold', costs 5p a turn; the player is returned 50p if more than 8 balls out of a total of 10 drop in a specified slot. The chance of any one ball dropping is c. Determine the chance of winning in a given turn, and for $c = 0{\cdot}65$ calculate the mean profit made by the machine on 500 turns.

Evaluate the proportion of losing turns in which the player comes within one or two balls of winning ($c = 0{\cdot}65$). *(C)*

2 Forces Acting at a Point

We are conscious of having to use muscular effort to move and lift bodies, and we speak of the efforts exerted as 'forces'. We are also aware that sufficient exertion will move a body and that no exertion produces no movement. Therefore, the efforts we loosely call forces are associated with motion and vice versa, and we take as a formal definition of a force 'that which produces motion', so that motion is a consequence of there being a force acting. [The exact relation between force and motion is provided by Newton's second law of motion (§5.1), which states that the force acting on a moving particle is proportional to its acceleration.]

When a body is dropped from rest above the earth, it moves (that is, it falls) and, therefore, there is a force acting on it. This is the force due to gravity and is commonly called the weight of the body.

A body which is at rest is said to be in equilibrium. Statics is effectively the study of the equilibrium of a body. Heavy structures such as the roof of a house are kept in position by using struts and trues. Equilibrium (that is, non-collapse of the roof) means that the total net force exerted at any point is zero and, therefore, forces exerted by the supports counterbalance the known force of gravity (the weight of the roof). It is important to be able to calculate the forces in these supports in order to make sure that they are strong enough for the job in hand. The basic object of the study of statics is to be able to carry out such calculations. To do this it is necessary to quantify and clarify our notion of force more precisely.

2.1 Characteristics of forces

If a small body is attached to a string and the other end of the string is pulled, then the body will move in the direction of the string. Furthermore, the more effort that is exerted on the string, the more rapid the motion. Therefore, since we are assuming force and motion to be related, there is associated with the force acting on a small body both a magnitude and a direction. These are the characteristics of a vector and we take as our basic assumption that force acting at a point is a vector. The vector character of forces also follows from the fact that force is proportional to acceleration. The resultant of two forces acting at a point is defined to be the vector sum of the two separate forces, and the resultant of n forces acting at a point is the vector sum of the n separate forces. The resultant can also be defined as the one force which has the same effect as all the separate forces. It is illustrated experimentally in elementary physics courses that this definition and the vector definition are consistent with each other.

Statics is largely concerned with the interaction of the forces occurring at points of contact of various bodies, and we make the fundamental assumption (Newton's third law, §5.1) that the force exerted by body A on body B at a point of contact is equal in magnitude but opposite in direction to that exerted by body B on body A.

A body which is small enough to be treated as a point is said to be in equilibrium if the resultant of all the forces acting on it is zero. By Newton's law its acceleration would then be zero.

2.2 Measurement of force

We now have enough information to enable us to use experiments to produce a system of measuring force. To do this, we imagine a small body suspended by a piece of elastic string (or a spiral spring) and measure the amount the string is extended when the body hangs freely, and so is in equilibrium. The total force acting is zero and, therefore, the string exerts an upward force equal in magnitude to the downward force of gravity. So, the extension is a measure of the weight of the body. We next imagine a second identical body attached to the first, and notice that the extension has doubled. The force of gravity would, by the law of vector addition, also be twice the original value, and hence there may be a linear relation between force and string extension. This would be confirmed by adding more particles similar to the original ones. Therefore, the extension of a string (or, similarly, the compression and extension of a spiral spring) could be used to produce a method for measuring the magnitude of forces. This could be done by defining the unit of force to be that which produces some particular extension. This is not actually the way the unit of force is defined in practice, but it is important for a deeper understanding of mechanics to realise that force can be measured independently of motion. The line of action of a force is defined to be that line, parallel to the force, which passes through a point at which the force is applied.

We shall take as our unit of force the absolute unit of force which is defined, for any unit system, as the force which produces unit acceleration in unit mass. An interpretation of mass will be given later but for the time being we shall just take it to be some scalar quantity associated with a body. (It is often referred to as the quantity of matter in a body but this is not really a meaningful expression.) We shall use the S.I. system where the units of mass, length and time are the kilogram (kg), the metre (m) and the second (s) respectively. In this system the absolute unit of force is the newton (N).

The advantage of using the absolute system is that if a force $\mathbf{F}$ is acting on a particle of mass m producing an acceleration $\mathbf{a}$ then Newton's second law becomes

$$\mathbf{F} = m\mathbf{a}.$$

A particle falling freely is observed to have a downward acceleration of $9{\cdot}8\,\mathrm{ms}^{-2}$ so that the force $\mathbf{F}$ on it is given by

$$\mathbf{F} = m\mathbf{g} \qquad 2.1$$

where $\mathbf{g}$ is a vector, with positive component in the vertically downward direction and of magnitude $g = |\mathbf{g}| = 9{\cdot}8\ \mathrm{ms}^{-2}$. The vector $\mathbf{g}$ is generally referred to as the acceleration due to gravity.

The magnitude of the force of gravity acting on a particle is known as the weight of the particle. So, in the absolute system, the weight of a particle of mass m is mg.

2.3 Forces in the physical world

In setting up models of practical situations, mathematical idealisations of the various commonly occurring types of forces have to be used. We now describe some of the simplest of these.

Forces in a light inextensible string

Fig. 2.1(a) shows a string AB attached at A to a small body P and pulled by a force of magnitude F at B in the direction AB. We know that P will move along AB and, therefore, the string at A exerts a force on P in the direction of AB. Such a force directed from one end to the other is called tension. Someone pulling at B would experience a force acting on them in the direction of B to A. Therefore, a string exerts a pull (tension) at its extremities as shown in Fig. 2.1(b).

If we take any point C between A and B, then the part AC will exert some force on the part BC and this force will be in the direction CA. The law of action and reaction shows that the part BC will exert a force on AC of equal magnitude but in the direction CB. The situation is as shown in Fig. 2.1(c).

Therefore, a string sustains a tension at all points along its length. A light inextensible string is defined to be such that its length remains constant and the tension is constant throughout its length. This is a consequence of lightness. The tension in a light inextensible string is also assumed to be the same whether it is straight or passing round a smooth

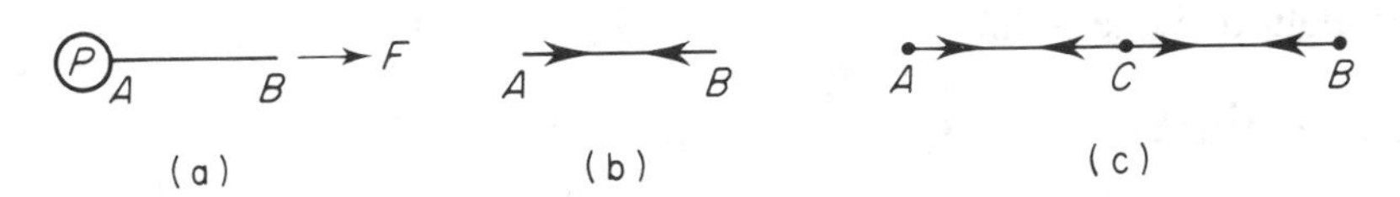

Fig. 2.1

pulley or a smooth peg. In the latter cases, the forces exerted by the string on the peg or pulley are as shown in Fig. 2.2. If the magnitude of the tension is T then, treating the peg as a point, the total force on the peg is $2T$ downwards. [The assumption that the tension is constant for that part of the string in contact with the peg (pulley) is not valid when the latter is rough.]

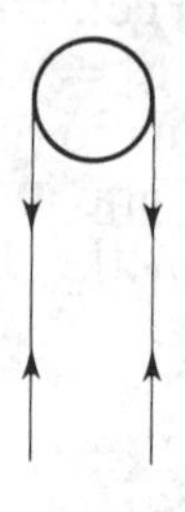

Fig. 2.2

Forces in a light extensible string

A light extensible (elastic) string possesses some of the properties of a light inextensible string in that it can only sustain a tension (that is, it cannot push) and the tension is constant along its length. It differs in that, as the adjective extensible suggests, the length of the string is not constant.

For most elastic strings (and certainly any that we shall consider) there is an experimentally determined relationship between tension and extension, known as *Hooke's law*. This states that when an elastic string of unstretched or natural length l is extended by an amount x then the magnitude of the tension T in the string is given by

$$T = \frac{\lambda x}{l} \qquad 2.2$$

where λ is an experimentally determinable constant known as the *modulus of elasticity* of the particular string, while λ/l is often called the *stiffness*.

Forces in light rods.

A light rod is a rigid connection which can maintain either a tension (that is, it pulls) or a thrust (that is, it pushes). A thrust is effectively a force directed out of the rod at both ends. Both tension and thrust are the same throughout its length.

Light elastic springs

Light elastic springs have all the properties of light elastic strings and, in addition, can sustain either a tension or a thrust.

Reaction of a smooth surface

If a small body A is in contact with a surface S as shown in Fig. 2.3(a) then there will be some force exerted by S on A. A smooth surface is defined to be such that its reaction on A would be normal to S at the point of contact and in the direction from the surface to A – that is, the surface exerts a push but not a pull.

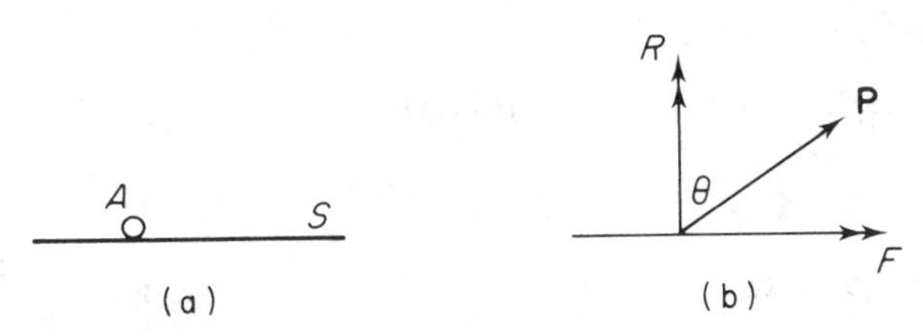

Fig. 2.3

Reaction of a rough surface

A rough surface is effectively one such that its reaction on any body is not necessarily normal to itself, as shown in Fig. 2.3(b). The reaction **P** is inclined at an angle θ to the normal to S. Its components along and perpendicular to S are denoted by F and R respectively. In this and other cases where a force (or any vector) and its components are shown on the same diagram, the latter are indicated by double arrows to avoid giving the impression that both the force and its components are acting. The component F along S is known as the force of friction, and by Newton's law is associated with motion along S.

The behaviour of the force of friction is rather complicated, but from observations we can state the following:

(a) The force of friction acts in the sense so as to prevent motion of a body.

(b) If there is a possibility of A moving along S due to the application of various forces then the force of friction will 'adjust itself' to the minimum value necessary to prevent motion.

(c) The ratio of the magnitude of the friction force to that of the reaction cannot exceed a value greater than a particular constant, which varies from surface to surface. This constant is usually denoted by μ and is called the *coefficient of friction*. In the above example, therefore,

$$\frac{|F|}{R} \leqslant \mu. \qquad 2.3$$

If $|F|/R$ would have to exceed μ to prevent motion, then equilibrium is not possible and there is a relative slipping. When $|F|$ equals μR, the friction is said to be limiting and the bodies are on the point of slipping.

Another quantity associated with friction is the angle of friction, normally denoted by λ and defined by

$$\mu = \tan \lambda.$$

The magnitude of the components of **P** are the lengths of the sides of the rectangle whose sides are parallel and perpendicular to S and whose diagonal is of length P (*Pure Mathematics for Advanced Level*, PMA, p. 169). Therefore,

$$R = P \cos \theta$$

and

$$|F| = P\,|\sin \theta|$$

and equation 2.3 gives

$$|\tan \theta| \leqslant \mu = \tan \lambda.$$

Therefore, another way of viewing the behaviour of the action of rough surfaces is to say that the angle between the reaction and the perpendicular to the surface cannot exceed λ.

Summary:

(a) Light inextensible strings can only sustain a tension which is constant throughout their lengths. In problems, the tension at both ends of any straight portion of string should be shown.

(b) Light extensible strings have all the properties of light inextensible strings except that they are not of fixed length. They also satisfy Hooke's law (equation 2.2).

(c) Light rods can sustain either a constant thrust or tension throughout their lengths. It is normally useful in problems to show in a diagram the forces exerted at the ends of such rods. This means having to assume that, in a particular case, the force is a thrust or a tension. If the wrong assumption is made, all that will happen is that the final answer for the force will be negative, showing that the assumption was unsound. It is impossible to obtain a negative tension in a string.

(d) Light springs have all the properties of light extensible strings except that they can sustain both tensions and thrusts.

(e) Smooth surfaces exert a reaction normal to, and away from, themselves. The reaction of a smooth surface cannot be negative.

(f) The reaction of a rough surface is not normally perpendicular to itself, but is inclined to the normal at an angle less than the angle of friction. The force of friction is the least possible one necessary to maintain equilibrium.

In friction problems, it is generally best not to assume that a system is in limiting equilibrium, but to calculate the normal and tangential force and then apply the condition $|F| \leqslant \mu R$. If a problem states the equilibrium is limiting, then it is safe to use $|F| = \mu R$, but it is then necessary to see in which direction the motion would take place so that F is shown in the opposite direction.

2.4 Several forces acting at a point

We are now in a position to be able to solve statical problems where several forces act at a point. In general, problems are likely to fall into one of two classes:

(i) those in which the separate forces are given explicitly and the resultant is to be found;

(ii) those in which there are known and unknown forces acting at a point and either the forces are in equilibrium or, occasionally, the resultant is a given force.

Both of these types are really quite similar and just require addition of forces at a point – that is, the addition of vectors. The first type simply requires straightforward addition, whilst the second requires addition and then a comparison of the result with the given resultant (zero for equilibrium) to give equations for the unknown forces.

Adding vectors can be done using the triangle or parallelogram rules (PMA, p. 74) or the ratio theorem (PMA, p. 82), but this can become complicated. It is generally far easier to add the components in two perpendicular directions together. This involves three separate steps:

(i) Finding the components of a given force;

(ii) Adding the separate components to get the components of the total force;

(iii) Constructing the total force from its two components.

Step (ii) is merely algebra, and we shall concentrate on the first and third steps. When starting to solve problems, it is useful to set down clearly the unit vectors in the two perpendicular directions chosen. This is done in the following few problems, where **i** and **j** denote two perpendicular unit vectors. As one gets more familiar with solving problems, it is not as necessary to set down these vectors.

Finding the components of a force in two perpendicular directions

When there are several forces acting, the safest way to proceed is to find the component of each force in two directions which are parallel to the basic directions chosen (but not necessarily in the same sense) and which both make acute angles with the direction of the force considered. The magnitude of the components of a vector **P** in two perpendicular directions are the lengths of the sides of a rectangle whose sides are parallel to the given directions and whose diagonal is represented by **P**. Therefore, from Fig. 2.4(a) the components of **P** in the direction of **i** and **j** are $P\cos\theta$ and $P\sin\theta$ respectively, and

$$\mathbf{P} = (P\cos\theta)\mathbf{i} + (P\sin\theta)\mathbf{j}.$$

The corresponding quantities for the force **Q** are shown in Fig. 2.4(b), but

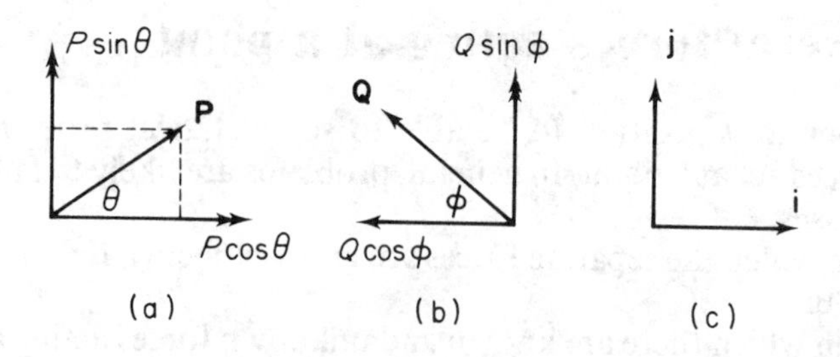

Fig. 2.4

since $Q \cos \phi$ is in the direction opposite to **i**

$$\mathbf{Q} = -(Q \cos \phi)\mathbf{i} + (Q \sin \phi)\mathbf{j}.$$

In this way, the components of each force can be found, taking care to obtain the correct sign for each component.

Determining a force given two perpendicular components

If the components of the force **R** in the directions of the unit vectors **i** and **j** are X and Y respectively then it is sufficient to write

$$\mathbf{R} = X\mathbf{i} + Y\mathbf{j},$$

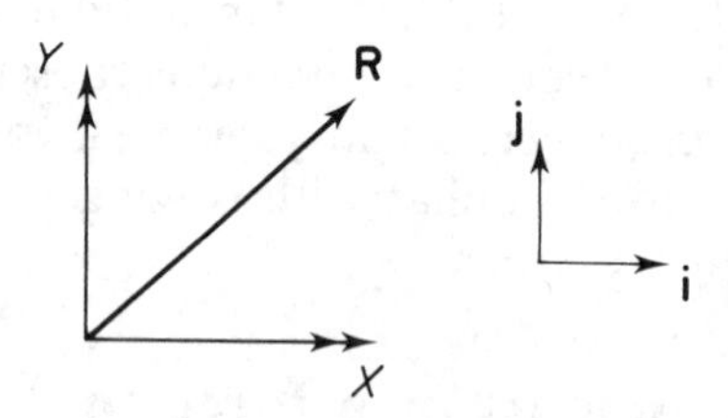

Fig. 2.5

provided that **i** and **j** have been clearly defined. Sometimes, however, the magnitude and direction of a force are asked for specifically. The magnitude of **R** is by definition $\sqrt{(X^2 + Y^2)}$ and it acts at an angle θ to **i**, where

$$\tan \theta = \frac{Y}{X}.$$

EXAMPLE 1 *Find the resultant of the forces of magnitude* 3 N *and* 4 N *acting as shown in Fig.* 2.6(*a*).

Using the method described above we find that the forces have components as

3 N
60°
4 N
4 N
4 N
(a)
j
i
(3 sin 60°) N
3 N
60°
(3 cos 60°) N
(b)

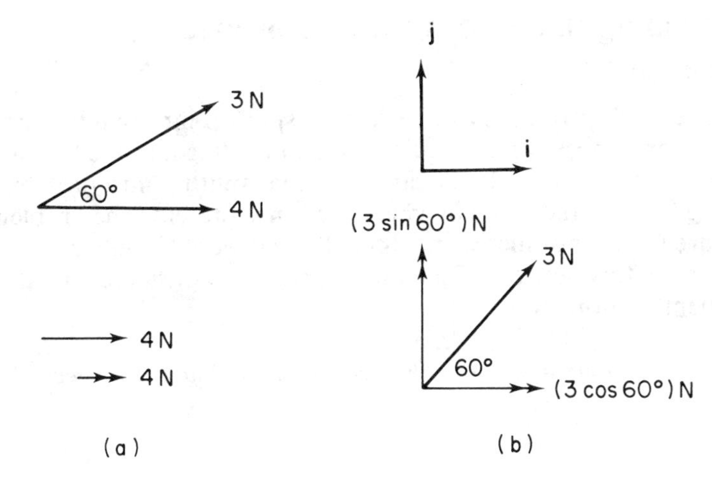

Fig. 2.6

shown in Fig. 2.6(b). Adding the components, the resultant **R** is given by

$$\mathbf{R} = (4 + 3\cos 60^\circ)\mathbf{i} + 3\sin 60^\circ\,\mathbf{j}$$
$$= \frac{11}{2}\mathbf{i} + \frac{3\sqrt{3}}{2}\mathbf{j}.$$

The magnitude of the resultant is $(37)^{\frac{1}{2}}$ N and it acts at an angle $\tan^{-1}(3\sqrt{3}/11)$ to **i**.

EXAMPLE 2 *Find the resultant of the forces shown in Fig. 2.7(a).*

Each force can be split into components as shown in Fig. 2.7(b). The first force has a negative component in the direction of **i**, whilst the third force has a negative component in the direction of **j**. So, the resultant **R** N is given by

$$\mathbf{R} = (3\cos 30^\circ + 5 - 7\cos 60^\circ)\mathbf{i} + (7\sin 60^\circ - 3\sin 30^\circ)\mathbf{j}$$
$$= \left(\frac{3\sqrt{3}}{2} + \frac{3}{2}\right)\mathbf{i} + \left(\frac{7\sqrt{3}}{2} - \frac{3}{2}\right)\mathbf{j}.$$

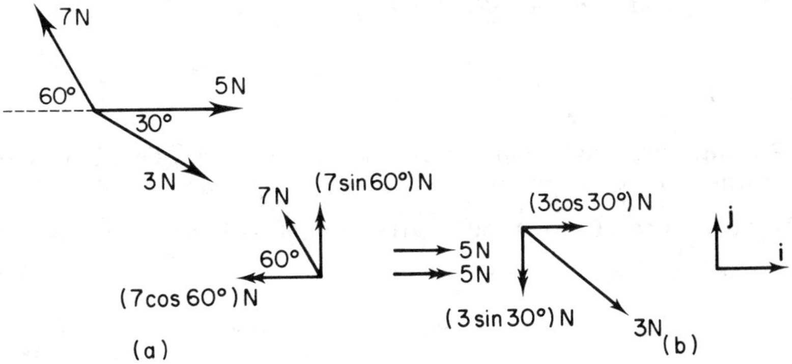

7N
60°
5N
30°
3N
(a)
7N
(7sin 60°) N
60°
(7cos 60°) N
5N
5N
(3cos 30°) N
(3 sin 30°) N
3N
j
i
(b)

Fig. 2.7

Determining unknown forces to produce a given resultant

In essence, such problems can be further split up into purely numerical problems, and those which require some modelling and the use of given physical conditions. The latter can be further split up into problems not involving friction and those in which friction is present. The problems, as stated, are in order of increasing difficulty and we shall only consider the first type in this section. The other types are considered in the two subsequent sections.

EXAMPLE 3 *Find R such that the forces shown in Fig. 2.8 are in equilibrium.*

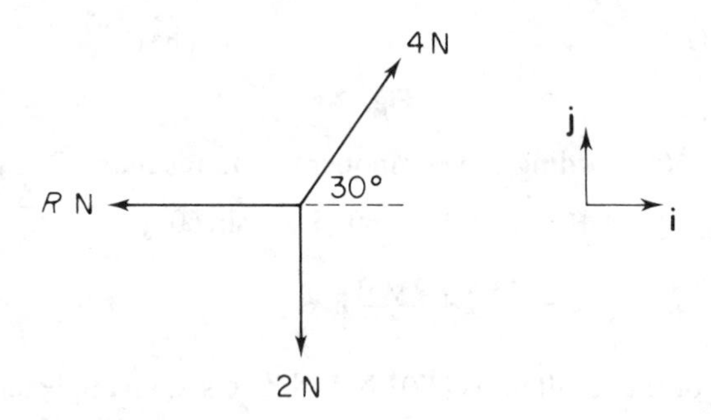

Fig. 2.8

The force of magnitude 4 N has components ($4\cos 30°$) N and ($4\sin 30°$) N to the right and up the page respectively. The unknown force has a component of magnitude R N to the left and the third force has a component of magnitude 2 N down the page. Paying attention to the senses of the various forces, the resultant is given by

$$\mathbf{R} = [(2\sqrt{3} - R)\mathbf{i} + (4\sin 30° - 2)\mathbf{j}]\,\text{N}.$$

For equilibrium both components have to be zero. The $\mathbf{j}$ component vanishes automatically and, therefore, $R = \mathbf{2\sqrt{3}}$.

EXAMPLE 4 *Find P and Q such that the system shown in Fig. 2.9(a) is in equilibrium.*

The separate forces have components as shown in Fig. 2.9(b). On taking account of the senses of the components, the resultant is equal to

$$[(P\cos 60° + 3\cos 30° + 6\cos 60° - Q)\mathbf{i} + (P\sin 60° + 3\sin 30° - 6\sin 60°)\mathbf{j}]\,\text{N}.$$

Both components have to be zero so

$$P = 2Q - 6 - 3\sqrt{3},\ P\sqrt{3} = 6\sqrt{3} - 3$$

This gives

$$P = \mathbf{6 - \sqrt{3}},\ Q = \mathbf{6 + \sqrt{3}}.$$

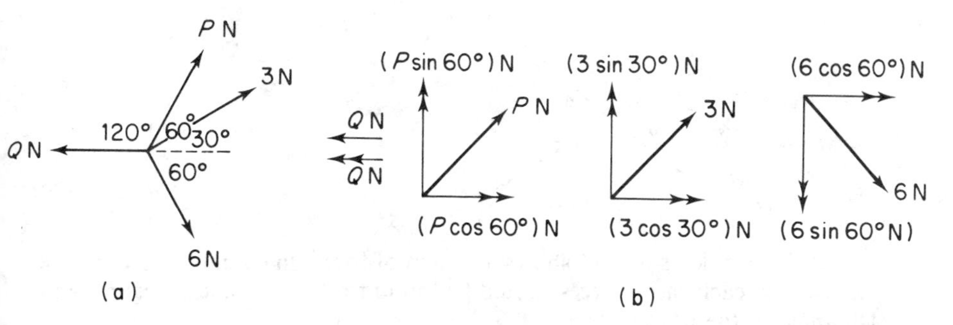

Fig. 2.9

As familiarity with using components develops it will not be necessary to split up the forces separately into components as shown here. Explicit use of **i** and **j** can also be avoided – for example, in the above the result could have been obtained by setting the components to the left (or upwards) equal to those to the right (or downwards).

EXERCISE 2.4

In each of the problems 1 to 5, the unit vectors **i** and **j** are defined to be across the page (from left to right) and up the page, respectively. Find the components, referred to the vectors **i** and **j**, of the resultants of the force systems shown. Find also the magnitude of the resultant and, taking the anti-clockwise sense to be positive, the angle the resultant makes with **i**. Forces should be given to three significant figures and angles to the nearest degree.

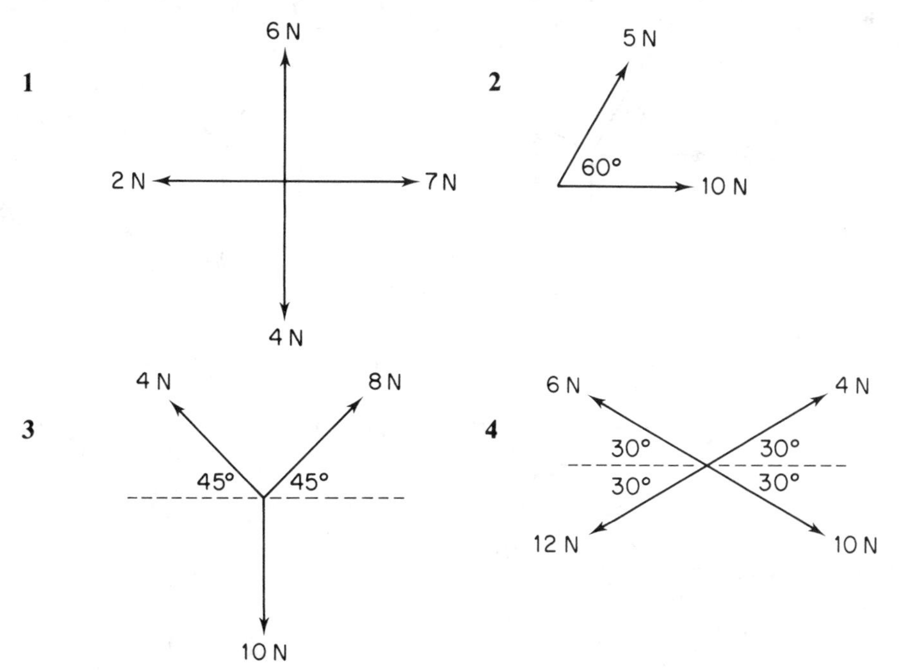

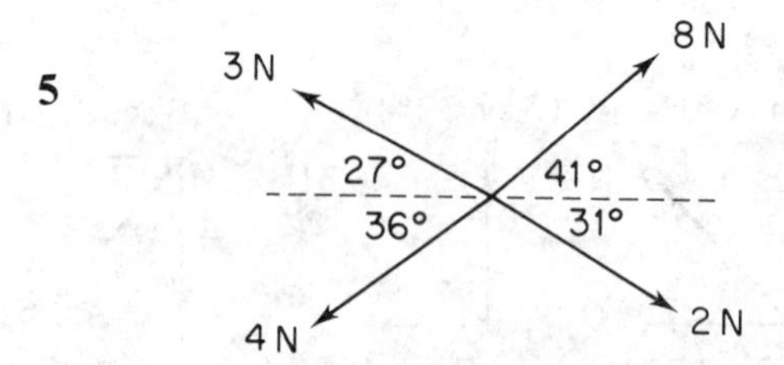

Each of the problems 6 to 11 shows a system of forces in equilibrium. Find the unknowns in each case. Forces should be found to three significant figures and each angle to the nearest degree.

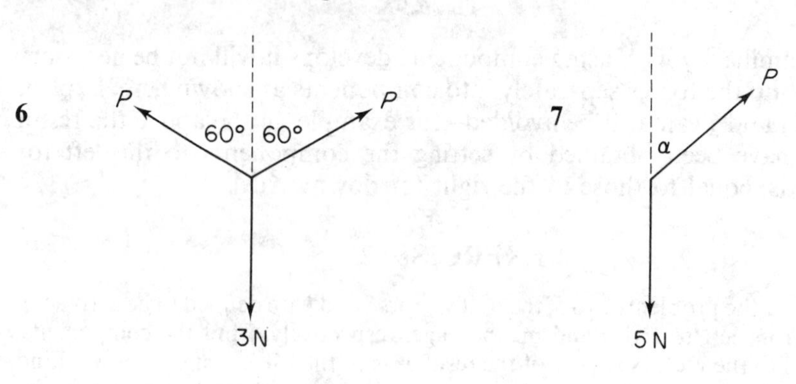

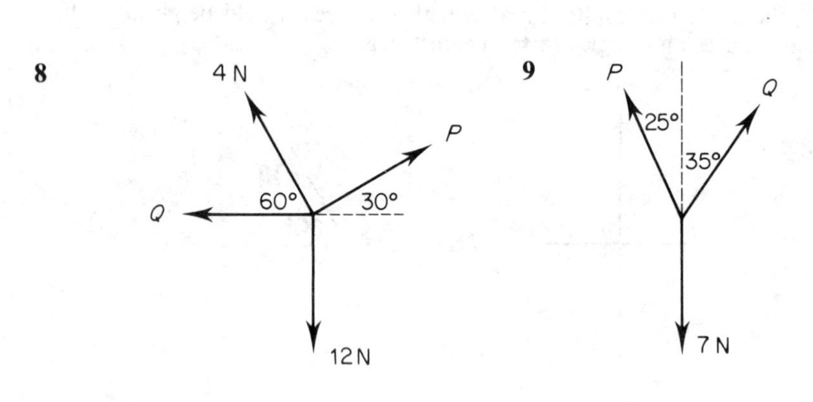

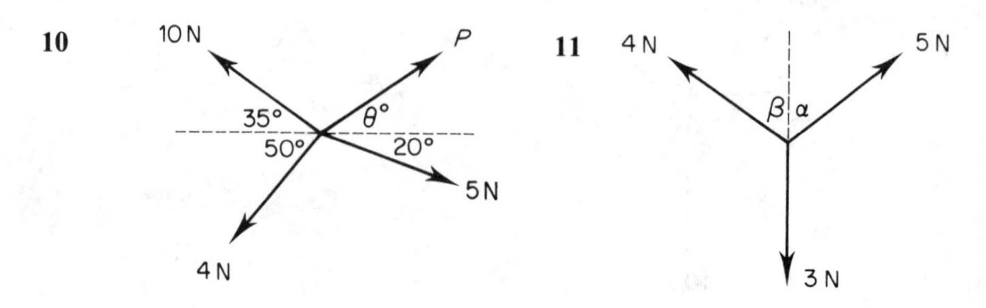

2.5 Problems involving physical modelling

The previous problems have all been relatively straightforward in that the various forces were prescribed exactly. In practical situations this is not so and it will be necessary to determine, from the statement of the problem, the nature of the forces acting. Most problems involve contact between bodies and the type of force depends on the nature of the contact.

The first step is to use the basic definitions to obtain all the forces acting at each point, and then to mark these forces clearly on a diagram. For strings, it is particularly important to mark the tensions at both ends of each straight part of each string. In drawing diagrams for particular problems it can be confusing to underline or use bold type for vector quantities. It generally suffices to use an arrow which is clearly associated with a letter or a number. This may denote force, or a component of a force, and the direction is indicated by the arrow as in Fig. 2.10. It should be remembered that the tension is the same throughout a given string but need not be the same in two strings tied to the same point.

For two bodies A and B in contact, it should be remembered that the force of A on B is equal and opposite to that of B on A. Once the model has been set up, the calculation is performed as shown in the previous section. Usually forces are required which produce equilibrium and, therefore, the unknown forces have to be such as to give a zero resultant. The reader should by now be reasonably confident about calculating resultants and can possibly stop using the basic vectors **i** and **j** to calculate explicit components. The equilibrium can be found, without use of **i** and **j**, by equating the components acting to the left (down) to the components acting to the right (up). This process of equating components is referred to as resolving in the two separate directions. The horizontal and vertical directions are not always the best ones in which to resolve and many problems are greatly simplified by resolving in other directions. For example, if there are several forces acting on a particle in equilibrium with two of the forces being unknown, then resolving perpendicular to one of the unknown forces automatically eliminates that force and allows the other one to be found immediately.

For the particular case where there are only three separate forces acting at a point and they are in equilibrium, it is possible to circumvent the use of components and solve the problem directly by using the law of addition of vectors. Though this appears to be a simpler way it can involve trickier manipulation occasionally. However, as it is sometimes useful, we shall describe it here. The basic result is usually referred to as Lami's theorem.

Lami's theorem

If three forces acting at a point are in equilibrium, then each is proportional to the sine of the angle between the other two.

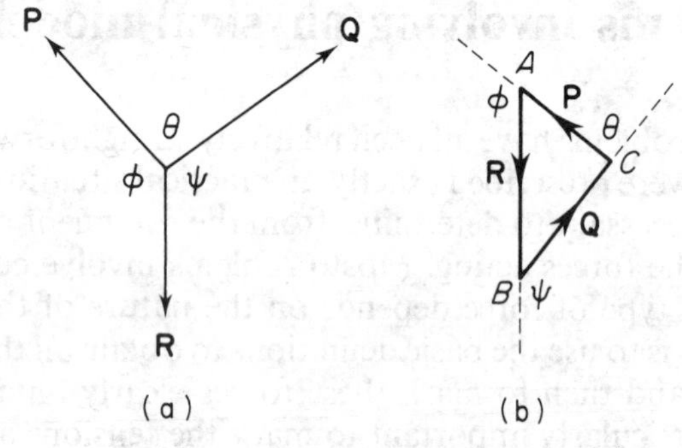

Fig. 2.10

Proof

If the three forces shown in Fig. 2.10 (a) are in equilibrium, then any one of them is minus the resultant of the other two or, equivalently

$$\mathbf{P}+\mathbf{Q}+\mathbf{R}=\mathbf{0}.$$

The vectors **P**, **Q** and **R** can, therefore, be represented diagrammatically as shown in Fig. 2.11(b). Applying the sine rule to triangle ABC

$$\frac{P}{\sin\psi}=\frac{Q}{\sin\phi}=\frac{R}{\sin\theta},$$

which is Lami's theorem.

EXAMPLE 1 *A particle of weight W is suspended in equilibrium from the end B of a light inextensible string AB. Find the tension in the string and the force necessary to hold the string at A.*

The forces acting at A and B are shown in Fig. 2.11. It is assumed that the force at A is of magnitude F at an angle θ to the vertical. The points A and B are both in equilibrium. At B the only forces acting are the tension and the weight and therefore

$$T=W.$$

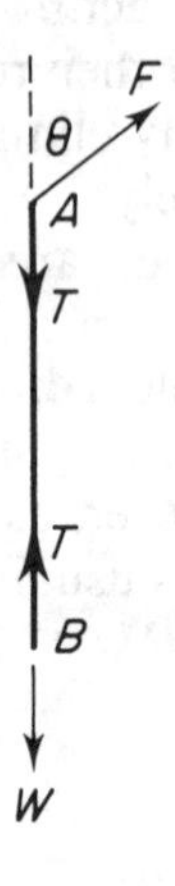

Fig. 2.11

The only horizontal force at A is $F \sin\theta$ and so for equilibrium

$$F \sin\theta = 0.$$

Considering the vertical components gives

$$F \cos\theta = T.$$

Clearly F cannot be zero otherwise we would have W equal to zero. Hence, θ has to be 0 or 180°. The latter would give $T = -W$ which is impossible. Therefore, $\theta = 0$ and

$$\boldsymbol{F = T = W}.$$

The above solution is rather long-winded since we assumed that the force at A could be in any direction. Common sense decrees that the force has to be applied upwards. It is often relatively obvious in which direction a particular force acts. However, where there is doubt the most general assumption should be made, and careful application of the equilibrium conditions will yield the correct direction.

EXAMPLE 2 *A small smooth ring R of mass m is threaded on a light inextensible string of length* $2l$. *The ends of the string are attached to two points in a horizontal line and at a distance* $2a$ $(a < l)$ *apart. The system is in equilibrium in a vertical plane. Find the tension in the string.*

The general situation is as shown in Fig. 2.12.

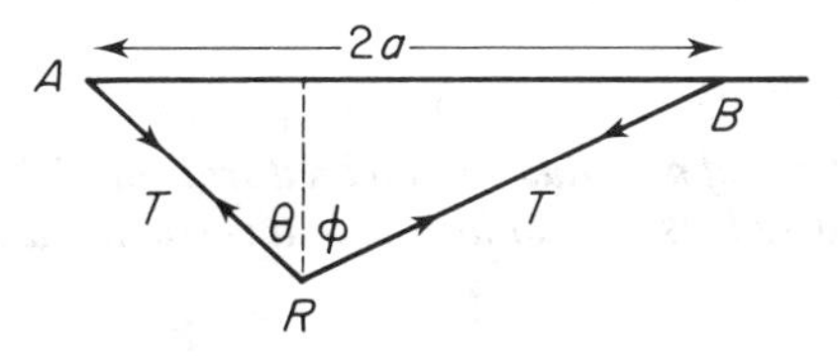

Fig. 2.12

There is no reason to assume immediately that both strings are inclined at the same angle to the horizontal, though symmetry suggests this. Considering the horizontal component of the force acting on R gives

$$T \sin\theta = T \sin\phi$$

So $\theta = \phi$ and $RA = RB$.

The force of gravity is mg and, hence, resolving vertically

$$2T \cos\theta = mg.$$

The depth of R below AB is, by Pythagoras, $\sqrt{(l^2 - a^2)}$ and, therefore,

$$\boldsymbol{T = \frac{mg}{2\cos\theta} = \frac{mgl}{2(l^2 - a^2)^{\frac{1}{2}}}}.$$

EXAMPLE 3 *Two light inextensible strings are each tied to a particle of mass m, and their other ends are attached to two points in a horizontal line so that the particle is in equilibrium with the strings inclined at angles of* 30° *and* 60° *to the horizontal. Find the tensions in the strings.*

Resolving horizontally we have

$$T_2 \cos 60^\circ = T_1 \cos 30^\circ$$

that is, $$T_2 = T_1\sqrt{3}.$$

Resolving vertically gives

$$T_1 \sin 30^\circ + T_2 \sin 60^\circ = W$$

that is, $T_1 = \dfrac{W}{2}$, and therefore $T_2 = W\sqrt{3}/2$.

A slightly quicker method would have been to resolve perpendicularly to the string with tension T_1. This gives $T_2 = W \sin 60^\circ = W\sqrt{3}/2$.

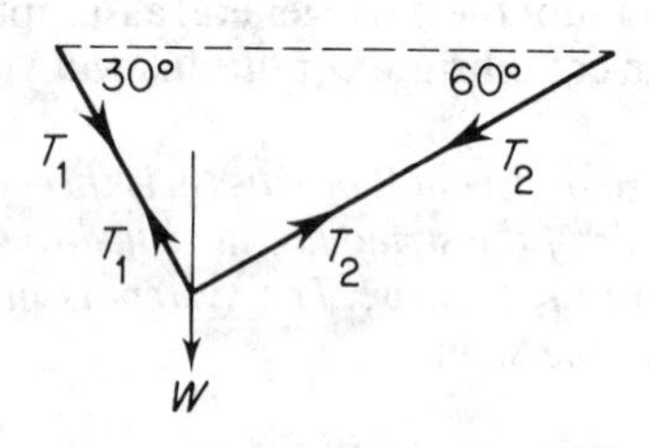

Fig. 2.13

EXAMPLE 4 *A particle of weight W is suspended from the end of a light extensible (elastic) string, of modulus 4 W and natural length l. Find the extension of the string when the particle is in equilibrium in a vertical plane with one end of the string held fixed.*

So far as the equilibrium condition is concerned, this example is exactly the same as the previous one, with $T = W$. For an elastic string, however, the extension x is related to the tension T by Hooke's law (equation 2.2) which gives

$$T = W = \frac{4Wx}{l}.$$

Therefore $$x = \frac{1}{4}l.$$

EXAMPLE 5 *Two small particles, each of weight W, are attached to the ends of a light inextensible string. The string passes over a small smooth peg and the particles are in equilibrium in a vertical plane. Find the force that has to be applied at the peg in order to maintain equilibrium.*

The forces acting at the peg and the particles are as shown in Fig. 2.14. The strings do not exert a horizontal force on the peg and, therefore, the force applied at the peg has no horizontal component. Equilibrium of the particles gives, as before, $T = W$ and equilibrium of the peg gives

$$F = 2T = 2W.$$

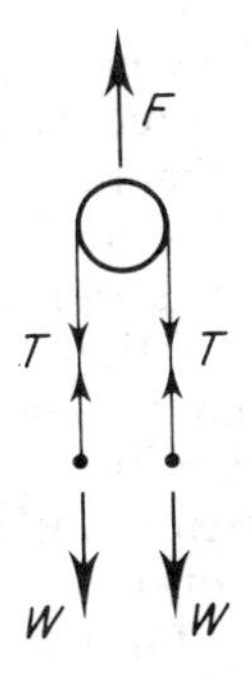

Fig. 2.14

EXERCISE 2.5

In numerical examples, forces should be found to three significant figures and g should be taken as $10\,\text{ms}^{-2}$.

Questions 1 to 3 refer to a body of weight W N suspended from a fixed point by a light elastic string of natural length a m, elastic modulus λ N. The extension is denoted by x m.

1 Find x, given $W = 20$, $a = 2$, $\lambda = 100$.

2 Find λ, given $W = 40$, $x = 0.1$, $a = 4$.

3 Find a, given $W = 40$, $x = 0.2$, $\lambda = 200$.

Questions 4 and 5 refer to a particle of mass 2 kg suspended by a light inextensible string, the other end of which is attached to a fixed point.

4 The particle is acted on by a horizontal force so that it is in equilibrium with the string inclined at an angle of 45° to the downward vertical. Find the force.

5 The particle is maintained in equilibrium with the string inclined at an angle of 30° to the downward vertical by a force acting perpendicular to the string. Find the force and the tension in the string.

6 A light elastic string AB, of natural length 1·5 m and modulus 150 N, has the end A fixed and a heavy particle attached to the end B. A horizontal force of magnitude P is then applied at B so that the system is in equilibrium with AB taut and inclined at an angle $\tan^{-1}(4/3)$ to the downward vertical and at a depth of 1·2 m below A. Find the tension in the string and the value of P.

7 A particle of mass 2 kg is suspended by two light inextensible strings from two fixed points on the same horizontal level. The strings are inclined at angles 20° and 40°, respectively, to the horizontal. Find the tensions in the strings.

8 A particle of mass 1 kg is suspended by two light inextensible strings of lengths 6 m and 8 m from two points on the same horizontal level and at a distance of 10 m apart. Find the tensions in the strings.

9 Two strings, attached to a particle P of mass M, pass over two smooth pegs at the same level and hang vertically in equilibrium with masses m and $2m$ at their ends. Given that the strings at P are perpendicular to each other, find M/m.

10 A ring R of mass m slides on a smooth vertical wire. A light inextensible string attached to it passes over a small smooth peg P and a particle of mass $4m$ is attached to the other end of the string. Find the inclination of PR to the vertical when the system is in equilibrium.

11 A smooth ring C of mass m slides on a light inextensible string whose ends A and B are fixed at two points on the same level. A horizontal force of magnitude F is applied at C so that the ring is in equilibrium vertically below B with AC inclined at an angle θ to the vertical. Find F.

12 A smooth ring R of mass $4m$ is threaded on a smooth wire in the form of a circle of radius a and fixed in a vertical plane. One end of a light inextensible string is attached to the ring and the other fixed at A, the highest point of the wire. The ring is in equilibrium, with AR inclined at an angle θ to the downward vertical. Find the tension in the string.

13 In the configuration of question 12, the string is no longer fixed at A but passes over a small smooth peg at A, and a particle of mass $8mg/5$ is attached to the free end of the string and allowed to hang freely in equilibrium. Find the length of AR.

2.6 Problems involving friction

The same basic principles have to be applied in order to solve problems involving friction, the additional difficulty now being that the ratio of the tangential to normal force is restricted (or, equivalently, the reaction must lie within a certain angular sector of the vertical). In such problems it is best, as has been mentioned earlier, to assume arbitrary values of the friction force (F) and the normal reaction (R) and obtain all the statical equations. It is after this stage that the condition on $|F|/R$ should be imposed. If a question specifically asks for consideration of limiting equilibrium, it is then in order to set $|F| = \mu R$.

EXAMPLE 1 *A particle P of mass $4m$ rests on a rough horizontal table, with coefficient of friction $1/6$. The particle is attached by a light inextensible string to a second particle Q. The string passes over a smooth pulley at the edge of the table and Q hangs in equilibrium. Find the greatest possible mass of Q.*

We assume that Q has mass M and that the forces acting at the various points are as shown in Fig. 2.15. We have assumed that the force of friction is acting to the left since motion, if it occurs, will be to the right.

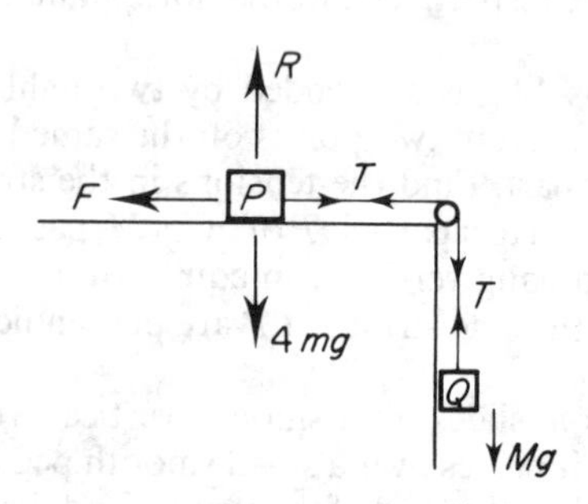

Fig. 2.15

Resolving horizontally and vertically for P gives

$$R = 4\,mg$$
$$F = T$$

Resolving vertically for Q gives

$$T = Mg.$$

Therefore

$$F = Mg.$$

If we had assumed that the friction force was acting to the right, we would have obtained a negative value showing that it was acting to the left. Applying the condition on the force of friction gives

$$\frac{M}{4m} \leqslant \frac{1}{6}$$

so that $\boldsymbol{M \leqslant 2m/3}$.

EXAMPLE 2 *A heavy particle of weight W is placed on a rough plane inclined at an angle α to the horizontal. The coefficient of friction between the plane and the particle is μ. Show that equilibrium is not possible unless* $\tan\alpha \leqslant \mu$. *Find, when* $\tan\alpha > \mu$, *the least value of the magnitude of the force acting up the line of greatest slope of the plane which will maintain equilibrium.*

The forces acting on the particle are as shown in Fig. 2.16. In problems involving inclined planes, it is slightly more convenient to resolve along and perpendicular to the plane, and the components of the force of gravity in these directions are shown. Resolving in these directions gives

$$F = W\sin\alpha, \quad R = W\cos\alpha.$$

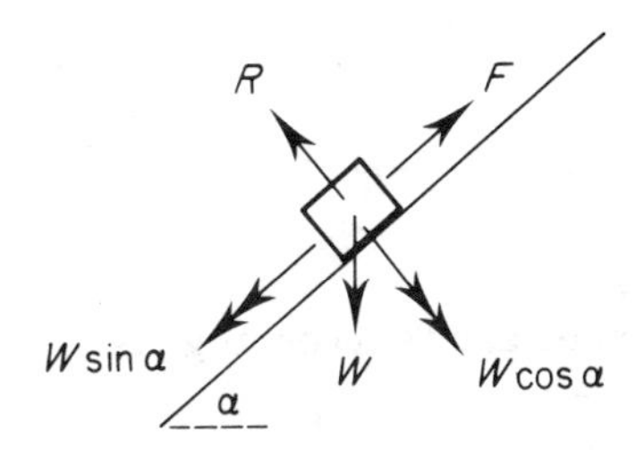

Fig. 2.16

Therefore

$$\frac{F}{R} = \tan\alpha$$

and therefore $\boldsymbol{\tan\alpha \leqslant \mu}$ **for equilibrium**.

If an additional force of magnitude P acts up the plane, we have

$$F + P = W\sin\alpha$$

so that
$$\frac{F}{R} = \tan\alpha - \frac{P}{W\cos\alpha}.$$

Therefore, for equilibrium,
$$\tan\alpha - \frac{P}{W\cos\alpha} \leqslant \mu$$

so that
$$P \geqslant W(\sin\alpha - \mu\cos\alpha).$$

The least value of P is therefore $\boldsymbol{W(\sin\alpha - \mu\cos\alpha)}$.

Problems of this type can often be solved just as easily by Lami's theorem, as illustrated in the following example.

EXAMPLE 3 *Find, for the case described in the previous problem, the greatest value of the force that can be applied up the line of greatest slope without disturbing equilibrium.*

As we intend to solve this problem using Lami's theorem, the reaction of the plane

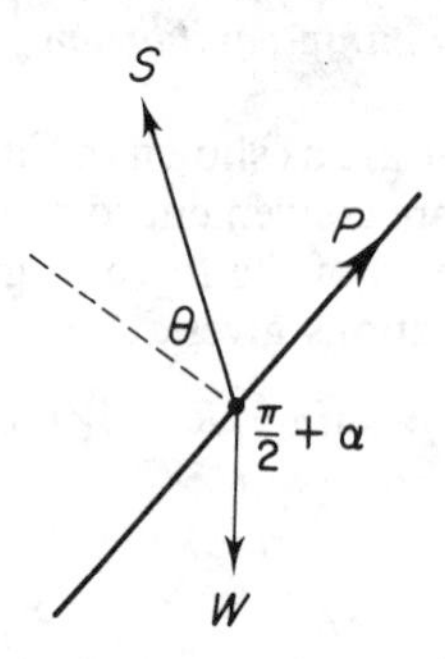

Fig. 2.17

is not split into components. Lami's theorem gives
$$\frac{P}{\sin(\pi + \theta - \alpha)} = \frac{W}{\sin\left(\dfrac{\pi}{2} - \theta\right)} = \frac{S}{\sin\left(\dfrac{\pi}{2} + \alpha\right)}$$

that is,
$$\frac{P}{\sin(\alpha - \theta)} = \frac{W}{\cos\theta} = \frac{S}{\cos\alpha}.$$

Therefore
$$P = W[\sin\alpha - \cos\alpha\tan\theta].$$

It follows that
$$-\mu \leqslant \tan\theta \leqslant \mu$$

and therefore
$$W(\sin\alpha - \mu\cos\alpha) \leqslant \boldsymbol{P} \leqslant \boldsymbol{W(\sin\alpha + \mu\cos\alpha)}.$$

The maximum value of P is given by the right hand part of this inequality.

EXAMPLE 4 *Two small rough rings A and B, of weights 2W and W respectively, slide on a fixed, rough, horizontal rod. The coefficient of friction between each rod and the ring is μ. A light inextensible string of length 2a is threaded through a smooth ring of weight W and its ends are attached to A and B. The whole rests in equilibrium in a vertical plane. Find the greatest distance apart of A and B.*

The general situation is as shown in Fig. 2.18. The fact that both strings are equally inclined to the horizontal follows as shown in example 2, §2.5. For equilibrium at A, B and C

$$R = 2W + T\cos\theta, \quad T\sin\theta = F_1,$$
$$S = W + T\cos\theta, \quad T\sin\theta = F_2,$$
$$2T\cos\theta = W.$$

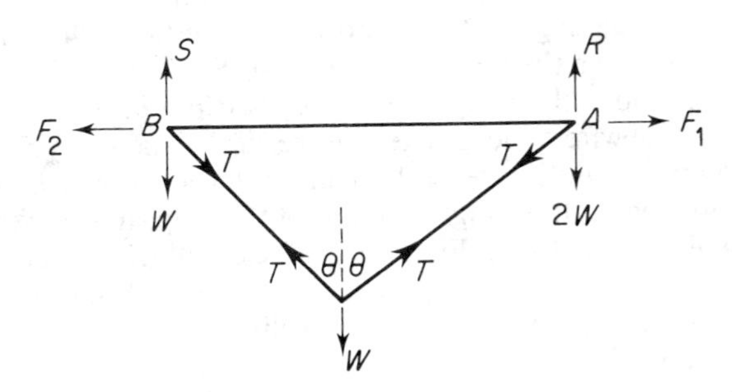

Fig. 2.18

Therefore $$F_1 = \frac{W}{2}\tan\theta = F_2 = F \text{ (say).}$$

$$R = \frac{5W}{2}, \quad S = \frac{3W}{2}.$$

Since $S < R$, it follows that for equilibrium

$$\frac{F}{S} \leqslant \mu$$

that is, $\tan\theta \leqslant 3\mu$. [This means that slipping first takes place at B].

The distance between A and B is $2a\sin\theta$. Since $\tan\theta \leqslant 3\mu$, $\sin\theta \leqslant 3\mu/(9\mu^2+1)^{\frac{1}{2}}$ and so

$$\mathbf{AB \leqslant \frac{6a\mu}{(9\mu^2+1)^{\frac{1}{2}}}}.$$

EXERCISE 2.6

In numerical examples, g should be taken as $10\,\text{ms}^{-2}$ and answers given correct to three significant figures.

1 A particle of mass 4 kg is in equilibrium on a rough horizontal plane, the coefficient of friction being 0·5. Find the least force which, acting (i) horizontally, (ii) at an angle of 30° to the downward vertical, would just move the body along the plane.

2 A particle of mass 2·5 kg, at rest on a rough horizontal plane, can just be moved by a horizontal force of 10 N. Find the coefficient of friction.

3 A particle of mass 2 kg is placed on a rough plane inclined at an angle $\sin^{-1} 3/5$ to the horizontal. The coefficient of friction between the plane and the particle is 0·25. Find the least force acting along a line of greatest slope of the plane required (i) to prevent the particle from sliding down, (ii) to pull it up the plane.

4 A particle of mass 5 kg is on the point of sliding down a rough inclined plane when a force 10 N is applied up the plane along a line of greatest slope. When the force is increased to 20 N the particle is on the point of moving up the plane. Find the coefficient of friction.

5 A body of mass m is to be moved along a rough horizontal plane by a force applied at an acute angle θ to the downward vertical. Show that, if $\theta < \lambda$, where $\mu = \tan \lambda$, then the body will not move. Show, for $\theta > \lambda$, that the least force that will move the body is $mg \sin \lambda \operatorname{cosec} (\theta - \lambda)$.

6 A circular loop of wire, which is fixed in a vertical plane, has a ring R of mass m threaded on it. A light inextensible string fastened to the ring passes over a small smooth peg at A, the highest point of the loop, and carries a particle of mass $\frac{1}{2}m$, which hangs freely. Find the coefficient of friction between the ring and the wire given that the ring is in limiting equilibrium on the point of slipping down the loop when AR is inclined at an angle of 30° to the downward vertical.

7 A particle of mass m can be just supported on a rough inclined plane by a horizontal force amg. It can also be just supported by a force bmg, acting up the plane along a line of greatest slope. Find, in terms of a and b, the cosine of the angle of friction.

MISCELLANEOUS EXERCISE 2

1 Two coplanar forces **P** and **Q** have magnitudes 15 and 7 units respectively and the magnitude of their resultant is 20 units. Draw a diagram showing the vectors representing the forces **P** and **Q** and their resultant.

Calculate the angle between the forces **P** and **Q** to the nearest degree.

The force **Q** is replaced by a force **S** equal to $-k.\mathbf{Q}$, where k is a positive constant. The resultant of the forces **P** and **S** is at right angles to the force **P**. Calculate

(a) the magnitude of the force **S**,
(b) the value of the constant k. *(L)*

2 A particle is in equilibrium under the action of three forces **P**, **Q** and **R**. Given that $\mathbf{P} = (3\mathbf{i} + 5\mathbf{j})$, $\mathbf{Q} = (-2\mathbf{i} + 6\mathbf{j})$, calculate

(a) the magnitude of **R**,
(b) the tangent of the acute angle made by the line of action of **R** with the positive x-axis. *(L)*

3 The forces $\mathbf{F}_1$ and $\mathbf{F}_2$, where $\mathbf{F}_1 = (\mathbf{i} + 3\mathbf{j})$ and $\mathbf{F}_2 = (2\mathbf{i} + \mathbf{j})$, both act through the point with position vector $(\mathbf{i} + \mathbf{j})$. Find the resultant of the two forces and show that its magnitude is 5. Obtain, in both vector and Cartesian form, equations for the line of action of this resultant. Determine the cosine of the angle between the direction of the forces $\mathbf{F}_1$ and $\mathbf{F}_2$. *(L)*

4 Referred to O as origin, the position vectors of the points A and B are $(5\mathbf{i}+12\mathbf{j})\,m$ and $(16\mathbf{i}+12\mathbf{j})\,m$ respectively.
(i) Find, in terms of $\mathbf{i}$ and $\mathbf{j}$, the unit vectors along $\overrightarrow{OA}$ and $\overrightarrow{OB}$.
Force $\mathbf{F}_1$ of magnitude 91 N acts along OA and force $\mathbf{F}_2$ of magnitude 80 N acts along OB.
(ii) Express $\mathbf{F}_1$ and $\mathbf{F}_2$ in terms of $\mathbf{i}$ and $\mathbf{j}$ and hence calculate the magnitude of their resultant.
(iii) Show that the line of action of this resultant passes through the point C whose position vector is $(9\mathbf{i}+12\mathbf{j})m$. *(AEB 1981)*

5 (i) A force of magnitude 10 N parallel to the vector $4\mathbf{i}+3\mathbf{j}$ is the resultant of two forces parallel respectively to the vectors $2\mathbf{i}+\mathbf{j}$, $\mathbf{i}+\mathbf{j}$. Find the magnitudes of these two forces. *(L)*

6 Forces of magnitude P and Q act along lines OA and OB respectively, and their resultant is a force of magnitude P; if the magnitude of the force along OA is changed to $2P$ the resultant is again a force of magnitude P. Find
(i) Q in terms of P,
(ii) the angle between OA and OB,
(iii) the angles which the two resultants make with OA. *(W)*

7 If any number of forces, acting at a point, can be represented in magnitude and direction by the sides of a polygon taken in order, prove that the forces are in equilibrium.

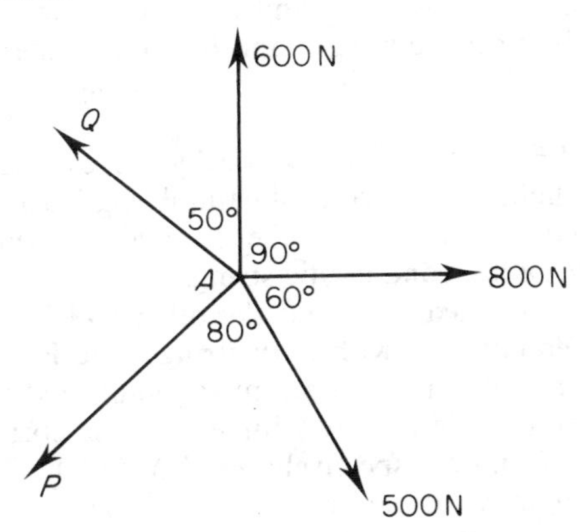

(i) The five coplanar forces acting at the point A are in equilibrium. Find graphically, or otherwise, the values of P and Q to the nearest 5 N.
(ii) If $P = 555$ N and $Q = 815$ N, find graphically, or otherwise, the magnitude and the direction of the resultant of the five forces. *(AEB 1973)*

8 With reference to Cartesian axes Ox, Oy, the vertices of a square are $A(a, a)$, $B(-a, a)$, $C(-a, -a)$, $D(a, -a)$. A particle $P(x, y)$ is subject to forces acting along the lines joining P to the four vertices.
(i) In the case when the forces are $k\overrightarrow{PA}$, $k\overrightarrow{BP}$, $k\overrightarrow{PC}$, $k\overrightarrow{DP}$ show that P is in equilibrium.
(ii) In the case when the forces are $k\overrightarrow{PA}$, $k\overrightarrow{PB}$, $k\overrightarrow{PC}$, $k\overrightarrow{PD}$ show, by using components or otherwise, that the resultant force is $4k\overrightarrow{PO}$. *(JMB)*

9 Show that the resultant of forces $\overrightarrow{PA}$ and $\overrightarrow{PB}$ is $2\overrightarrow{PM}$, where M is the midpoint of AB.

If A, B, C are fixed non-collinear points and P is a variable point such that the resultant of the forces $\overrightarrow{PA}$ and $\overrightarrow{PB}$ passes through C, find the locus of P.

If K, L, M, N are fixed points, no three of which are collinear, and the resultant of the forces $\overrightarrow{PK}, \overrightarrow{PL}, \overrightarrow{PM}$ passes through N, find the locus of P. *(O&C)*

10 The resultant of three forces $\mathbf{F}, \mathbf{G}, \mathbf{H}$ is $\mathbf{R}$, the resultant of $\mathbf{F}, \mathbf{G}, -\mathbf{H}$ is $\mathbf{S}$ and the resultant of $\mathbf{F}, -\mathbf{G}, \mathbf{H}$ is $\mathbf{T}$. Prove that if $\mathbf{S}$ is perpendicular to both $\mathbf{R}$ and $\mathbf{T}$ then $\mathbf{S}$ is also perpendicular to $\mathbf{G}$. *(C)*

11 The resultant of two concurrent forces $\mathbf{F}$ and $\mathbf{G}$ is $\mathbf{H}$. If $F = H$ and the angle between $\mathbf{F}$ and $\mathbf{H}$ is θ, prove that

(i) $G = 2F \sin \frac{1}{2}\theta$,

(ii) the angle between $\mathbf{F}$ and $\mathbf{G}$ is $\frac{1}{2}(\pi + \theta)$. *(C)*

12 Two forces $\mathbf{P}$ and $\mathbf{Q}$ act at O. $\mathbf{R}$ is the resultant of $\mathbf{P}$ and $\mathbf{Q}$, and $\mathbf{S}$ is the resultant of $\mathbf{P}$ and $-\mathbf{Q}$. Show that

(i) $R^2 - S^2 = 4\mathbf{P}.\mathbf{Q}$,

(ii) $R^2 + S^2 = 2(P^2 + Q^2)$,

(iii) the angle between $\mathbf{R}$ and $\mathbf{P}$ is $\cos^{-1}\left(\dfrac{3R^2 + S^2 - 4Q^2}{4RP}\right)$. *(C)*

13 One end of a light string is fixed at a point A. The string passes through a small smooth ring B fixed at the same level as A. A particle, of mass m, hangs freely from the other end of the string. A smooth ring C, also of mass m, is free to slide on the string between A and B and the system is in equilibrium. Show that AC and BC are both inclined at an angle $\pi/3$ to the vertical. *(L)*

14 The points A and B are fixed with A vertically above B and the distance AB is $3a$. One end of a light elastic string, of natural length a and modulus $2\,mg$, is attached to A and the other end of the string is attached to a particle P of mass m. One end of a second elastic string, also of natural length a but of modulus $\frac{1}{2}\,mg$, is attached to B and the other end is attached to P. The particle rests in equilibrium with both strings taut. Find the distance AP.

The particle is now held in equilibrium at a point C, where $AC = BC$ and C is distant $2a$ from the line AB, by a force $\mathbf{F}$ whose horizontal and vertical components are X and Y respectively. Find X and Y. *(L)*

15 The ends of a light inextensible string are attached to two fixed points A and F at the same level. Particles with masses $2m, m, m, 2m$ are attached to the string at points B, C, D, E respectively, where $AB = BC = CD = DE = EF$. When the system rests in equilibrium, the sections AB and EF are each inclined to the horizontal at an angle of $60°$. Find the tensions in each of the sections of the string. Show that BC is inclined to the horizontal at an angle of $30°$. *(L)*

16 AB is a light inelastic string. The end A is attached to a fixed point on a smooth straight wire inclined at an angle $\pi/4$ to the vertical. The end B is fixed to a small bead of mass m which is free to slide on the wire. A particle of mass m is fixed to the midpoint C of the string AB and the system hangs in equilibrium. Prove that the tension in AC is $\dfrac{2\sqrt{5}}{3}\,mg$ and find the reaction between the bead and the wire. *(O)*

17 A particle of mass m is on a rough horizontal plane. A force with its line of action making an angle θ with the plane is applied to the particle. Show that, if this force is just sufficient to pull the particle along the plane, the magnitude, P, of the force is $[mg \sin \lambda]/\cos(\theta - \lambda)$, where $\tan \lambda$ is the coefficient of friction. State the least value of P.

The particle is now placed on the same plane which is tilted at an angle α to the horizontal. A force of magnitude $mg \sin \lambda$ acting along the line of greatest slope is just sufficient to move the particle up the plane. Show that

$$\sin(\lambda + \alpha) = \sin \lambda \cos \lambda.$$ (*L*)

18 Two particles of the same mass are connected by an inextensible string. One particle lies on a rough plane inclined at an angle θ to the horizontal and the other hangs freely. The string connecting them passes over a smooth pulley which is above the particles and which separates the string into a part parallel to the inclined plane and a vertical part. Show that the system will move when released from rest if the coefficient of friction between the plane and the particle is less than $\sec \theta - \tan \theta$. (*W*)

19 A particle of weight W rests on a rough horizontal plane, the coefficient of friction being $\frac{12}{5}$. The particle is acted on by a force W inclined to the vertical at an angle $\beta = \tan^{-1}(\frac{4}{3})$ and by a variable force X inclined to the vertical at an angle $\alpha = \tan^{-1}(\frac{12}{5})$. These two forces act in the same vertical plane, as indicated in the diagram. Determine whether, as X increases from zero, the particle slips before it is lifted from the plane. (*C*)

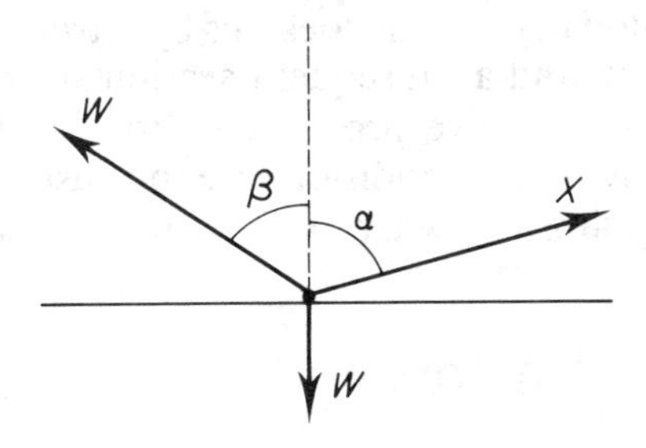

3 Action of Forces on a Finite Body

The previous chapter showed how the problems of equilibrium of small particles can be handled by assuming only that force is a vector. In practice, however, it is clearly not possible to treat everything as a small particle, and we will now start to look at the effect of forces acting on finite bodies. If a heavy stick was placed on a table and we applied roughly the same force (that is, a force of the same magnitude and direction) at different points, we would see that the effect varied with the point at which the force was applied. It would also be noticed that the stick would not only tend to shift bodily but also would tend to turn. Therefore, in dealing with finite bodies, it is necessary to quantify in some way the turning effect of the force. The following section does this by introducing the moment of a force about a point, and a subsequent section shows that by using the moment it is possible to give precise conditions which determine the equilibrium of a body. These conditions are then used to examine in detail the problems of equilibrium when parallel forces act on a body.

3.1 Moment of a force

The moment about a point O of a force $\mathbf{F}$ acting through a point P is defined to be a vector, perpendicular to the plane containing O, P and $\mathbf{F}$ and of magnitude F times the perpendicular distance of the line of action of $\mathbf{F}$ from O. In Fig. 3.1, the magnitude of the moment of $\mathbf{F}$ about O is

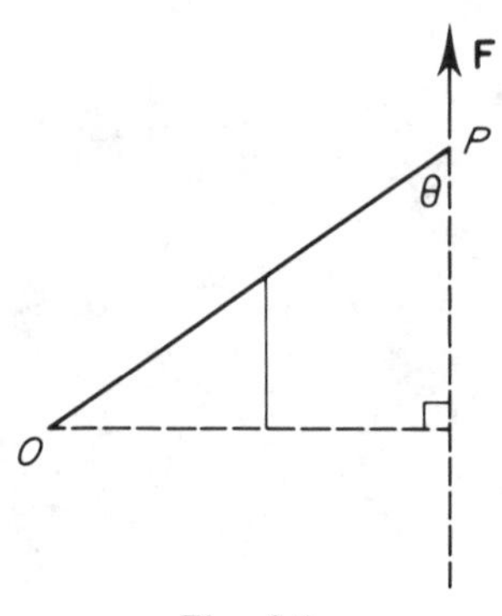

Fig. 3.1

therefore $F.\,OP\sin\theta$. This is clearly a measure of the turning power of **F** because we know from practical experience that a small force exerted on a large spanner can be more effective than a larger force on a small spanner.

The above definition is not precise as it stands because there are two directions, differing by 180°, in which a vector can be perpendicular to a plane. This ambiguity is removed by imagining a right-hand (normal) screw placed perpendicular to any given plane and a screw-driver applied to turn the screw in a counter-clockwise direction. The screw then moves in the direction out of the plane and this direction is defined as the positive sense out of the plane. We shall denote the unit vector in this direction by **k**. (The positive sense out of this page is towards the reader.)

Having defined the positive sense for any plane, we can now obtain an exact definition of the moment. If a force **F** at P would tend to turn a light rod along OP in a counter-clockwise direction then the moment **M** is given by

$$\mathbf{M} = F\,|OP\sin\theta|\,\mathbf{k}.$$

If the force **F** would produce a clockwise rotation then

$$\mathbf{M} = -F\,|OP\sin\theta|\,\mathbf{k}.$$

At this stage, we shall only be looking at planar problems, and it turns out that for these it is possible to avoid treating moment as a vector. It is not possible to do this for problems where forces are acting in different planes, and we have chosen to define moment precisely in order to avoid confusion later – it can be bewildering to find that a quantity that has previously been treated as a scalar is really a vector!

We can avoid using the vector notation for planar problems, because all the moments will be parallel to **k** and so we can just drop the **k**. Therefore, we can treat the moment of a force about a point as

$+|\text{Force}| \times |\text{perpendicular distance}|$ when counter-clockwise rotation would be produced,

$-|\text{Force}| \times |\text{perpendicular distance}|$ when clockwise rotation would be produced.

Strictly speaking, the moment, thus defined, is the component of **M** in the direction of **k**, and it is sometimes called the moment about the axis **k**.

In order to calculate moments, it is sometimes useful to have an expression available for the moment in terms of the coordinates of the points and the components of the force. The following shows how such an expression is obtained.

The coordinates of P will be taken to be (x_0, y_0) and the x and y components of **F** will be denoted by

$$X = F\cos\theta, \quad Y = F\sin\theta,$$

where θ is the angle between the positive x-direction and the positive sense

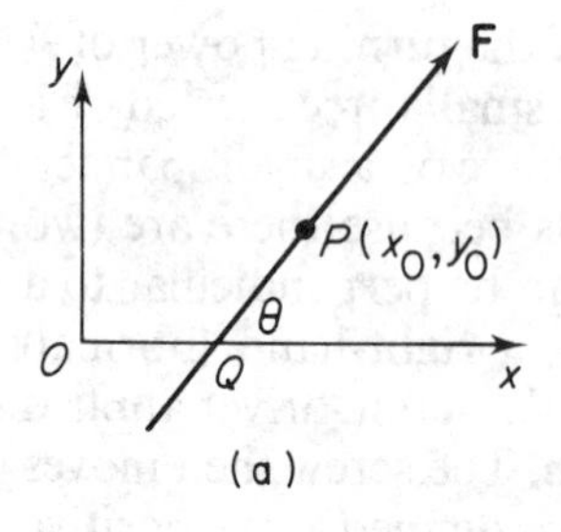

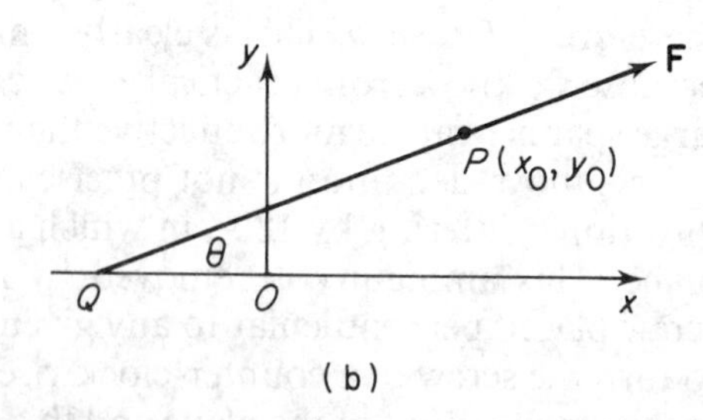

Fig. 3.2

of **F**. The point Q denotes the point of intersection of the line of action of **F** with the x-axis. We assume first that the situation is as depicted in Fig. 3.2(a), with $0 < \theta < 180°$ and Q to the right of O.

The equation of the line of action of **F** is

$$\frac{x - x_0}{y - y_0} = \cot\theta$$

so that the x-coordinate of Q is $x_0 - y_0 \cot\theta$. For the configuration shown, the moment M is positive and given by

$$M = F\,.\,OQ\,.\sin\theta = F\,(x_0 - y_0 \cot\theta)\sin\theta$$
$$= Yx_0 - X\,y_0 \qquad 3.1$$

For the configuration shown in Fig. 3.2(b), with $0 < \theta < 180°$, the moment M is negative and is given by

$$M = -F\,.\,OQ\,.\sin\theta.$$

In this case the x-coordinate of Q is negative, so

$$OQ = -(x_0 - y_0 \cot\theta)$$

and

$$M = Yx_0 - X\,y_0.$$

Carrying out similar calculations for $\theta \geqslant 180°$ produces the same expression. Similarly, the moment M' about the point (h, k) is given by

$$M' = Y(x_0 - h) - X\,(y_0 - k).$$

The right-hand side of equation 3.1 can also be interpreted as the sum of the moments about O of the Cartesian components of **F**.

If we have forces $(\mathbf{F}_1, \ldots, \mathbf{F}_n)$ with components of magnitude $(X_1, \ldots, X_n)$ and $(Y_1, \ldots, Y_n)$ parallel to the x- and y-axes then the moment of the resultant is

$$(Y_1 + Y_2 + \ldots + Y_n)x_0 - (X_1 + X_2 + \ldots + X_n)\,y_0$$
$$= (x_0 Y_1 - y_0 X_1) + \ldots + (x_0 Y_n - y_0 X_n).$$

The right-hand side is the sum of the moments of the separate forces and we can, therefore, deduce the following important theorem, often known as *Varignon's theorem.*

Theorem For forces acting at a point, the moment about any point of the resultant is the sum of the moments, about that point, of the separate forces.

We define the moment of a system of forces acting through different points to be the sum of the moments of the separate forces. In the S.I. system, the unit of moment is the Newton metre (Nm).

EXAMPLE 1 *Find the moment about the origin of a force of magnitude* 4 N *acting through the point* (2 m, 0) *in a direction inclined at an angle of* 45° *to both the positive x and y directions.*

The situation is as shown in Fig. 3.3. The perpendicular distance from the origin to the line of action of the force is

$$(2 \cos 45°)\,\text{m} = \sqrt{2}\,\text{m}.$$

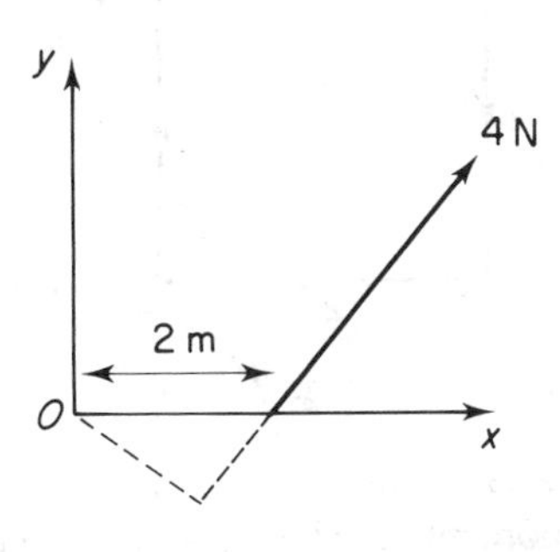

Fig. 3.3

The moment, therefore, has magnitude $4\sqrt{2}$ Nm and, as the force would produce a counter-clockwise rotation, the moment is $\mathbf{4\sqrt{2}}$ **Nm.**

An alternative method of calculating the moment would have been to determine the components of the force parallel to Ox and Oy and calculate the moments separately. The x component passes through O and hence has no moment, whilst the y component is $2\sqrt{2}$ N and therefore has moment $4\sqrt{2}$ Nm.

EXAMPLE 2 *Find the moment about the point Q* (3, 5) *of the force defined in the previous example.*

One way of calculating the moment is to work out the perpendicular distance of Q from the line of action of the force. The geometry necessary for this can, however, be avoided by evaluating the components of the force and then calculating the

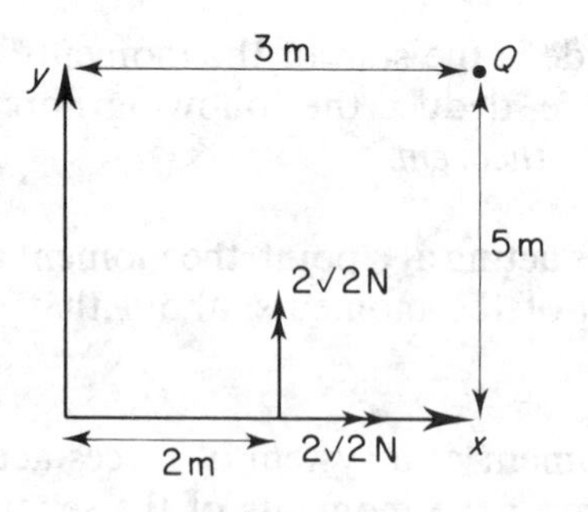

Fig. 3.4

moment. The components are as shown in Fig. 3.4 and the moment is

$$[5.2\sqrt{2}-2\sqrt{2}(3-2)]\,\text{Nm} = \mathbf{8\sqrt{2}\,Nm}.$$

EXAMPLE 3 *Forces of magnitude* 4 N, 3 N, 1 N *and* 2 N *act in the senses shown in Fig.* 3.5 *along the sides BA, BC, DC and DA respectively of a square ABCD. The square is of side* 2 m. *Find the moment of the system about A and about C.*

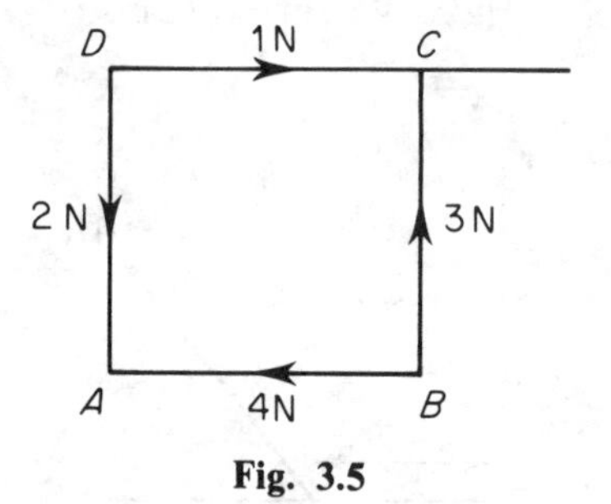

Fig. 3.5

The forces which pass through A give zero moment about A. The force along BC has a counter-clockwise moment of 3·2 Nm, whilst that along DC has a clockwise moment of 1·2 Nm. The total moment about A is therefore **4 Nm**.

The forces passing through C have no moment about C. The force along BA has a clockwise moment of 4·2 Nm whilst that along DA has a counter-clockwise moment of 2·2 Nm. The moment about C is therefore **−4 Nm**.

EXERCISE 3.1

In questions 1 to 4, forces are shown acting at various points on a straight line and the distances shown are measured in metres. Find the moments of the systems about the points A and B.

1 **2**

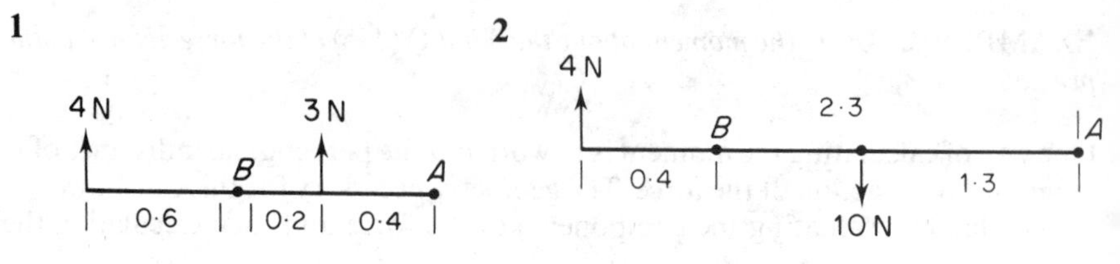

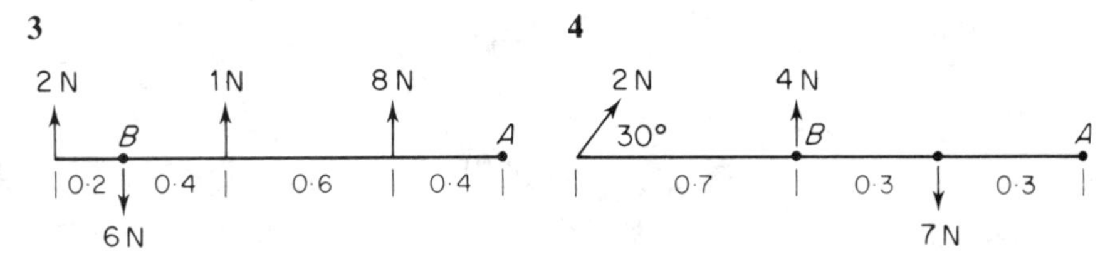

In questions 5 to 7, the moments of the force system described are to be found about the point with position vector $(a\mathbf{i}+b\mathbf{j})$ m.

5 $(2\mathbf{i}+3\mathbf{j})$ N acting at $(\mathbf{i}+\mathbf{j})$ m, $(4\mathbf{i}+5\mathbf{j})$ N acting at $(3\mathbf{i}-\mathbf{j})$ m, $a = b = 0$.

6 $(5\mathbf{i}-3\mathbf{j})$ N acting at $(-4\mathbf{i}+\mathbf{j})$ m, $(-2\mathbf{i}-\mathbf{j})$ N acting at $(7\mathbf{i}-\mathbf{j})$ m, $(11\mathbf{i}-2\mathbf{j})$ N acting at $(4\mathbf{i}-7\mathbf{j})$ m, $a = 2$, $b = 1$.

7 $(3\mathbf{i}+\mathbf{j})$ N acting at $(2\mathbf{i}-3\mathbf{j})$ m, $(-7\mathbf{i}+2\mathbf{j})$ N acting at $(4\mathbf{i}+\mathbf{j})$ m, $(5\mathbf{i}-9\mathbf{j})$ N acting at $(-4\mathbf{i}-\mathbf{j})$ m, $a = -3$, $b = 2$.

In questions 8 to 11, find the moments of the systems shown about the points A and B.

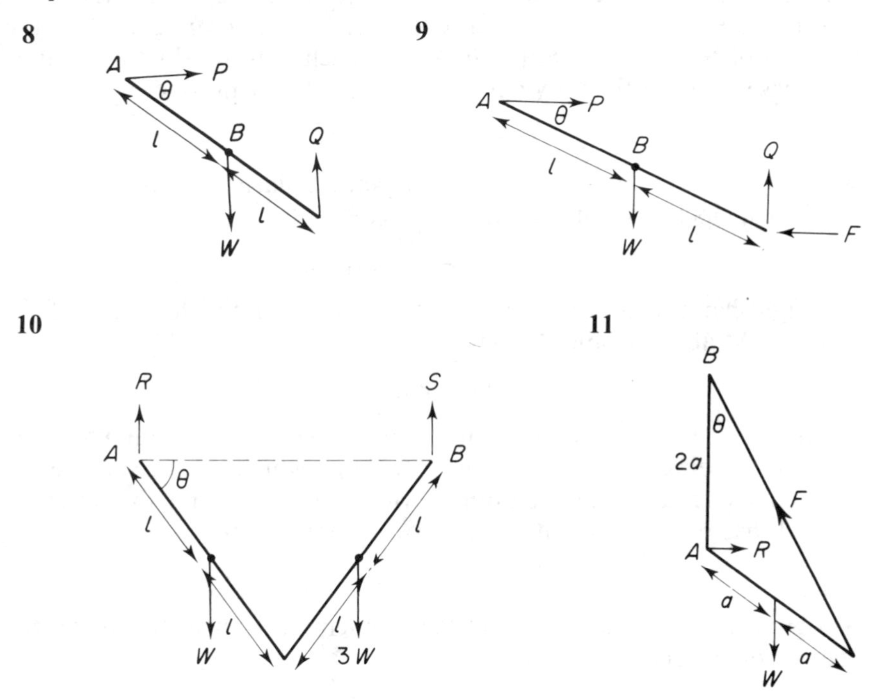

3.2 Alternative definition of moment

(This section should be omitted by those with no knowledge of the vector product.)

Readers with a working knowledge of the vector product may have noticed a similarity between the definition of moment and that of the vector product. For the benefit of those readers, we show how the vector product can be used to define moment of force about a point. In Fig. 3.6,

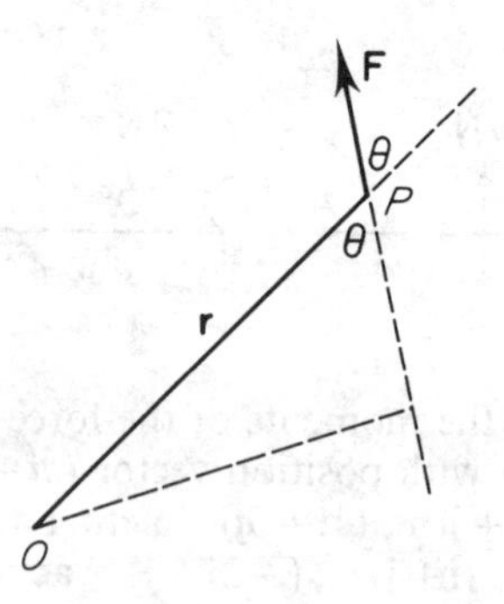

Fig. 3.6

the force $\mathbf{F}$ is shown acting at point P with position vector $\mathbf{r}$ relative to Q. The angle between the positive senses of $\mathbf{F}$ and $\mathbf{r}$ is θ. The magnitude of the moment of $\mathbf{F}$ about O is $Fr|\sin\theta|$. If $\mathbf{F}$ is such as to produce a counter-clockwise rotation then, by definition of the vector product,

$$\mathbf{r} \times \mathbf{F} = Fr|\sin\theta|\,\mathbf{k},$$

where $\mathbf{k}$ is a unit vector out of the paper. Similarly, if $\mathbf{F}$ is such as to produce a clockwise rotation then

$$\mathbf{r} \times \mathbf{F} = -Fr|\sin\theta|\,\mathbf{k}.$$

The right-hand sides are precisely those obtained in the definition of the moment $\mathbf{M}$ about O and, therefore,

$$\mathbf{M} = \mathbf{r} \times \mathbf{F}.$$

The vector $\mathbf{r}$ is the position vector, relative to the point about which the moment is being taken, of a point P through which the force passes. If moments were taken about a point A with position vector $\mathbf{a}$, then, relative to A, P has position vector $\mathbf{r}-\mathbf{a}$ so the moment is $(\mathbf{r}-\mathbf{a}) \times \mathbf{F}$.

Note

Varignon's theorem is now merely a statement of the distributive law for the vector product.

3.3 Equilibrium conditions for finite bodies

We now have sufficient apparatus to be able to obtain conditions for the equilibrium of finite rigid bodies. We take as our definition of such a body, a collection of a large (generally infinite) number of particles which are rigidly connected to each other. Since the particles are all rigidly connected, equilibrium (that is, non-movement) of the body is equivalent to equilibrium of each separate particle. We already know how to deal with the equilibrium of separate particles, and we can use this information to set up conditions for the equilibrium of a rigid body. The derivation of the conditions for a large (infinite) number of particles is rather complicated and will not be given. Section 3.7 contains a simplified derivation for three particles rigidly connected in order to give some idea of how the general derivation is carried out.

It is not really necessary to be able to understand the proof and the model as used may not necessarily be the correct one. The important thing is that the conditions obtained do give correct results in practical situations, though it is obviously useful to have some idea of the assumptions which led to them. The basic model assumes:

(a) a rigid body consists of interacting particles;
(b) the interactive forces between each pair of particles satisfies the law of action and reaction, and acts along the line joining the particles;
(c) as well as the interactive forces between particles, additional applied forces will be acting on some particles.

Having made these assumptions, it can be deduced that necessary and sufficient conditions for a rigid body to be in equilibrium are that:

(i) the vector sum of the applied forces acting on the body is zero;
(ii) the total moment of the applied forces about any point is zero or, equivalently, the clockwise moment is equal to the counter-clockwise moment.

These conditions yield three scalar equations – two 'component' equations and one 'moment' equation. It is shown in §13.5 that both component equations can be replaced by moment equations, and that they can also be replaced by two further moment equations provided that the three points about which the moments are taken are not collinear. Therefore the following equilibrium conditions are equivalent to conditions (i) and (ii).

(iii) the component of the vector sum of the applied force in any one direction and the sum of the moments about two separate points must be zero;
(iv) the moment of the applied forces about three non-collinear points must be zero.

Conditions (iii) and (iv) are rarely used in practice.

3.4 Equilibrium of a body under parallel forces

The above conditions are sufficient to determine the equilibrium of any rigid body, and here they are applied to problems involving parallel forces. As in all equilibrium problems, the first step is to draw a diagram, showing the forces at their point of action, and then to apply the basic conditions. For parallel forces there is only one component of the vector sum so that (i) is easy to apply. There is some choice in the point about which the moment is taken. It does not matter what point is chosen but it can simplify matters to take moments about a point at which an unknown force is acting. This force will not then occur in the moment equation.

EXAMPLE 1 *A light beam AB of length* 10 m *has a load weight of* 100 N *attached to its mid-point, O. The beam rests horizontally in equilibrium on two smooth pegs CD with* $AC = 2$ m, $DB = 4$ m *and with loads of weights* 10 N *and* 2 N *attached at the ends A and B respectively. Determine the reactions at C and D.*

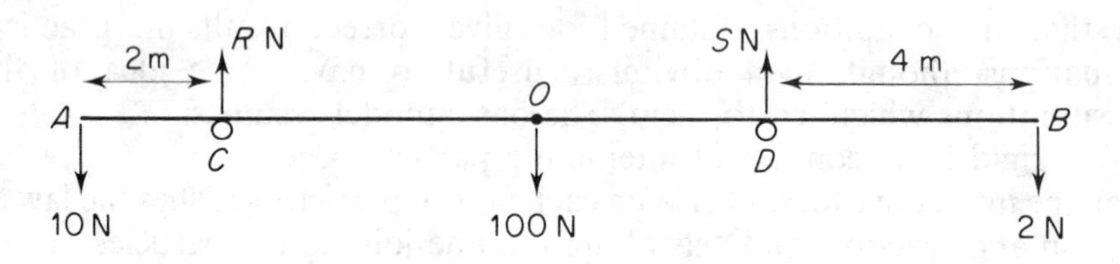

Fig. 3.7

Since the pegs are smooth the reactions are perpendicular to the pegs, as shown. The total vertical component of the forces is $(R+S-112)$N in the upward direction and, therefore,

$$R+S = 112\,\text{N}.$$

We can eliminate R by taking moments about C. Equating the clockwise moment about C to the counter-clockwise one gives

$$2.CB + 100.CO = 10.AC + 4S$$

that is,

$$16 + 300 = 20 + 4S$$

$$\mathbf{S = 74,\ R = 38.}$$

EXAMPLE 2 *Determine, for the previous example, the maximum weight that can be placed at B without disturbing equilibrium.*

We assume the required weight is W N and replace 2 N by W N. This gives

$$R+S = 110 + W.$$

Taking moments about C now gives

$$8W + 300 = 20 + 4S.$$

So

$$S = 2W + 70,$$

$$R = 40 - W.$$

However, we are not required to find R and S but to find when equilibrium is broken. Effectively equilibrium is broken when the configuration shown is not possible, and this will happen if R or S is negative because the pegs cannot exert a pull. Therefore, the maximum possible value of W is that which gives $R = 0$ that is

$$\mathbf{W = 40.}$$

When $R = 0$ the peg need not be there or, equivalently, the beam is just leaving the peg. This agrees with practical experience, for as the load on B increases the beam turns about D. We could have shortened the problem slightly by thinking about what happens physically in this situation and then looking for W such that $R = 0$. However, it is not possible to do this clearly in all problems, and it is usually safer to find out all the forces and then to establish the conditions which make them physically sensible.

EXAMPLE 3 *Two light rods AB and BC are rigidly joined at B so that they are perpendicular to each other. AB is of length* 2 m *and BC is of length* 4 m, *and weights* 2 *W and W are attached to A and C respectively. The configuration is suspended in equilibrium by a light string at B. Find the inclination of BC to the vertical.*
The situation is as depicted in Fig. 3.8, with AB and BC at angles θ and ϕ to the horizontal, and $\theta + \phi = 90°$. Taking moments about B gives

$$2W.2\cos\phi = W.4\cos\theta$$

so $\tan\theta = 1$ and $\theta = \mathbf{45°}$.

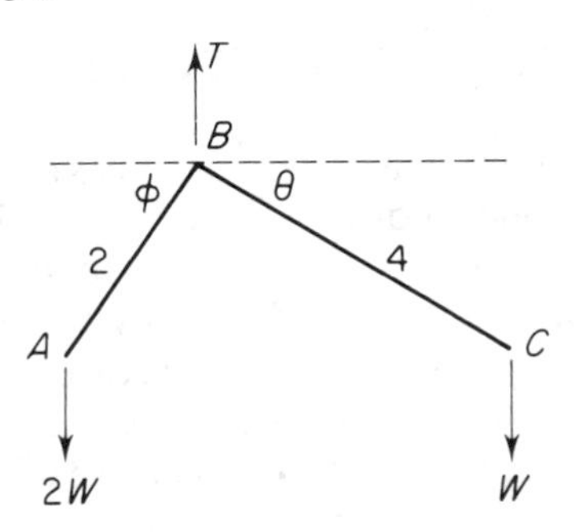

Fig. 3.8

EXERCISE 3.4

Questions 1 to 4 involve a light rod AB fixed at points C and D and acted on by the forces shown, the distances being measured in metres. Find the forces acting at the points C and D, taking the direction up the page as positive.

1
2
3

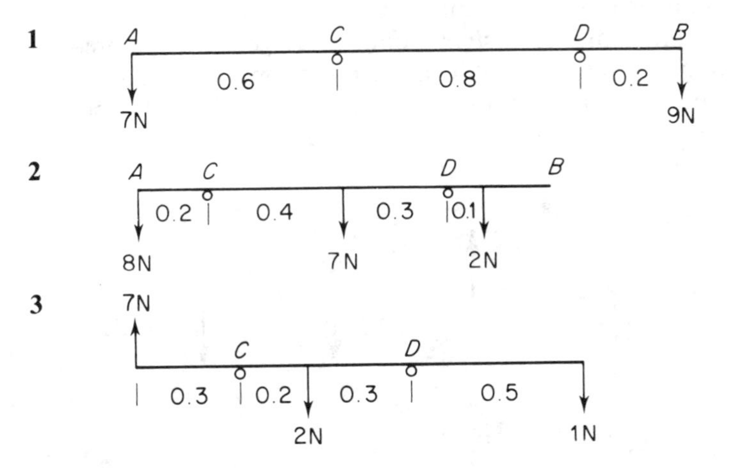

4

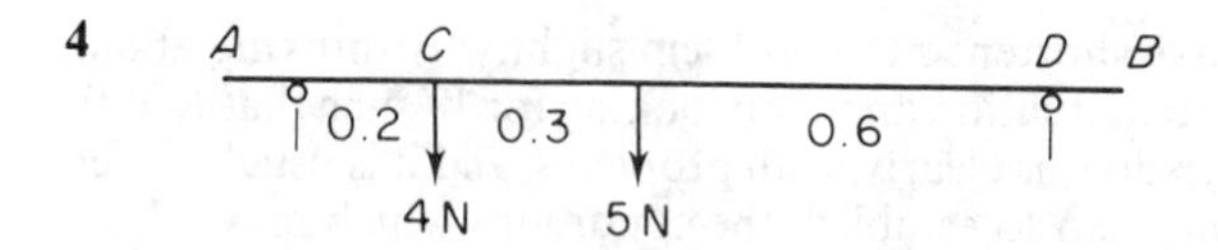

5 A light rod AB of length 2 m rests on smooth pegs C and D, where $AC = 0{\cdot}5$ m and $BD = 0{\cdot}3$ m. A force of 48 N is applied vertically downwards at the mid-point of AB. Find the forces at the pegs. Find also the least force applied vertically downwards at A sufficient to disturb the equilibrium.

6 A light square lamina $ABCD$ is free to turn about its centre in a vertical plane. A particle of mass m is attached at A. Find the mass of the particle that has to be attached at B so that the square can be in equilibrium, with AB inclined at an angle of 30° above the horizontal and A lower than B.

3.5 Resultant of parallel forces

The concept of the resultant of a system of forces acting on a rigid body is not a particularly easy one, and it will not be considered in general until Chapter 13. In fact, it is not even necessary to solve relatively difficult problems, but it is important to study briefly the particular case of the resultant of a system of parallel forces. The reason for this is that a rigid body is a system of particles, and the force of gravity acts on each particle. For practical purposes, the direction of the force of gravity is the same at all points of a relatively small body and, therefore, its effect is that of a system of parallel forces.

The resultant of a system of parallel forces whose vector sum is not zero is defined to be the force equal to the vector sum and acting along a line such that its moment about any point is the sum of the moments of the separate forces about the same point. So, the resultant is equivalent to the simplest force system which has the same effect as the original system.

The situation will in general be as shown in Fig. 3.9, with **R** denoting the resultant. The moment of the resultant about a point O of itself is zero and so the total moment of the system about O is zero. Therefore, an alternative and more useful way of locating points on the resultant is to find points about which the total moment is zero or, equivalently, points about which the clockwise moment is equal to the counter-clockwise one.

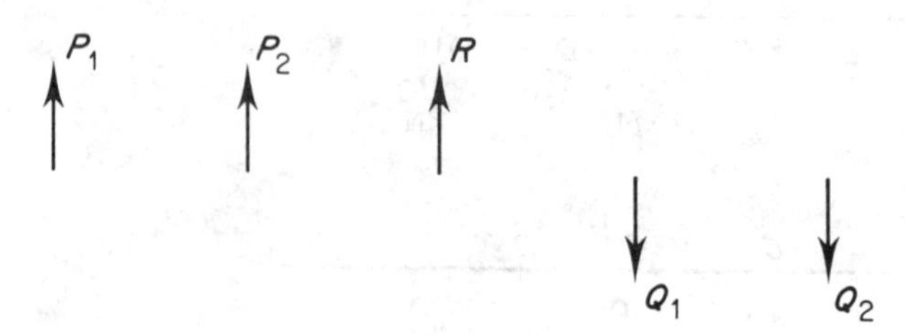

Fig. 3.9

These definitions will be used in Chapter 12 to locate the point at which the force of gravity acts on a body. This point is known as the centre of gravity.

EXAMPLE 1 *Find the resultant of the system of forces shown in Fig.* 3.10 *acting on the light beam AB* (*which is* 10 m *long*) *and the point on AB through which it acts.*

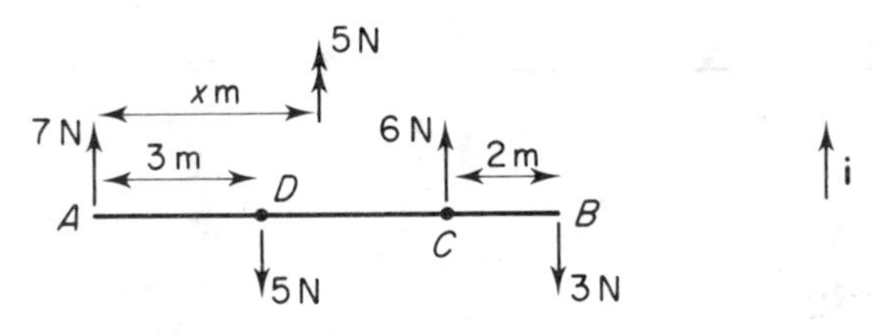

Fig. 3.10

The forces acting are $7\mathbf{i}$ N at A, $-5\mathbf{i}$ N at D, $6\mathbf{i}$ N at C and $-3\mathbf{i}$ N at B, where $\mathbf{i}$ is the unit vector shown. The resultant is therefore

$$(7\mathbf{i}-5\mathbf{i}+6\mathbf{i}-3\mathbf{i})\,\text{N} = 5\mathbf{i}\,\text{N}.$$

There is actually no need to use vector notation here, and the resultant can be found equivalently by noting that the sum of the components 'up' the page is 13 N whilst that down is 8 N.

If the resultant is assumed to be acting at a distance x m from A as shown, then equating the moment of the resultant about A to the sum of the moments of the other forces gives

$$5x = -5.3+6.8-3.10$$

$$x = \frac{3}{5}.$$

EXERCISE 3.5

In the following questions, find the resultant of the forces shown. Its position should be given relative to O, using the convention that an upwards force is positive. Distances are all measured in metres.

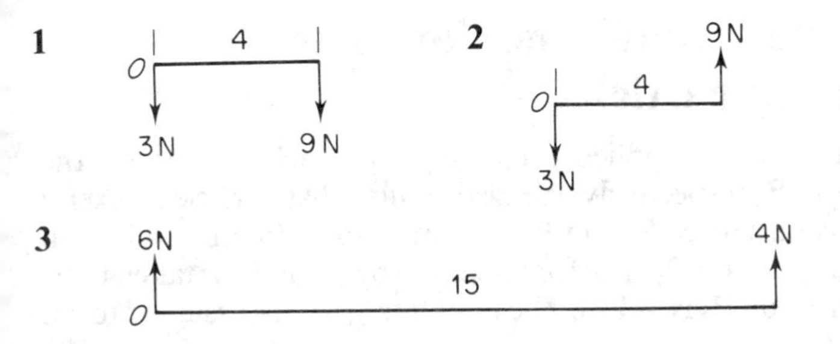

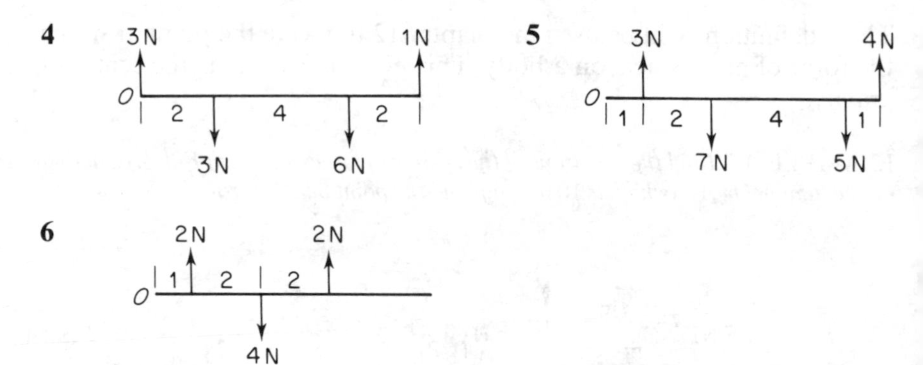

3.6 Couples

There is a particular system of parallel forces for which a resultant cannot be defined as above. This is a system where the vector sum of the forces is zero, such as that shown in Fig. 3.11. An example of this kind of force system, with parallel forces of equal magnitude but opposite senses, is provided by the action of unscrewing a stopper. The effort is entirely a rotating one, and such a system is referred to as a couple. The moment about any point O is $F(p_2 - p_1)$ where p_2 and p_1 are the perpendicular distances from 0 to the forces. Therefore, the moment is equal to Fd where d is the perpendicular distance between the lines of action of the forces. This is independent of the point about which the moment was taken, and so a couple is completely defined by its moment.

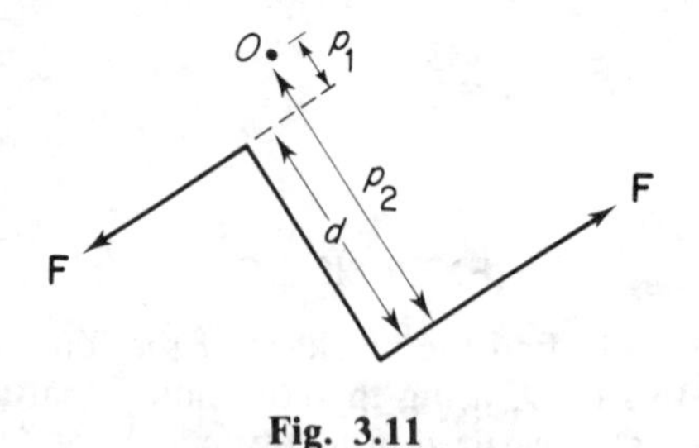

Fig. 3.11

3.7 Equilibrium conditions for three interacting particles

We assume that there are applied forces $\mathbf{F}_1$, $\mathbf{F}_2$ and $\mathbf{F}_3$ acting on the particles P_1, P_2 and P_3 respectively. We also assume that particle P_1 exerts a force $\mathbf{F}_{12}$ on P_2 and a force $\mathbf{F}_{13}$ on P_3, so that P_2 exerts a force $-\mathbf{F}_{12}$ on P_1 and P_3 a force $-\mathbf{F}_{13}$ on P_1. The force exerted by P_2 on P_3 is taken to be $\mathbf{F}_{23}$ whilst that of P_3 on P_2 is $-\mathbf{F}_{23}$. The interacting forces, assumed to act along the line joining the corresponding particles, are shown in Fig. 3.12. The equilibrium of the separate particles gives

$$\mathbf{F}_1 - \mathbf{F}_{12} - \mathbf{F}_{13} = 0,$$
$$\mathbf{F}_2 + \mathbf{F}_{12} - \mathbf{F}_{23} = 0,$$
$$\mathbf{F}_3 + \mathbf{F}_{23} + \mathbf{F}_{13} = 0.$$

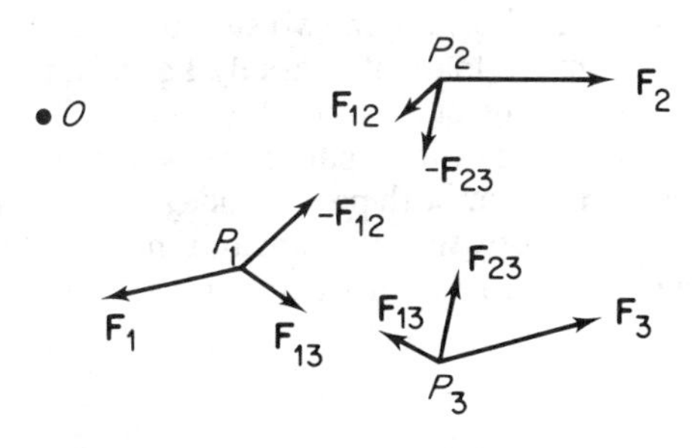

Fig. 3.12

These equations are simply vector identities and adding them gives

$$\mathbf{F}_1 + \mathbf{F}_2 + \mathbf{F}_3 = 0$$

which states that the vector sum of the forces is zero.

We can now take moments about any point O for the system of forces acting on the three separate forces. Assuming the separate particles are in equilibrium then by definition the total moment is zero (the net force on each particle is zero). Therefore,

$$\begin{aligned}&\text{Moment about } O \text{ of } (\mathbf{F}_1 - \mathbf{F}_{12} - \mathbf{F}_{13}) \text{ at } P_1 + \\ &\text{Moment about } O \text{ of } (\mathbf{F}_2 + \mathbf{F}_{12} - \mathbf{F}_{23}) \text{ at } P_2 + \\ &\text{Moment about } O \text{ of } (\mathbf{F}_3 + \mathbf{F}_{23} + \mathbf{F}_{13}) \text{ at } P_3 = 0.\end{aligned}$$

The forces $-\mathbf{F}_{13}$ at P_1 and $\mathbf{F}_{13}$ at P_3 are along P_1P_3 and are the same perpendicular distance from O. They will produce rotations in opposite senses and, therefore, the sum of their moments is zero. The same is true for the other pairs of forces interactions. Therefore

$$\text{Moment about } O \text{ of } (\mathbf{F}_1 \text{ at } P_1 + \mathbf{F}_2 \text{ at } P_2 + \mathbf{F}_3 \text{ at } P_3) = 0.$$

The choice of point O was entirely arbitrary and, therefore, we conclude that the sum of the moments of the applied forces about *any* point is zero.

MISCELLANEOUS EXERCISE 3

1 A light straight rod $ABGCD$ with a load applied at G rests horizontally in equilibrium on two smooth supports at B and C. The lengths AB, BC and CD

are 1·0 m, 2·0 m and 1·5 m respectively. The rod just starts to tilt when either a load weighing 100 N is attached at A, or a load weighing 40 N is attached at D. Calculate
(a) the load at G,
(b) the length AG,
(c) the reactions on the rod at each support when loads weighing 100 N and 40 N are attached at A and D respectively at the same time. (*L*)

2 A light horizontal beam ACB, of length 2 m and carrying a load of 100 N at its midpoint rests in equilibrium on two supports, one at A and the other at C, where $AC = 1{\cdot}7$ m. A load of mass 12 kg is hung from B. Find the magnitude, in N, of the reaction at each support. (*L*)

3 A light rod AB of length $2a$ with a particle of weight W at its midpoint rests horizontally between two smooth pegs P and Q as shown in the figure, where $AP = 5a$ and $QB = a$. Copy this diagram and show on it all the forces acting on the rod. Calculate the reactions between the rod and the pegs.

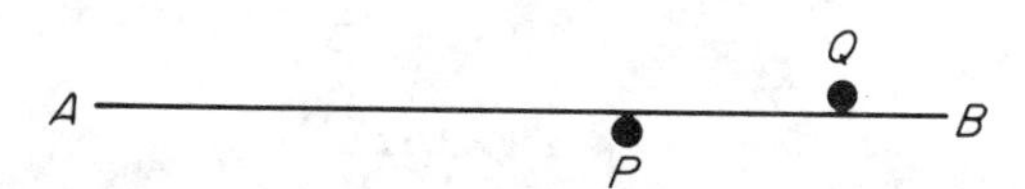

A particle of weight $4W$ is hung from A and the greatest force that the peg Q can sustain is $17W$. Calculate the greatest vertical force that can be applied at B
(a) upwards,
(b) downwards,
if equilibrium is not to be broken. (*L*)

4 A straight light beam $ABCD$, where $AB = 1$ m, $BC = 2{\cdot}5$ m and $CD = 0{\cdot}5$ m, rests horizontally on supports at B and C. The beam carries a particle of mass 10 kg at its midpoint. Calculate the magnitude, in N, of the force exerted by the beam on each support. (*L*)

5 A light beam rests, in a horizontal position, on smooth supports at its points of trisection and a heavy particle is placed at a point on it. When a mass of 1 kg is placed on one end, the beam just tilts. When a mass of 1 kg is placed on the other end, the pressures on the supports are equal. Find the mass of the particle and the ratio in which the length of the beam is divided by the particle. (*O*)

6 A light plank $ABCD$ of length $6a$ and carrying a weight W at its midpoint rests on two supports B and C; $AB = BC = CD = 2a$. A man of weight $4W$ wishes to stand on the plank at A, for which purpose he places a counterbalancing weight Y at a point P between C and D at a distance x from D. Show that the least value of Y required is

$$\frac{7Wa}{4a-x}.$$

Show that, if the board is also to be in equilibrium when he is *not* standing on it, then the greatest value of Y that may be used is

$$\frac{Wa}{2a-x}.$$

(*JMB*)

7 A see-saw consists of a light straight beam of length 3 m which is hinged at its centre C. When a boy of mass 30 kg sits at one end and a girl of mass 25 kg sits at the other end, the see-saw remains horizontal. Calculate the magnitude of the frictional couple which is acting around C to prevent the see-saw tilting, giving your answer in Nm. *(L)*

8 A light rod AB, of length $2a$, is rigidly attached at its midpoint C to one end of a light rod CD of length $4a$, the two rods being at right angles. Particles of mass $2m$, $6m$ and $10m$ are attached to A, B and D respectively and the system is free to rotate in a vertical plane about E, the midpoint of CD. The angle between EC and the horizontal is denoted by θ as shown in the figure.

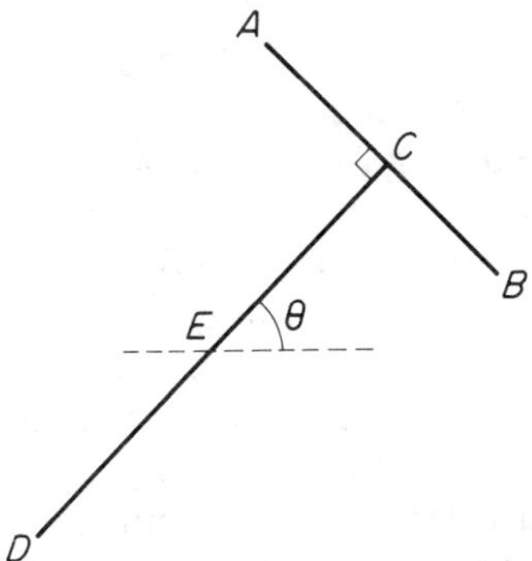

Calculate the moment about E of the gravitational forces acting on the system and hence determine the values of θ which correspond to equilibrium positions.

The mass $10m$ at D is now removed. Find the value of a new mass which, when attached at D, is such that the system rests in equilibrium with CD horizontal.

This new mass is now also removed and replaced by a light elastic string of natural length a. One end of this string is attached to D and the other end to a point at a distance $3a$ directly below the new position of C so that the system remains in equilibrium with CD horizontal. Find the modulus of elasticity of this string. *(AEB 1982)*

9 Weights W, $5W$, mW, $2W$, nW, $3W$ are attached to points A, B, C, D, E, F respectively equally spaced around the circumference of a uniform circular disc of radius a. The disc is free to rotate in a vertical plane about a fixed horizontal axis through its centre O. When E is vertically below O, the disc is kept at rest by a couple $\frac{1}{2}Wa\sqrt{3}$ acting in the sense ABC in the plane of the disc. When D is vertically below O, the disc is kept at rest by a similar couple of the same magnitude but acting in the sense CBA. Find the values of m and n. *(L)*

10 A light rigid beam AB, of length $3a$, carries a weight nW at the point R and rests on supports P and Q at the same level, where $AP = PQ = QB = a$. When a load of weight W is hung from A, the beam is on the point of tilting about P. Find the distance AR.

When an additional load of weight W_1 is hung from B, the forces exerted on the supports at P and Q are equal. Find W_1 in terms of n and W.

If a couple, of moment L and acting in the vertical plane through AB, is now applied to the loaded beam, the reaction at P is increased in the ratio 3:2. Show that

$$L = \tfrac{1}{3}(n+1)\,Wa. \qquad (JMB)$$

4 Kinematics of Rectilinear Motion

Particle dynamics is concerned effectively with finding the position of a particle given the force acting on it. The latter, by Newton's first law of motion, determines the acceleration of the particle, and in §4.1 acceleration is defined precisely for rectilinear motion. In this chapter we shall be concerned with methods of determining the position of a particle, free to move on a line, given the acceleration. This is a purely mathematical problem. The particular case of constant acceleration is considered in §4.3, and rectilinear motion under gravity, which is a special case of motion with constant acceleration, is examined in §4.4. The methods are extended in §4.5 to problems where the acceleration is a given explicit function of time. The chapter concludes with §4.6, where numerical methods of solving kinematic problems are discussed.

4.1 Basic definitions

When trying to solve any problem involving the rectilinear motion (that is, motion in a straight line) of a particle, it is essential at the outset to choose a particular direction to be the positive direction and to refer everything to this direction. The choice of reference direction does not matter; the important thing is to adhere to the same reference direction throughout a given problem. Failing to do this is the greatest source of error in problems, particularly in setting up equations of motion. For simplicity we shall consider motion along the x-axis and take the positive direction to be that of increasing x.

Displacement

The position of a particle at any time is determined by its x-coordinate, and in kinematics this coordinate is termed the *displacement* of the particle. Of course, the displacement x can be positive or negative depending on whether the particle lies to the right or the left of the origin. The distance of the particle from O is $|x|$. By definition, the displacement of a particle specifies its position uniquely, whereas the distance from O does not do this because it does not identify the side of the origin on which the particle lies.

Velocity

The velocity v in the positive direction is defined to be the rate of change of displacement with respect to time, that is,

$$v = \frac{\mathrm{d}x}{\mathrm{d}t} = \dot{x},$$

(using the convention that superscript dots imply differentiations with respect to time t). The velocity can again be either positive or negative and there is no direct dependence between the signs of x and $\dot{x}$. For example, $x = 1 - t^2$ is positive for $0 < t < 1$, whereas the velocity, which is $-2t$, is negative for $0 < t < 1$. All that is implied by a negative velocity is that the motion is in the opposite direction to the reference one. The speed is the magnitude of the velocity, that is, speed is equal to $|\dot{x}|$. Velocity and speed as defined can vary with time. It is often useful to determine an average speed, and this is defined as total distance/total time.

Acceleration

The acceleration in the positive direction is defined as the rate of change with respect to time of the velocity in the positive direction, that is,

$$a = \frac{\mathrm{d}v}{\mathrm{d}t} = \dot{v} = \frac{\mathrm{d}^2x}{\mathrm{d}t^2} = \ddot{x},$$

and again this can be positive or negative. When there is no possibility of misunderstanding, the phrase 'in the positive direction' is omitted after velocity and acceleration, but it should be remembered that it is always implied. If a particle is said to be moving with retardation r, then conventionally this means that the acceleration in the reference direction is equal to $-r$.

The acceleration as defined can be found directly if v is a given function of t, but it is often useful to have an alternative method of calculating the acceleration when v is given as an explicit function of x. We have that

$$a = \frac{\mathrm{d}v}{\mathrm{d}t} = \frac{\mathrm{d}v}{\mathrm{d}x}\frac{\mathrm{d}x}{\mathrm{d}t},$$

or

$$a = \frac{\mathrm{d}v}{\mathrm{d}t} = v\frac{\mathrm{d}v}{\mathrm{d}x}. \qquad 4.1$$

Equation 4.1 allows a to be calculated when v is given in terms of x.

4.2 Types of problems to be solved

The object of particle dynamics is the determination of the position of a particle at any time when the forces acting on the particle are known.

Newton's second law of motion, which is discussed in the following chapter, states that the acceleration of a particle is proportional to the resultant of the forces acting on it. Therefore, the force effectively gives the acceleration $\ddot{x}$, and it is necessary to find x from a known $\ddot{x}$.

In the most general case, $\ddot{x}$ can depend on all three of t, x and $\dot{x}$, and finding x from $\ddot{x}$ requires the solution of a differential equation. Solving this equation will involve two integrations and, hence, there will be two arbitrary constants in the solution found. In order to determine x completely these constants have to be found, and this can be done by giving the values of x and $\dot{x}$ at some time (or x at two different times).

The approach used in the subsequent sections is first to find the solution of a given differential equation in a form involving two arbitrary constants. This is referred to as the general solution of the equation. Second, further information available is then used to find these constants. It should always be remembered that any general expression obtained for x should involve two arbitrary constants, and a corresponding expression obtained for v should involve one such constant.

4.3 Constant acceleration

The simplest possible case of constant acceleration is when the acceleration a is zero, so that

$$\ddot{x} = 0.$$

Integrating this equation once with respect to t gives

$$\dot{x} = v,$$

where v is a constant which can be found if $\dot{x}$ is known at any time. The particle is then said to be moving with uniform velocity v. A further integration gives

$$x = vt + c,$$

where c is another constant. The constant c will be determined uniquely if x is known for one value of t. If, for simplicity, we assume that $x = 0$ when $t = 0$, then $x = vt$. We also see that in this case the average speed is equal to the actual speed $|v|$. If $\dot{x}$ were not given at any time, the constants v and c could still be found if x were given for two values of t.

The problem of a particle moving with a constant non-zero acceleration a in the positive direction can be solved almost as simply. In this case

$$\ddot{x} = a, \tag{4.2}$$

and integrating once gives

$$\dot{x} = at + c, \tag{4.3}$$

where c is a constant. If it is assumed that u denotes the value of $\dot{x}$ when

$t = 0$, substitution into equation 4.3 gives $c = u$. A further integration gives

$$x = \tfrac{1}{2}at^2 + ut + b,$$

where b is another constant. If x_0 denotes the value of x when $t = 0$, then we see that $b = x_0$. The above results can be written in a standard form by setting $\dot{x} = v$ and writing s for the change in displacement in time t (that is, $s = x - x_0$). We then have

$$v = u + at, \qquad 4.4$$

$$s = ut + \tfrac{1}{2}at^2, \qquad 4.5$$

where v is sometimes referred to as the final velocity (that is, the velocity at time t) and u as the initial velocity. Eliminating t between equations 4.4 and 4.5 gives

$$v^2 = u^2 + 2as. \qquad 4.6$$

Equation 4.6 can also be obtained independently of the other two by using equation 4.1,

$$\ddot{x} = v\frac{dv}{dx}.$$

This can be re-written as

$$v\frac{dv}{dx} = \frac{d}{dx}\left(\frac{1}{2}v^2\right) = a.$$

Integration with respect to x now gives

$$v^2 = 2ax + c,$$

where c is a constant. At $t = 0$ we have $v^2 = u^2$ and $x = x_0$, so

$$c = u^2 - 2ax_0$$

and hence $$v^2 = u^2 + 2a(x - x_0).$$

Substituting s for $x - x_0$ gives equation 4.6.

Using equation 4.4, equation 4.5 can be written as

$$s = \tfrac{1}{2}(u + v)t. \qquad 4.7$$

This alternative form is sometimes useful in particular problems.

Equations 4.4 to 4.7 are sufficient to enable all problems involving motion with constant acceleration to be solved, and they should be committed to memory. Before describing the use of these equations in solving problems, we shall give them a graphical interpretation which is very useful in solving particular types of problems.

The graph of equation 4.4 is illustrated in Fig. 4.1. It is a graph of v against t, with v being the y-coordinate and t corresponding to the x-coordinate, and is a straight line of gradient a. The area of the region above the t-axis and under the line, and between $t = 0$ and $t = T$, is the area of the trapezium $OABC$ and this is $\frac{1}{2}(u+v)T$ which, by equation 4.7, is equal to s. This is a special case of the general result (shown in §4.6), that the area under the graph of v against t is equal to the distance covered. Effectively, all the information supplied in equations 4.4 and 4.5 is contained in Fig. 4.1 when the area $OABC$ is interpreted as s. Such a graph can be a useful, compact way of setting down the conditions in a given problem.

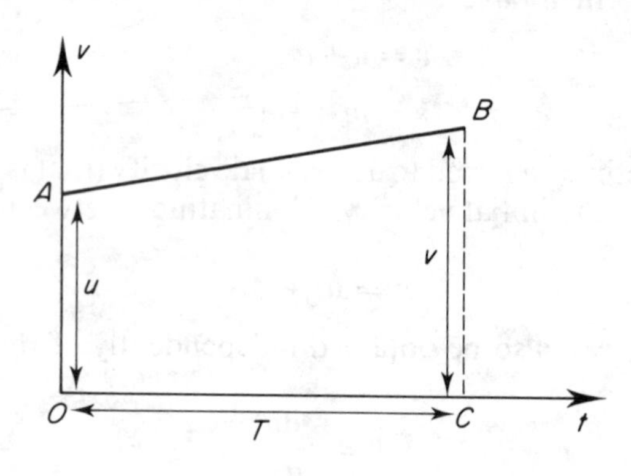

Fig. 4.1

Problem solving

The simplest kind of problem for motion under constant acceleration is the one in which u and a are given, and values of s and v are required for particular values of t. These problems are solved by substitution. In solving problems it is important that all the quantities in equations 4.4 to 4.7 are evaluated in the same unit system. In the following examples, we shall use the S.I. unit system and denote the initial and final speeds by $u\,\text{ms}^{-1}$ and $v\,\text{ms}^{-1}$ respectively, the acceleration by $a\,\text{ms}^{-2}$ and the time by t s. The quantities u, v, a and t are, therefore, dimensionless numbers which satisfy equations 4.4 to 4.7.

EXAMPLE 1 *Initially, a particle P moving with uniform acceleration* $4\,\text{ms}^{-2}$ *has a velocity of* $2\,\text{ms}^{-1}$. *Find its velocity after* 3 s *and the distance moved in the first* 4 s *of the motion.*

This is the simplest kind of problem, since both $u(=2)$ and $a(=4)$ are given. The velocity after 3 s is found by setting $t=3$ in equation 4.4, giving

$$v = 2+12 = 14,$$

so the velocity is $\mathbf{14\,ms^{-1}}$. The distance travelled in the first 4 s is found by substituting $t=4$ into equation 4.5, giving

$$s = 2.4+\tfrac{1}{2}4.16 = \mathbf{40}.$$

A slightly harder class of problem arises when u and a are not given directly, but sufficient information is available to find them. In solving this kind of problem, the best method is to list the unknowns and then find them systematically by choosing whichever of equations 4.4 to 4.7 contains only one unknown. This equation then gives that unknown. First try to find u and t. Then s and v can be found for all values of t by using equations 4.4 to 4.7.

EXAMPLE 2 *The velocity of a particle P moving with a uniform acceleration of* $4\,\text{ms}^{-2}$ *increases from* $2\,\text{ms}^{-1}$ *to* $4\,\text{ms}^{-1}$ *as P moves from A to B. Find the distance between A and B.*

In this case $v = 4$, $u = 2$, $a = 4$, s is required and this suggests using equation 4.6. Making the appropriate substitutions gives

$$16 = 4 + 8s,$$

so that $s = 1{\cdot}5$. The points A and B are therefore **1·5 m** apart.

This problem could also have been solved by substituting in equation 4.4 to find the total time

$$4 = 2 + at,$$

giving $t = 0{\cdot}5$, and then using equation 4·7 to obtain the value of s.

EXAMPLE 3 *The displacement from its original position of a particle moving with uniform acceleration is* 2 m *after* 2 s *and* 12 m *after* 4 s. *Find the displacement* 6 s *after the start of the motion.*

Neither a nor u is given, but values of s are given for two values of t. Substituting into equation 4.5 for these two values gives

$$2 = 2u + 2a, \qquad 12 = 4u + 8a,$$

so that $a = 2$ and $u = -1$. The displacement after 6 s is

$$s = (-1)6 + \tfrac{1}{2}2.36 = 30,$$

and therefore the required displacement is **30 m**.

Probably the most complicated problems involving constant acceleration are those in which the acceleration is constant for a particular period but then switches to another constant value for a different period. This kind of problem can occur, for example, in the motion of a train which accelerates from rest to a steady speed, keeps that steady speed for a while, and then retards to come to rest. In such problems, equations 4.4 to 4.7 have to be applied systematically for each period, and the information given in a question used to find all the unknowns. It is in these problems, where the given information can be complicated, that the graphical method described above is most useful. The given information can be displayed compactly on a diagram and elementary geometry used to complete the question. This approach is particularly useful for 'rest-to-rest' problems. The graph of v against t will be a series of zig-zag lines, with the parts of the motion where the acceleration is zero being represented by segments parallel to the t-axis.

EXAMPLE 4 *Starting from rest, a train moves with uniform acceleration and attains a maximum velocity of* 20 ms^{-1} *in* 50 s. *It runs at this velocity for* 35 s *and then comes to rest with uniform retardation in* 40 s. *Find the total distance travelled.*

Fig. 4.2 shows the information set out on a (v, t) graph.
The distance is the total area of the figure, which is **1600 m**.

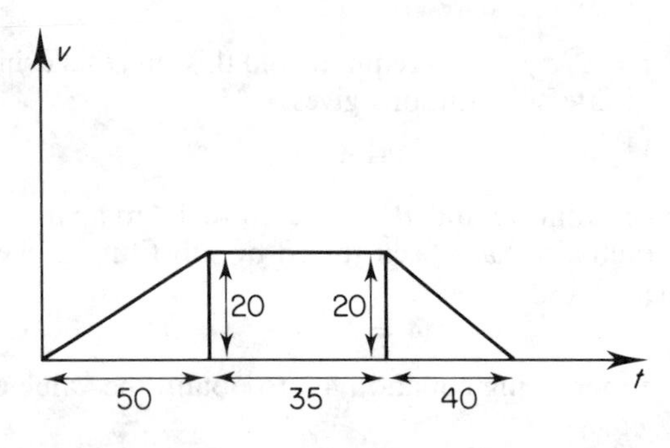

Fig. 4.2

EXAMPLE 5 *Over a* 100 m *length, a runner accelerates uniformly from* 6 ms^{-1} *to* 10 ms^{-1} *and then maintains the latter speed over the remaining length. If the total time for the* 100 m *distance is* 11 s, *find the acceleration.*

This is an example of a change in acceleration from an unknown value to zero and we therefore try to solve it using a (v, t) graph. Fig. 4.3 shows all the data set out on a graph.

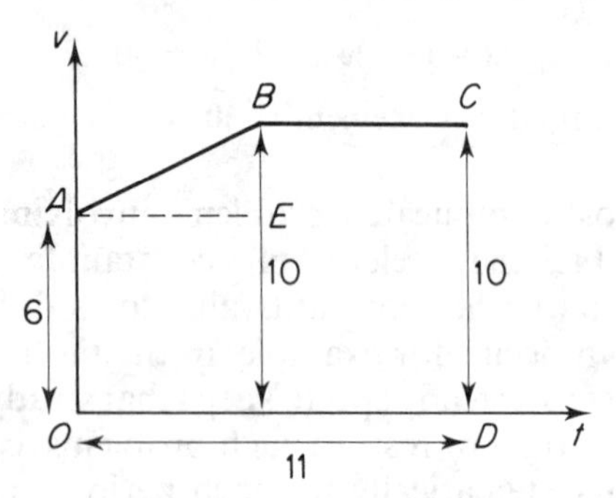

Fig. 4.3

The area of the figure is 100 and the gradient of AB is the unknown acceleration a ms^{-2}. We see that

$$\frac{4}{AE} = a$$

so that $AE = 4/a$. The total area is

$$\tfrac{1}{2}(6+10)\,AE + 10(11 - AE)$$

and since this is equal to 100, $AE = 5$. Therefore, the acceleration is $\mathbf{0{\cdot}8\,ms^{-2}}$.

EXERCISE 4.3

Questions 1 to 11 refer to a particle moving on a straight line with constant acceleration $a\,\text{ms}^{-2}$, so that at time t s the velocity and the displacement of the particle from a fixed point O are given by $v\,\text{ms}^{-1}$ and s m, the initial velocity of the particle being $u\,\text{ms}^{-1}$. The acceleration, velocities and displacement are with respect to the same reference direction.

1 $a = 4$, $v = 12$, $u = 4$; find t.
2 $a = -5$, $u = 3$, $v = -12$; find t and s.
3 $u = 6$, $t = 2$, $s = 20$; find v.
4 $u = -2$, $a = 3$, $t = 5$; find s.
5 $a = 4$, $t = 4$, $s = 42$; find u.
6 $v = 11$, $u = 7$, $t = 2$; find a.
7 $v = 8$, $u = 6$, $s = 7$; find a.
8 $s = 25$, $u = 4$, $v = 11$; find t.
9 $s = 70$, $u = 4$, $t = 5$; find a.
10 $s = 8$, $u = 9$, $a = -5$; find the possible values of t.
11 $s = 25$ when $v = 13$, $s = 52$ when $t = 4$; find the possible values of u and a.

Questions 12 to 14 refer to particles P and Q free to move on adjacent parallel lines; O is a fixed point assumed to be common to the lines.

12 At time $t = 0$ s, the particle P starts from rest with a constant acceleration of $2\,\text{ms}^{-2}$ and at time $t = 2$ s, the particle Q passes through O moving towards P with a constant speed of $9\,\text{ms}^{-1}$. Show that they are level with each other at two points and find the distance between these points.

13 At time $t = 0$ s, P passes through O moving towards a point B with a speed of $7\,\text{ms}^{-1}$ and thereafter moves with a constant acceleration of $2\,\text{ms}^{-2}$ directed towards B. At the same time, Q passes through the point B, which is 114 m away from O, with a velocity directed towards P of $3\,\text{ms}^{-1}$ and thereafter Q has acceleration $1\,\text{ms}^{-2}$ in the direction BO. Find when P and Q are level with each other.

14 At time $t = 0$ s, P passes through O with a speed of $4\,\text{ms}^{-1}$ and thereafter moves with a constant acceleration of $2\,\text{ms}^{-2}$. At time $t = 2$ s Q passes through the point B, 100 m away from O, with a velocity directed towards P of $17\,\text{ms}^{-1}$. Thereafter Q has an acceleration of $4\,\text{ms}^{-2}$ in the sense BO. Find when P and Q are at a distance of 60 m apart.

15 An underground train covers 576 m from rest to rest in 60 s. At first, it has a constant acceleration of $0.5\,\text{ms}^{-2}$, then moves uniformly and finally has a constant retardation of $1\,\text{ms}^{-2}$. Find the time taken for each stage of the journey.

16 A train approaching a station travels two successive distances of 0·25 km in 10 s and 20 s respectively. Assuming the retardation to be uniform, find the further time taken before coming to rest.

17 A train starting from rest in uniformly accelerated during the first 80 s of its journey in which it covers 600 m. It then runs at a constant speed until it is brought to rest in a distance of 750 m by applying a constant retardation. Find the maximum speed of the train and the magnitude of the retardation.

4.4 Vertical motion under gravity

It is an observed fact that particles free to move in a vertical direction near the earth have the same constant acceleration, denoted by $g(\approx 9{\cdot}8\,\text{ms}^{-2})$ downwards. Therefore, all problems involving such motion can be solved by the methods described above. If s is measured upwards from the point of projection then the previous formulae hold with $a = -g$, whilst if s is measured downwards a has to be replaced by g. It does not matter whether the upwards or downwards direction is taken as positive. Normally it is more sensible to measure s upwards for particles projected upwards, and downwards for those released from rest. Anything projected up will, of course, come down and in such cases, taking the positive direction of s to be upwards for the complete motion avoids sign errors.

A particle projected up with speed u will, by equation 4.6, have a speed of $(u^2 - 2gs)^{\frac{1}{2}}$ when at a height s. Therefore, the maximum height h reached will be when the speed is zero, so that

$$u^2 = 2gh. \qquad 4.8$$

If a particle is dropped from rest then its speed v at a depth s below the point of projection is given by $v^2 = 2gs$. Therefore, a particle projected upwards with a speed u will rise a distance $u^2/2g$, and its speed on the downward path at the point of projection (that is, at a depth $u^2/2g$ below the point at which it stops) is given by $V^2 = 2g(u^2/2g)$. So, the velocity when it returns to the point of projection is equal in magnitude but opposite in direction to the original velocity of projection.

EXAMPLE 1 *A stone is thrown vertically up with speed* 20 ms^{-1} *from the top of a building* 20 m *high. After what time and with what speed will it strike the ground? (Take g as* 10 ms^{-2}.)

The upward displacement s m at time t s is

$$s = 20t - 5t^2,$$

and we are required to find the time when the stone is a distance 20 m below its original position, that is, when $s = -20$. Therefore,

$$-20 = 20t - 5t^2,$$

that is,

$$t^2 - 4t - 4 = 0.$$

The positive root of this equation is $2 + \sqrt{8}$, so the time taken is **4·83 s**. The speed v when $s = -20$ can be found by substitution in equation 4.6 with $a = -10$, giving

$$v^2 = 400 + 400$$

and the speed is, therefore, $10\sqrt{8}$ ms^{-1} or approximately **28·3 ms^{-1}**.

EXAMPLE 2 *A particle is projected vertically upwards with a speed of* 10 ms^{-1}. *Find the maximum height reached and the time taken before the particle returns to the point of projection. (Take g as* 10 ms^{-2}.)

The maximum height is given from equation 4.8 as **5 m**, and the displacement s m from the point of projection at time t s is

$$s = 10t - 5t^2.$$

It returns to the projection point when $s = 0$, that is, after **2 s**.

EXERCISE 4.4

Take $g = 10$ ms^{-2} throughout this exercise. Questions 1 to 5 refer to a particle projected vertically upwards from a point O with speed u ms^{-2}.

1 $u = 20$; find the height reached and the time taken to reach the ground again.

2 The maximum height reached is 45 m. Find the speed of projection and the time taken to first reach a height of 40 m above O.

3 $u = 45$; find the times at which (a) the particle speed is 35 ms^{-1}, (b) the particle is at a height 90 m above O.

4 It is found that when the particle is 35 m above O it takes 6 s to reach that point again. Find u.

5 $u = 25$; find the time taken for the particle to reach 30 m below O.

Questions 6 and 7 refer to a particle projected vertically downwards from O with speed u ms^{-2}.

6 $u = 5$; find the speed when the particle has dropped 10 m and the time taken to reach this position.

7 $u = 4$; find the distance fallen in 9 s.

8 A particle is projected upwards from a point O with speed u, and at time T later, a second particle is projected upwards from O with the same speed. Find the time that passes before they meet.

4.5 Acceleration depending only on time

Problems involving accelerations depending only on time can be solved in a very similar way to those involving constant acceleration. We illustrate the general method by taking the case when $a = t$. We now have to solve

$$a = \frac{dv}{dt} = t, \qquad 4.9$$

and integrating this gives

$$\frac{dx}{dt} = v = \tfrac{1}{2}t^2 + c,$$

where c is a constant. Integrating this equation again gives

$$x = \tfrac{1}{6}t^3 + ct + b,$$

where b is a second constant. This general solution contains two arbitrary constants which can be found if x and $\dot{x}$ are given for one value of t, or x is given for two values of t. The algebra of finding the constants can be slightly simplified when v and x are given for a specific value of $t(t_0)$ by integrating both sides of equation 4.9 between t_0 and t. For example, if $v = u$ at $t = t_0$ we obtain

$$v - u = \tfrac{1}{2}(t^2 - t_0^2).$$

For arbitrary accelerations the process is the same:

(i) Integrate with respect to t to find $\dot{x}$, remembering to introduce an arbitrary constant or to integrate between limits if v is given for some value of t.
(ii) Integrate the expression for $\dot{x}$ to find x, remembering to introduce a second constant or to integrate between appropriate limits if possible. Then find the constants using the given initial conditions.

In more complicated problems, the form of the acceleration may differ from one time interval to another. In such cases, the general solutions should be found for each of the time intervals separately and the constants determined from the given conditions. The values of x and v at the end of one time interval may be needed to work out the appropriate constants in the succeeding interval.

EXAMPLE 1 *A particle moving under an acceleration of* t^2 ms^{-2} *at time* t s *has a velocity of* 1 ms^{-1} *when* $t = 3$ *and its displacement from a given point is* 6 m *when* $t = 12$. *Find its displacement from the given point when* $t = 6$.

The displacement x m at time t s satisfies the equation

$$\ddot{x} = \frac{dv}{dt} = t^2,$$

the integral of which is

$$\dot{x} = \tfrac{1}{3}t^3 + c,$$

where c is a constant. A second integration gives

$$x = \tfrac{1}{12}t^4 + ct + b,$$

where b is a further constant. Since $\dot{x} = 1$ at $t = 3$,

$$1 = 9 + c.$$

Also, $x = 6$ when $t = 12$, so

$$6 = 1728 + 12c + b.$$

Therefore $c = -8$ and $b = -1626$. When $t = 6$,

$$x = 108 - 48 - 1626 = -1566.$$

So, the displacement is **−1566 m.**

Some of the algebra could have been avoided by integrating the first equation between 3 and t giving

$$v - 1 = \frac{t^3}{3} - 9,$$

that is,
$$v = \dot{x} = \frac{t^3}{3} - 8.$$

Integrating this between 12 and t gives

$$x - 6 = \tfrac{1}{12}(t^4 - 12^4) - 8(t - 12),$$

which gives the same result as before.

EXAMPLE 2 *A particle starts from rest with acceleration* $(2 + 6t)$ ms^{-2} *at time* t s *and after* 2 s *the acceleration is maintained at a constant value of* 14 ms^{-2}. *Find the distance covered in the first* 5 s *of the motion.*

This motion has to be considered in two parts. For the first 2 s, the displacement x m at time t s satisfies

$$\ddot{x} = 2 + 6t.$$

Integrating with respect to t gives

$$\dot{x} = 2t + 3t^2 + b,$$

where b is a constant. A second integration with respect to t gives

$$x = t^2 + t^3 + bt + c,$$

where c is a further constant. At $t = 0$, $x = \dot{x} = 0$, and so $c = b = 0$.

After 2 s the constant acceleration formulae with $a = 14$ hold, but some care is needed in interpreting them. Remember that s, t and u in equation 4.5 refer to changes from some initial position, but in this problem a only becomes constant after 2 s. Hence equation 4.5 applies with s, u and t referred to the position where the acceleration first becomes constant. We have, with this interpretation,

$$s = ut + \frac{14t^2}{2}.$$

Here u is the velocity at the start of constant acceleration, that is, $\dot{x}$ at $t = 2$. Using the equation for $\dot{x}$ above, u is equal to

$$2.2 + 3.2^2 = 16.$$

A total time of 5 s means 3 s seconds at constant acceleration, so the displacement in the second part of the motion is

$$16.3 + \tfrac{14}{2}.9 = 111.$$

To this must be added x evaluated at $t = 2$, which is 12, so the total distance moved is **123 m**.

EXERCISE 4.5

The following questions refer to a particle moving along Ox, where x m denotes the displacement from O of the particle at time t s, and u ms^{-1} and a ms^{-2} denote the velocity and acceleration in the sense of increasing x, at time t s.

1 $u = 3t^3 + 4t^2 + 1$; find a.
2 $x = 7t^4 + 2t^3 + 5$; find a.
3 $a = 24t^2 + 18t + 2$, $u = 2$ and $x = 0$ when $t = 0$; find x.
4 $a = 20t^3 + 12t^2$, $x = 1$ when $t = 0$, $u = 6$ when $t = 1$; find x.
5 $a = e^{-t}$, $x = 2$ and $u = 4$ when $t = 0$; find x.
6 $a = 6t$ when $0 \leqslant t \leqslant 1$, $a = 6$ when $t \geqslant 1$, $u = 1$ and $x = 0$ when $t = 0$; find x for $t = 1$ and $t = 2$.
7 $a = 24t^2 + 6$ when $0 \leqslant t \leqslant 1$, $a = 30t$, $t \geqslant 1$, $u = 2$ and $x = 1$ when $t = 0$; find x for $t > 1$.
8 $a = |\sin t|$, $u = 2$ and $x = 0$ when $t = 0$; find x when $t = 2\pi$.
9 $a = |\cos t|$, $u = 3$ and $x = -1$ when $t = 0$; find x when $t = 2\pi$.
10 $u^2 = 2x^2 + 1$; find a in terms of x.

4.6 Graphical or numerical methods

In many cases the acceleration is not a simple function of time which can be integrated easily, and a graphical or numerical method has to be used to find x from $\ddot{x}$. The word 'graphical' is used to describe non-exact methods of solving kinematic problems, and does not mean that scale drawings are used. Most of the graphical methods of kinematics require the measurement of the area under a curve and though counting squares is one method of estimating area, it is not really necessary when pocket calculators are so easily available. Determining the area under a curve is equivalent to carrying out numerical integration and to a large extent, though not entirely, this is the most important aspect of non-exact methods of solution. Nowadays, the graphical aspect is merely the use of a graph to represent the information given.

There are several different graphical methods and we consider first those which are the direct analogues of the exact methods for determining $\ddot{x}$ from x described in §4.5. The two relevant graphs are known as the (a, t) and (v, t) graphs, the first letter always referring to the ordinate y and the second to the coordinate corresponding to x. In practice, when dealing with observed data (a, t) graphs are seldom used because acceleration is difficult to measure, though modern accelerometers can be very accurate. It is much easier to measure speed as in, for example, police radar checks.

Fig. 4.4 illustrates a possible (a, t) graph. The area of the region enclosed by $t = t_0$, $t = T$, the curve and the line $a = 0$ is, by the definition of an integral, equal to

$$\int_{t_0}^{T} a \, \mathrm{d}t = \int_{t_0}^{T} \frac{\mathrm{d}v}{\mathrm{d}t} \, \mathrm{d}t,$$

that is,

$$\text{area} = \text{velocity at time } T - \text{velocity at time } t_0.$$

Therefore, if the velocity is known for one value of t, it can be found for any other value by numerical integration. As mentioned earlier, this can be estimated, by counting squares but we shall concentrate on numerical

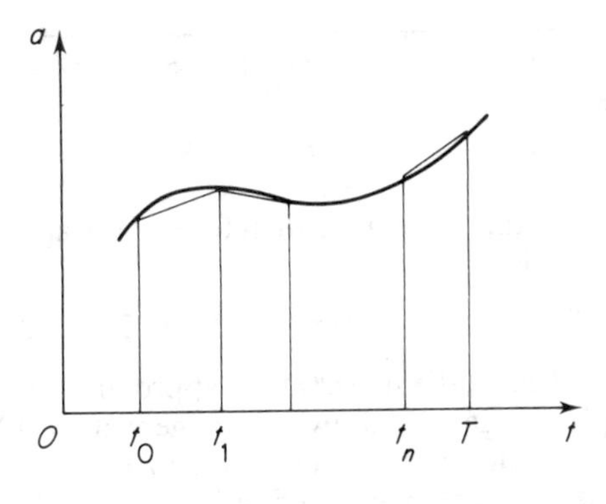

Fig. 4.4

methods. The simplest such method is to divide the segment between t_0 and T into small segments $t_0t_1, t_1t_2, \ldots, t_nT$ and let the values of a at time t_i be denoted by a_i and those at t_0 and T by a_0 and a_T. The area under the curve is thus effectively divided into small trapezia as shown in Fig. 4.4. The area of each trapezium can be worked out and then all of them added together to give an approximate value for the total area. For the special case when the segments have the same length h along the t-axis, the total area

$$= \tfrac{1}{2}(a_0 + a_1 + a_1 + a_2 + \ldots + a_{n-1} + a_n + a_n + a_T)h$$

$$= \frac{h}{2}[a_0 + a_T + 2 \times (\text{sum of ordinates at the inner points})].$$

Usually a more accurate estimate of the area can be found using Simpson's rule (*Pure Mathematics for Advanced Level* §20.6.)

If this process is continued, a graph of v against t or a (v, t) graph can be constructed. The area of the region under this graph between $t = t_0, t = T$ and the t-axis is

$$\int_{t_0}^{T} v\,\mathrm{d}t = \int_{t_0}^{T} \frac{\mathrm{d}x}{\mathrm{d}t}\,\mathrm{d}t,$$

that is

$$\text{area} = x \text{ at time } T - x \text{ at time } t_0.$$

Therefore, if x is known for one value of t, it can be found for all values of t. The gradient of the (v, t) graph is $\mathrm{d}v/\mathrm{d}t$, which is the acceleration, and hence measuring the gradient of the tangent to the (v, t) graph at a point gives the acceleration at that point. In this case it is far better to measure the gradient on a graph than to use a numerical method, as numerical methods of differentiation tend to be inaccurate.

EXAMPLE 1 *The table below gives simultaneous values of time and speed for a train. Estimate the distance travelled in the first* 200 s, *the acceleration after 2 minutes and the distance travelled in reaching a speed of* 22 ms^{-1}.

Time in s	0	50	100	150	200	250	300
Speed in ms^{-1}	0	10·8	16·0	19·2	21·3	22·6	23·5

The time intervals are the same and, therefore, the trapezoidal rule gives the distance as approximately

$$\tfrac{1}{2}[0+21{\cdot}3+2\times(10{\cdot}8+16{\cdot}0+19{\cdot}2)]\times 50 \text{ m}.$$

So the distance travelled in the first 200 s is approximately **2·83 km**.

Plotting the graph of v against t, drawing a tangent at $t = 120$ and measuring its gradient gives an estimate of the acceleration to be **0·067 ms^{-2}**. To find the distance travelled before reaching a speed of 22 ms^{-1} we need to know the time when this is reached and the graph gives $t = 238$. The distance is, therefore, that already calculated plus the area of the 'trapezium' formed by the curve and the lines $t = 200$ and $t = 238$. This latter area is approximately

$$0{\cdot}5\times 38\times(21{\cdot}3+22) = 822{\cdot}7$$

and hence the total distance covered is approximately **3·65 km**.

In experimental work, the easiest variables to observe are the displacements and the times; and the observations can be used to draw an (s, t) graph. The basic significance of such a graph is that the gradient at any point gives the velocity at that point. There is also another graph whose gradient can be used to give a relevant physical quantity, and this is the $(\tfrac{1}{2}v^2, s)$ graph. The gradient in this case is

$$\frac{\mathrm{d}}{\mathrm{d}s}(\tfrac{1}{2}v^2) = \frac{v\,\mathrm{d}v}{\mathrm{d}s}$$

and this, from equation 4.1, is the acceleration.

Other graphs which can be used to obtain numerical information by calculating areas are the $\left(\frac{1}{v}, s\right)$, $\left(\frac{1}{a}, v\right)$ and the (a, s) graphs.

$\left(\frac{1}{v}, s\right)$ graph

The area of the region under the curve is

$$\int\frac{\mathrm{d}s}{v} = \int \mathrm{d}t,$$

that is,

area = difference in times.

$\left(\frac{1}{a}, v\right)$ graph

The area of the region under the curve is

$$\int \frac{\mathrm{d}v}{a} = \int \mathrm{d}t$$

that is,

area = difference in times.

(a, s) graph
The area of the region under the curve is

$$\int a \mathrm{d}s = \int \frac{\mathrm{d}}{\mathrm{d}s}(\tfrac{1}{2}v^2)\mathrm{d}s$$

that is,

area = change in $\frac{1}{2}v^2$.

EXERCISES 4.6

The following questions refer to a particle moving along Ox, where x m denotes the displacement from O of the particle at time t s and u ms^{-1} and a ms^{-2} denote the velocity and acceleration, in the sense of increasing x, at time t s. Give the answers to question 1 correct to the nearest integer and to the other questions correct to one decimal place.

1 Corresponding values of a and t are given by

t	0	1	2	3	4
a	42	120	214	280	390

Given that $u = 1$ when $t = 0$ estimate, using the trapezoidal rule, u when $t = 1, 2, 3, 4$ and hence estimate the distances travelled in each of the first 4 s.

2 Corresponding values of x and a are given by

x	0	0·5	1	1·5	2	2·5	3
a	0	1·2	6	14·3	34	71	110

Estimate, using both the trapezoidal rule and Simpson's rule, u when $x = 3$ given that $u = 2$ when $x = 0$.

3 Corresponding values of x and u are given by

x	1	2	3	4
u	3·1	4·7	7·3	8·8

Estimate, using the trapezoidal rule, the time taken for the speed to increase from 3·1 ms^{-1} to 8·8 ms^{-1}.

4 Corresponding values of u and a are given by

u	1	2	3	4
a	1·8	3·1	4·3	5·5

Estimate, using the trapezoidal rule, the time taken for the speed to increase from 1 ms^{-1} to 4 ms^{-1}.

MISCELLANEOUS EXERCISE 4

1 A particle is uniformly accelerated from A to B, a distance of 96 m, and is then uniformly retarded from B to C, a distance of 30 m. The speeds of the particle at A and B are 6 ms^{-1} and u ms^{-1} respectively and the particle comes to rest at C. Express, in terms of u only, the times taken by the particle to move from A to B and from B to C.

Given that the total time taken by the particle to move from A to C is 18 seconds, find
(i) the value of u,
(ii) the acceleration and the retardation of the particle. (*AEB 1977*)

2 Due to track repairs a train retards uniformly, with retardation 1 ms^{-2}, from a speed of 40 ms^{-1} at A to a speed of 10 ms^{-1} at B. The train travels from B to C, a distance of 3·5 km, at a constant speed of 10 ms^{-1} and then accelerates uniformly, with acceleration 0·2 ms^{-2} so that its speed at D is 40 ms^{-1}. Sketch the velocity-time graph for the journey from A to D, and show that the distance from A to D is 8 km.

Show that the journey from A to D takes 330 s more than it would if the train travelled at a constant speed of 40 ms^{-1} from A to D. (*C*)

3 A lift travels vertically a distance of 22 m from rest at the basement to rest at the top floor. Initially the lift moves with constant acceleration x m/s^2 for a distance of 5 m; it continues with constant speed u m/s for 14 m; it is then brought to rest at the top floor by a constant retardation y m/s^2. State, in terms of u only, the times taken by the lift to cover the three stages of its journey.

If the total time that the lift is moving is 6 seconds, calculate
(i) the value of u,
(ii) the values of x and y. (*AEB 1975*)

4 The vertical descent of a lift-cage is undertaken in three stages. During the first stage the lift uniformly accelerates from rest at $5k$ m/s^2, during the second stage it moves at a constant speed of 10 m/s and during the third stage it uniformly retards at $2k$ m/s^2 until it comes to rest.
(i) Express, in terms of k, the times taken during the first and third stages of the descent.
(ii) Given that the total distance covered by the lift during the descent is 350 m and that this distance is covered in 40 s, calculate the value of k.
(*AEB 1980*)

5 A car A moves with constant acceleration f from rest up to its maximum speed U. Just as it starts it is overtaken by a car B moving with constant speed V. Given that A catches up with B in a distance a ($> U^2/2f$), show that A has been travelling with speed U for a time

$$(a/V)-(U/f).$$

Show also that

$$U^2V-2afU+2afV=0.$$

Hence find U in terms of a, f and V, explaining carefully how you decide which root to choose of the quadratic equation. (*L*)

6 A bus starts from rest and moves along a straight road with constant acceleration f until its speed is V; it then continues at constant speed V. When

the bus starts, a car is at a distance b behind the bus and is moving in the same direction with constant speed u. Find the distance of the car behind the bus at time t after the bus has started
(i) for $0 < t < V/f$ (ii) for $t > V/f$.

Show that the car cannot overtake the bus during the period $0 < t < V/f$ unless $u^2 > 2fb$.

Find the least distance between the car and the bus in the case when $u^2 < 2fb$ and $u < V$. State briefly what will happen if $u^2 < 2fb$ and $u > V$. (*JMB*)

7 Two tube stations are d m apart. An electric train travelling between the two stations is able to accelerate uniformly at $\alpha\,\mathrm{ms}^{-2}$ and brake uniformly at $2\alpha\,\mathrm{ms}^{-2}$. The train is subject to a maximum speed limit of $\sqrt{(\alpha d/3)}\,\mathrm{ms}^{-1}$ during the journey between the two stations. If the train completes the journey, from rest to rest, in the shortest possible time without exceeding the speed limit, find the time taken and show that the distance covered at maximum speed is $3d/4$ m. Sketch the velocity-time and the distance-time graphs, illustrating in each case the three different stages of the journey between the two stations. (*AEB 1970*)

8 Two trains A and B are travelling in the same direction on parallel tracks; A has a constant acceleration and B has a constant deceleration. At a given time they are abreast and their speeds are in the ratio $1:n$. After a further time t they are again abreast, having travelled a distance a, and their speeds are in the ratio $n:1$. Prove that the acceleration of A and the deceleration of B are each equal to

$$2a(n-1)/t^2(n+1).$$

Prove that A and B both pass the half-way mark with speed

$$a\sqrt{(2n^2+2)}/t(n+1). \qquad (O)$$

9 Two points A and B are 60 m apart on rough horizontal ground. A particle P is projected along the ground from A towards B with speed $4\,\mathrm{ms}^{-1}$ and moves under a constant frictional retardation of $\frac{1}{3}\,\mathrm{ms}^{-2}$. At the same instant a particle Q is projected along the ground from B towards A with speed $v\,\mathrm{ms}^{-1}$ and moves under a constant frictional retardation of $2\,\mathrm{ms}^{-2}$. By considering the distance travelled by each particle before coming to rest, show that the particles collide if $v \geqslant 12$.

If $v = 13$, show that the collision occurs 18 m from A.

Find the value of v if the collision occurs $\frac{40}{3}$ m from A. (*C*)

10 When the driver of a car sees an obstruction ahead, there is a delay of T s (the driver's reaction-time) before he takes any action. He then applies the brakes, and the car is subject to a constant retardation of $F\,\mathrm{ms}^{-2}$. If the driver sees an obstruction when travelling at a uniform speed of $12\,\mathrm{ms}^{-1}$, he can bring the car to rest in 20 m; at a uniform speed of $24\,\mathrm{ms}^{-1}$ the corresponding distance is 64 m. Calculate F and T.

The driver sees an obstruction when travelling at $18\,\mathrm{ms}^{-1}$ with a uniform acceleration of $3\,\mathrm{ms}^{-2}$. Show that he can bring the car to rest in 46 m.

When the car is travelling along a narrow road with a uniform speed of $30\,\mathrm{ms}^{-1}$, the driver sees a lorry X m ahead. The lorry is travelling in the same direction as the car with uniform speed $6\,\mathrm{ms}^{-1}$. Find the condition which X must satisfy if a collision is to be avoided. (*C*)

11 A car is travelling along a motorway at 110 km/h. At a point A the driver observes a sign warning of road works ahead and ordering a reduction of speed to v km/h. The driver immediately applies the brakes so that the speed of the car is uniformly retarded in t minutes to v km/h. The car maintains this speed until a 'road clear' sign is reached. The driver then accelerates uniformly for $3t$ minutes until the car reaches a speed of 110 km/h at a point B. The distance travelled at v km/h is $2\frac{1}{2}$ km, $AB = 6\frac{1}{2}$ km and the car takes 6 minutes to travel from A to B. Sketch the velocity-time graph and hence, or otherwise, show that
(i) $v(6-4t) = 150$, (ii) $(110+v)t = 120$.
Hence show that $v = 50$. *(AEB 1971)*

12 After passing a police radar check-point at 60 km/h, a sports car immediately began to retard uniformly until its speed was 40 km/h and it continued to move at this speed until it was passed by a police car 1 km from the check-point. This police car had started from rest at the check-point at the same instant as the sports car had passed the check-point and had then moved with constant acceleration until it had passed the sports car. Assuming that the time taken by the sports car in slowing down from 60 km/h to 40 km/h was equal to the time that it travelled at constant speed before being passed by the police car, find, by using a velocity-time sketch, or otherwise,
(i) the time taken by the police car to reach the sports car,
(ii) the speed of the police car at the instant when it passed the sports car,
(iii) the time, measured from the check-point, when the speeds of the two cars were equal. *(AEB 1972)*

13 At time $t = 0$ a particle A is projected vertically upwards from the point O with speed U. At time $t = U/(2g)$ a particle B is projected vertically upwards from O with speed $3U$. Show that the particles collide before A has reached its maximum height.

Find the height above O at which the particles collide.

If, instead of B, another particle C is projected vertically upwards from O at time $t = U/(2g)$ and collides with A at time $t = U/g$, find the speed of projection of C. *(L)*

14 Two free-falling raindrops leave the top of a cliff of height h such that the second one begins to fall when the first one has already fallen a distance s. Show that the distance between the drops when the first drop hits the ground is

$$2(hs)^{\frac{1}{2}} - s.$$

If the height of the cliff is 28 m and this final distance apart is 3 m, find the value of s to the nearest centimetre. Take g as 9.8 ms^{-2}. *(W)*

15 A particle moving along a straight line starts at time $t = 0$ seconds with a velocity 4 ms^{-1}. At any subsequent time t seconds the acceleration of the particle is $(6t-8)$ ms^{-2}. Find
(a) the distance the particle moves before first coming to instantaneous rest.
(b) the total time T seconds taken by the particle to return to the starting point,
(c) the greatest speed of the particle for $0 \leqslant t \leqslant T$. *(L)*

16 A particle moves in a straight line so that its speed is inversely proportional to $(t+1)$, where t is the time in seconds for which it has been moving. After

2 seconds, the particle has a retardation of $10/9 \text{ m/s}^2$. Calculate the distance moved in the first second of the motion. (*L*)

17 A particle moves in a straight line. At time t its displacement from a fixed point O on the line is x, its velocity is v, and its acceleration is $k \sin \omega t$, where k and ω are constant. The displacement, velocity and acceleration are all measured in the same direction. Initially the particle is at O and is moving with velocity u. Show that

$$v = \frac{k}{\omega}(1 - \cos \omega t) + u$$

and hence find x in terms of t, k, ω and u.
(i) Show that, if $u = 0$, the particle first comes to instantaneous rest after travelling a distance $2\pi k/\omega^2$. (*JMB*)

18 A particle, moving in a straight line, starts from rest at time $t = 0$. At time t seconds the velocity v m/s of the particle is given by

$$v = 3t(t-4) \quad \text{for} \quad 0 \leqslant t \leqslant 5,$$
$$v = 75/t \quad \text{for} \quad 5 \leqslant t \leqslant k,$$

where k is a constant.
(i) Sketch a velocity-time graph for the particle for $0 \leqslant t \leqslant k$.
(ii) Find the set of values of t for which the acceleration of the particle is positive.
(iii) Show that the *total* distance covered by the particle in the interval $0 \leqslant t \leqslant 5$ is 39 metres.
(iv) Given that the distance covered by the particle in the interval $5 \leqslant t \leqslant k$ is also 39 metres, find, to 2 significant figures, the value of k.

19 A car of mass 1000 kg moves along a horizontal road with acceleration proportional to the cube root of the time t seconds after starting from rest. When $t = 8$, the speed of the car is 8 m/s. Neglecting frictional resistances, calculate in kW, the rate at which the engine driving the car is working when $t = 27$. (*AEB 1979*)

20 A car of mass 1000 kg moves from rest on a straight level road. The engine exerts a tractive force P which varies with the time t, while the resistance to motion can be assumed constant at 50 newtons. The table below gives the values of P in newtons against the time t in seconds. Use Simpson's rule to estimate
(a) the speeds when $t = 4$ seconds and $t = 8$ seconds.
(b) the distance travelled in the first 8 seconds.

P	250	350	390	420	450
t	0	2	4	6	8

(*L*)

21 A train of mass 90 000 kg is accelerated from rest on a level track and reaches a maximum speed of 40 ms^{-1} in 2 minutes. The speed during the acceleration varies with time according to the table.

Speed (ms^{-1})	0	16	26	32	37	39	40
Time (s)	0	20	40	60	80	100	120

Use Simpson's rule to estimate the distance moved by the train in reaching the maximum speed.

The resistance to the motion from frictional forces is proportional to the speed. When the train is moving at its maximum speed the power exerted by the engine is 3 600 kW. Find (to the nearest kW) an approximation to the power being exerted by the engine 40 s after the start. (*JMB*)

22 Show how the time taken by a particle to travel a given distance can be found from the graph of the reciprocal of the speed against the distance covered by the particle.

The speeds of a car at distances from a check-point are noted and recorded in the following table:

Distance (km)	0	0·2	0·4	0·6	0·8	1·0	1·2
Speed (km/h)	20	37·6	48·8	58·8	66·6	74·0	80·0

Find, graphically or otherwise, the time taken by the car to travel 1·2 km from the check-point. (*AEB 1977*)

23 A car of mass 1 tonne was started from rest and given continuous acceleration for 35 seconds. Measurements of the resultant propulsive force were recorded and are shown in the following table correct to the nearest ten newtons.

Time/s	0	5	10	15	20	25	30	35
Force/N	1200	1020	870	740	620	500	390	290

Use the trapezium rule to calculate an estimate of the speed of the car in metres per second at the end of the 35 second period. (*JMB*)

5 Dynamics of Rectilinear Motion

In chapters 2 and 4, forces and motion have been studied separately. We now proceed to study these together and examine the effect of forces on bodies and the changes in motion produced by these forces. The study is based on Newton's three laws.

5.1 Newton's laws of motion

First law

This law states that every body continues in a state of rest or of uniform motion in a straight line unless compelled to change that state by a force. If a body changes from a state of uniform motion, that body has an acceleration and, therefore, Newton's first law can be interpreted as stating that a force acting on a body produces an acceleration and that any body which possesses an acceleration is being acted upon by a force.

It is observed experimentally that if a given force is applied to various bodies, the accelerations produced vary with the bodies considered. A further observation is that, whatever the magnitude of a particular force, the ratios of the magnitudes of the accelerations produced by the same force acting on the different bodies are constant. Therefore, there exists some independent relative property of the bodies which is demonstrated by this ratio. This property is referred to as the *mass* of a body.

Having recognised the existence of mass, the next step is to quantify it. The accelerations produced by the same force acting on different particles P, Q and R are measured and the magnitude of these accelerations are $|\mathbf{a}_P|$, $|\mathbf{a}_Q|$ and $|\mathbf{a}_R|$ respectively. The masses of P, Q and R, denoted by m_P, m_Q and m_R, are then defined so that

$$m_Q/m_P = |\mathbf{a}_P|/|\mathbf{a}_Q|, \; m_R/m_P = |\mathbf{a}_P|/|\mathbf{a}_R|.$$

If one of these particles (P say) is then chosen to have unit mass, the masses of the other particles can be found by measuring the accelerations. It can also be verified experimentally that, for mass defined in this way, the mass of the combined particle Q and R is the sum of the masses of Q and R.

Therefore, in principle a method of measuring mass can be established, and the definition of the unit of mass is independent of the choice of units of length, time and force.

Second law

Force, acceleration and mass are, therefore, all independently measurable. Many experiments have verified that force is proportional to the product of mass and acceleration. This statement expressed in symbols is

$$F = kma, \qquad 5.1$$

where k is a positive constant, F is the resolved part of the force in a particular reference direction and a is the acceleration in that direction. The unit system chosen is such that $k = 1$, so that one unit of force (the Newton, N) produces unit acceleration (1 ms^{-2}) when acting on unit mass (1 kg). In this system, equation 5.1 becomes

$$F = ma. \qquad 5.2$$

The vector nature of equation 5.2 is implicit in the fact that F and a refer to quantities measured with respect to the same direction. The same general relation, however, holds for non-rectilinear motion in which equation 5.2 becomes

$$\mathbf{F} = m\mathbf{a}. \qquad 5.3$$

The vector nature of Newton's second law is explicit in equation 5.3.

Third law

Newton's third law states that action and reaction are equal and opposite. This means that if a body B exerts a force $\mathbf{F}$ on a body A, then A exerts a force $-\mathbf{F}$ on B. A simple example of this is supplied by a man pushing against a wall; if the man exerts a force $\mathbf{F}$ on the wall, then the wall exerts a force $-\mathbf{F}$ on the man.

The third law also applies to bodies in motion. If an engine pushes a tender with a force $\mathbf{F}$ when both engine and tender are in motion, the tender exerts a force $-\mathbf{F}$ on the engine.

5.2 The force of gravity

A particle of mass m left to fall freely is observed to have a constant acceleration of magnitude 9·8 ms^{-2}. This acceleration is referred to as the acceleration due to gravity and is denoted by $\mathbf{g}$. We conclude from Newton's first law that there is a downward force acting on the particle, the force due to gravity. This force is called the weight of the particle. The magnitude of this weight, obtained from Newton's second law, is mg, where g is the magnitude of $\mathbf{g}$.

5.3 Problems involving constant forces

It is most important in setting up an equation of motion to make a clear sketch showing the position of a system at any time, and to choose a reference direction. All displacements must be measured from a fixed origin. All forces should be marked on the sketch and then the resultant force in the reference direction calculated. Equation 5.2 then gives the acceleration in that direction. This procedure is illustrated in the following simple examples.

EXAMPLE 1 *A stone of mass* $\frac{1}{2}$ kg *is falling through water. The buoyancy of the water provides an upward force of* 4 N *and the resistance provides a further upward force of* $\frac{1}{2}$ N. *Find the acceleration of the stone.*

We take the reference direction to be vertically downwards. The vertical forces acting are the force of gravity of $0{\cdot}5\,g$ downwards, the resistance and the buoyancy, and these forces are shown in Fig. 5.1. The total force in the downward direction is

$$(0{\cdot}5 \times 9{\cdot}8 - 4 - 0{\cdot}5)\,\text{N} = 0{\cdot}4\,\text{N}.$$

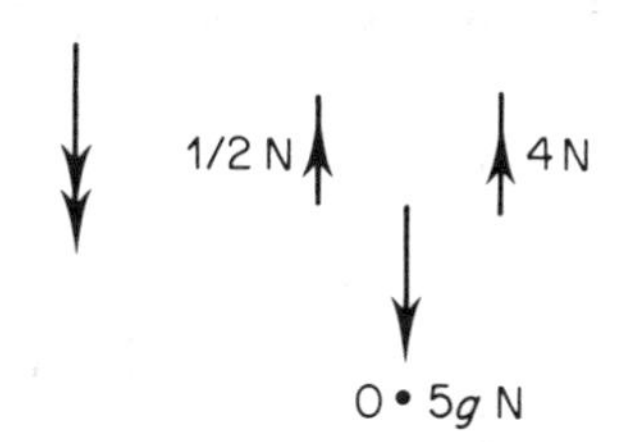

Fig. 5.1

Hence, the downward acceleration a is given by

$$0{\cdot}4 = 0{\cdot}5a$$

and the acceleration is $\mathbf{0{\cdot}8\,ms^{-2}}$.
If we were given the speed and the position of the stone at any instant, then we could use the constant acceleration (equations 4.4 to 4.7 (§4.3)) to determine the position of the stone completely.

EXAMPLE 2 *A bobsleigh is being pushed by two men, each of whom push with a force of* 150 N *in the same direction. The sleigh is on horizontal ground and while it is being pushed it moves with acceleration* 3 ms^{-2}. *The mass of the sleigh is* 80 kg. *Find the frictional resistance to the sleigh.*

We choose the reference direction to be that in which the two men are pushing. The only other force acting on the sleigh is the frictional resistance of magnitude

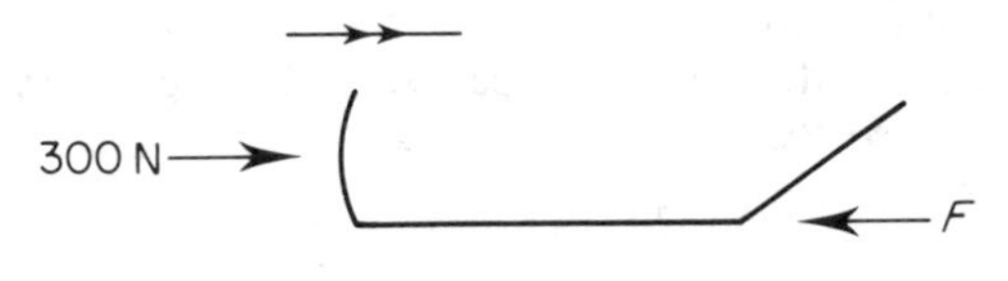

Fig. 5.2

F, which will be in the opposite direction to that in which the men are pushing. The total force in the reference direction is, therefore, $(300-F)$ N. Hence, by equation 5.2

$$300-F = 80 \times 3$$
$$F = \mathbf{60\,N}.$$

EXAMPLE 3 *A man of mass 70 kg is in a lift. Find the forces exerted by the floor on the man*
(*i*) *when the lift descends with an acceleration of* $0{\cdot}5\,\text{ms}^{-2}$,
(*ii*) *when on its downward journey, the lift is slowing up with a downward retardation of* $0{\cdot}5\,\text{ms}^{-2}$.

We take the downward direction as the reference one, and the forces are as shown in Fig. 5.3.
(i) The only forces acting on the man are the force of gravity down and the unknown reaction of the floor, of magnitude X N, acting up.
The total downward force in newtons is, therefore, $70 \times 9{\cdot}8 - X$.

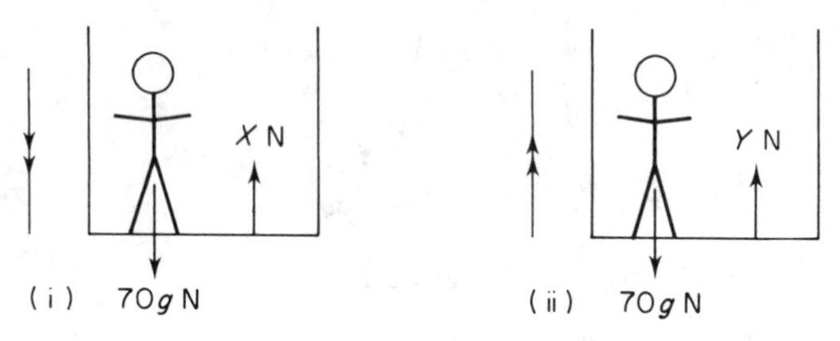

Fig. 5.3

Hence,

$$70 \times 9{\cdot}8 - X = 70 \times 0{\cdot}5$$
$$X = 651.$$

So the force exerted by the floor on the man is **651 N** which is less than the man's weight. This would be demonstrated in practice by a slight lightening effect on the legs of the man when the lift starts to accelerate downwards.
(ii) A downward retardation of $0{\cdot}5\,\text{ms}^{-2}$ means that the downward acceleration is $-0{\cdot}5\,\text{ms}^{-2}$ and, therefore, using the same reference direction with the upward force exerted on the man now being Y N,

$$70 \times 9{\cdot}8 - Y = -70 \times 0{\cdot}5$$
$$Y = 721.$$

In this case the force is **721 N** and there is an extra thrust on the legs as a lift slows down when going downwards. A downward retardation means effectively a positive acceleration upwards.

The second part could also have been solved by changing the reference direction to upwards so that

$$Y - 70 \times 9{\cdot}8 = 70 \times 0{\cdot}5$$

giving, of course, $Y = 721$, as before.

EXAMPLE 4 *An engine of mass* 50 tonnes *pushes a carriage of mass* 10 tonnes *with an acceleration of* $0{\cdot}4\,\text{ms}^{-2}$. *The resistance to the motion of the engine is* 1200 N *and to the motion of the carriage is* 800 N. *Find*
(*i*) *the force in the buffers exerted on the carriage,*
(*ii*) *the total driving force of the engine.* (*Note that* 1 tonne = 1000 kg.)

We must consider separately (i) the motion of the carriage and (ii) the motion of the engine. The reference direction is that of the acceleration and the forces are shown in Fig. 5.4.
(i) The forces on the carriage are the push of magnitude P N transmitted through the buffers, and the resistance opposing the motion. Hence,

$$P - 800 = 10\,000 \times 0{\cdot}4$$
$$P = 4\,800.$$

The force in the buffers is **4 800 N**.

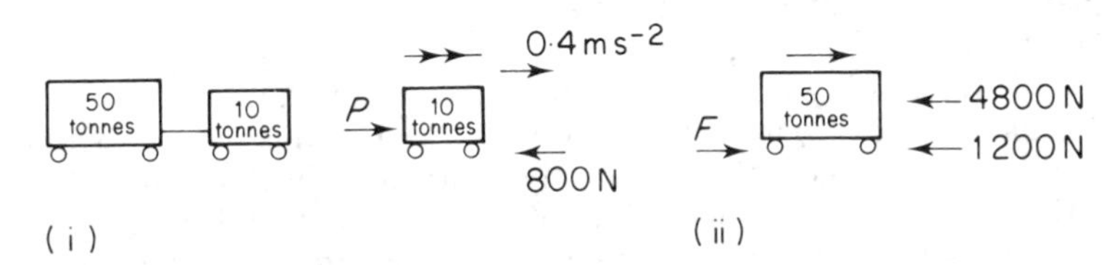

Fig. 5.4

(ii) We use Newton's third law in stating that the buffers push back on the engine with the force $-P$ N. Using the same reference direction, the forces on the engine are the driving force, F N transmitted at the points where the wheels are in contact with the track, the resistance and the force in the buffers. Hence,

$$F - 4\,800 - 1\,200 = 50\,000 \times 0{\cdot}4$$
$$F = 26\,000.$$

The driving force is **26 000 N**.

EXERCISE 5.3

1 A tug tows a barge of mass 10 000 kg with acceleration $0{\cdot}1\,\text{ms}^{-2}$. The water resistance to the movement of the barge is 1000 N. Given that the tow rope is horizontal, calculate the tension in this rope. (You may assume that the acceleration and velocity of the barge are in the direction of the tow rope.) (*L*)

2 A car of mass 800 kg moving along a level road has its speed reduced from $18\,\text{ms}^{-1}$ to $12\,\text{ms}^{-1}$ by a constant retarding force of 2000 N. Calculate, for the interval during which the car reduces speed,
(a) the time taken,
(b) the distance covered. (*L*)

3 A constant force acting vertically upwards on a body of mass 10 kg moves the body from rest to a height of 5 m above its starting point and gives it a speed of $6\,\text{ms}^{-1}$. Find the magnitude of the force.

4 A car is moving along a horizontal road against a constant resistance to motion of 1500 N. If the engine of the car is exerting a constant force of 5500 N and the car reaches a speed of 25 ms^{-1} from rest in 10 s, calculate
(a) the acceleration, in ms^{-2} during the first 10 s,
(b) the mass in kg of the car,
(c) the distance in m travelled by the car during the first 10 s. *(L)*

5 A truck is moving along a horizontal road at a steady speed of 30 ms^{-1}. Show that the magnitude of the uniform force necessary to bring the truck to rest in a distance of 40 m is three times the magnitude of the uniform force necessary to bring the truck to rest in a time of 8 s. *(L)*

6 A car of mass 800 kg tows a caravan of mass 600 kg along a horizontal road. Given that the driving force is 280 N and neglecting resistances, calculate the acceleration in ms^{-2} of the car.

7 A car of mass 800 kg is pulling a caravan of mass 600 kg by means of a tow bar along a straight horizontal road.The resistive forces opposing the motions of the car and the caravan are 100 N and 60 N respectively. Given that the car is accelerating at $\frac{1}{2}$ ms^{-2}, calculate in N
(a) the tension in the tow bar,
(b) the force being produced by the engine of the car. *(L)*

8 A lift accelerates at $\frac{1}{4}$ ms^{-2} to a maximum speed of 3 ms^{-1} and, after moving at this constant speed, it is uniformly retarded at $\frac{1}{2}$ ms^{-2}. A parcel of mass 10 kg is standing on the floor of the lift during this ascent from the basement to a penthouse. Find in N the magnitude of the force exerted by the floor of the lift on the parcel during each stage of the ascent. *(L)*

9 A tug, of mass 50 000 kg tows three barges in line astern along a canal. Each barge is of mass 10 000 kg and the resistance to the motion of each barge is 800 N. The resistance to the motion of the tug is 3 000 N. Calculate the tensions in the tow ropes between
(a) the tug and the first barge,
(b) the first and second barges,
(c) the second and third barges, when
(i) the tug and barges are moving at a uniform speed,
(ii) they are all accelerating at $\frac{1}{2}$ ms^{-2}.

10 A stone is sliding over the ice on a pond and it slides a distance of 200 m in coming to rest from a speed of 15 ms^{-1}. Assuming that the force of friction between the stone and the ice is constant, calculate the coefficient of friction.

5.4 Quasi-rectilinear problems

There are some problems in which motion is only possible along a line but in which the forces do not act completely along this line. For lack of a better word we shall refer to these as quasi-rectilinear problems. A simple example of such a problem is when a particle of mass m slides along a smooth plane inclined at an angle θ to the horizontal as illustrated in Fig. 5.5(a).

The forces acting on the particle are the reaction of the plane acting in a direction perpendicular to the plane and the force of gravity. Neither of

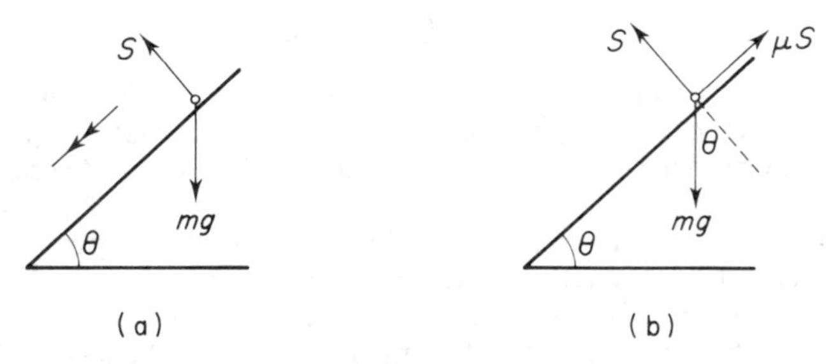

Fig. 5.5

these lies along the plane. The reference direction is chosen as downwards along a line of greatest slope of the plane.

Strictly, such a problem should be solved by using the vector equation $\mathbf{F} = m\mathbf{a}$ but, as the acceleration is along one line, the problem can be split into separate dynamical and statical ones. The total force in the direction perpendicular to the line of possible motion is zero, since the acceleration in that direction is zero. Therefore, we may obtain one equation by resolving in this direction. Taking the resolved part of the vector equation $\mathbf{F} = m\mathbf{a}$ along the direction of motion:

Resolved part of the total force in given direction = mass × acceleration in given direction.

Hence, all that has to be done is to find the sum of the resolved parts of all the forces along the line of motion and then treat the problem as a simple rectilinear one.

In the particular example of motion down an inclined plane, we choose the reference direction to be downwards along the line of greatest slope of the plane. The resolved part of the force of gravity is magnitude $mg \sin \theta$. As the plane is smooth, the reaction of the plane, of magnitude S, is normal to the plane and has no resolved part along the plane.
The downward acceleration, a, along the reference direction is given by

$$mg \sin \theta = ma$$
$$a = g \sin \theta.$$

Fig. 5.5(b) shows the situation when the plane is rough and the particle is sliding down the plane. Here, it is necessary to find the normal reaction to the plane in order to find the force of friction. Resolving in a direction at right angles to the plane as there is no resolved part of the acceleration in this direction,

$$S = mg \cos \theta.$$

Hence the frictional force is $\mu mg \cos \theta$.

In this case, $\quad mg \sin \theta - \mu mg \cos \theta = ma$

$$a = g(\sin \theta - \mu \cos \theta).$$

EXERCISE 5.4

1 A ring of mass 0·01 kg is threaded on a smooth fixed vertical wire. A force of 0·3 N acts on the ring and the line of action of the force makes an angle of 60° with the upward vertical. Find the magnitude of the upward acceleration of the ring.

2 A particle of mass m starts from rest and moves up a line of greatest slope of a plane inclined at an angle θ to the horizontal, where $\tan\theta = \frac{3}{4}$, under the influence of a force F of magnitude $3mg/2$ acting parallel to the plane. Given that the coefficient of friction between the particle and the plane is $\frac{1}{2}$, find the acceleration of the particle.
Calculate, in metres, the distance moved by the particle in the first 2 s of its motion. *(L)*

3 A particle of mass m is projected up a line of greatest slope of a plane inclined at an angle θ to the horizontal, where $\tan\theta = \frac{3}{4}$, with speed 12 ms^{-1}.
(i) If the plane is smooth, calculate how far up this line of greatest slope the particle will travel.
(ii) If the plane is rough, and the particle travels up the line of greatest slope for 6 m before coming to rest, calculate the coefficient of friction between the particle and the plane.

4 A particle of mass m slides down a line of greatest slope of a plane inclined at an angle θ to the horizontal, where $\tan\theta = \frac{3}{4}$. The coefficient of friction between the particle and the plane is $\frac{1}{4}$. Calculate the acceleration of the particle.

5 A particle of mass 0·1 kg is pushed up a line of greatest slope of a plane which is inclined at an angle θ to the horizontal, where $\tan\theta = \frac{3}{4}$, by a horizontal force of magnitude 1 N. If the plane is smooth, calculate the magnitude of the acceleration.

6 A ring of mass 0·01 kg is threaded on a smooth wire which is in the form of a straight line inclined at an angle of 30° with the horizontal. A vertical force of magnitude 0·2 N acts on the ring pulling it up the wire. Find the magnitude of the acceleration of the ring.

7 A particle of mass m is projected with speed U up a line of greatest slope of a plane which is inclined at an angle θ with the horizontal. The plane is rough and the coefficient of friction between the particle and the plane is μ. Show that the particle will travel a distance S up the line of greatest slope before coming to instantaneous rest, where

$$S = U^2/[2g(\sin\theta + \mu\cos\theta)].$$

If $\tan\theta > \mu$, show that the particle will return to the point from which it was projected with speed V given by

$$V^2 = \frac{U^2(\sin\theta - \mu\cos\theta)}{\sin\theta + \mu\cos\theta}.$$

5.5 Forces dependent on time

If the forces acting on a body moving along a straight line include variable ones, again it is vital to choose a clear reference direction and, hence, the direction of the acceleration and velocity will be specified.

The equation of motion $F = ma$ gives the acceleration in terms of the variable forces. As the acceleration is equal to d^2x/dt^2, a differential equation is thus obtained and the problem is a kinematic one as covered in Chapter 4.

EXAMPLE 1 *A particle of mass* 300 g *is subject to a force of magnitude F which acts for* 5 s *in the positive x-direction. The particle starts at rest. At time t* s, *the magnitude of the force is given by* $F = 3\sin(\frac{\pi t}{5})$ N, *for* $0 \leqslant t \leqslant 5$. *Find the velocity of the particle and the distance travelled when* $t = 5$ s.

The direction of reference is the positive x-direction, and the only force acting is F. Hence,

$$3\sin\left(\frac{\pi t}{5}\right) = 0{\cdot}3\,\frac{dv}{dt}.$$

Integrating both sides from 0 to 5 gives

$$\int_0^5 3\sin\left(\frac{\pi t}{5}\right) dt = 0{\cdot}3\,(v - u),$$

where v is the velocity after 5 s and u is the velocity at $t = 0$.

$$\left[-3 \times \frac{5}{\pi}\cos\frac{\pi t}{5}\right]_0^5 = 0{\cdot}3\,v$$

$$\frac{30}{\pi} = 0{\cdot}3\,v$$

$$v = \frac{100}{\pi}.$$

The velocity is approximately **31·8 ms^{-1}**.

To find the distance travelled, we need first to find the velocity as a function of the time.

$$0{\cdot}3\,v = \frac{-15}{\pi}\cos\left(\frac{\pi t}{5}\right) + c.$$

Since $v = 0$ when $t = 0$, $c = \dfrac{15}{\pi}$. Therefore,

$$0{\cdot}3\,v = \frac{15}{\pi}\left[1 - \cos\left(\frac{\pi t}{5}\right)\right].$$

Integrating both sides from 0 to 5,

$$0{\cdot}3\,s = \frac{15}{\pi}\left[t - \frac{5}{\pi}\sin\left(\frac{\pi t}{5}\right)\right]_0^5 = \frac{75}{\pi},$$

where s is the distance travelled in 5 s. So,

$$s = \frac{250}{\pi}$$

and the distance travelled is approximately **79·6 m**.

EXERCISE 5.5

1 A particle of mass m moves in a straight line under the action of a force acting along the same straight line, the magnitude of which at time t s is $m(2+6t)$. The particle is moving at 20 ms^{-1} when $t = 2$. Calculate the speed of the particle when $t = 0$.

2 A body of mass 2 kg starts from rest at O and moves along the x-axis under the action of a force acting along Ox, the magnitude of which at time t s is $(6t - t^2)$N. What is the speed of the body (i) after 3 s, (ii) after 9 s, from the start.

3 A particle of mass m starts from rest at the origin and moves in a straight line under the action of a force along this line, the magnitude of which is given at time t s by the formula $0{\cdot}2\, m \cos(\pi t/6)$. Find the velocity of the particle when $t = 3$, and the distance of the particle from the origin when $t = 2$.

4 A particle moves up a line of greatest slope of a smooth inclined plane, the angle θ made by this line of greatest slope with the horizontal being $\tan^{-1}(\frac{3}{4})$. There is a force acting on the particle up this line of greatest slope whose magnitude at time t s is given by $m(12-3t)$. Find the velocity acquired when starting from rest in (i) 2 s, (ii) t s. (iii) Find the distance travelled in 2 s from rest.

5.6 Impulse and momentum for rectilinear motion

Practically by its definition, a force changes the velocity of a particle when it acts on that particle. One measure of the effect of a force acting along a straight line for a time t is $\mathbf{v}-\mathbf{u}$, where $\mathbf{v}$ is the velocity at the end of the time t and $\mathbf{u}$ is the velocity just before the force is applied. We first evaluate this quantity for a constant force $\mathbf{F}$. This force, produces an acceleration $\mathbf{F}/m$ and, therefore, using the constant acceleration equation 4.4

$$\mathbf{v} = \mathbf{u} + \mathbf{a}t,$$

we obtain

$$m(\mathbf{v}-\mathbf{u}) = \mathbf{F}t. \qquad 5.4$$

This equation relates the change of velocity to the force and the time for which the force is applied.

Definition The product of mass and velocity is defined as the *linear momentum* of a body and so equation 5.4 relates change of momentum to the effect of the force. The product $\mathbf{F}t$ is defined as the *impulse* of the constant force $\mathbf{F}$ acting for time t. Thus equation 5.4 states

'Change in linear momentum = impulse'. 5.5

Therefore, knowledge of the impulse of a force is sufficient to determine the change in momentum.

Note that associated with rectilinear motion there is always a particular direction involved. It is important to emphasise that linear momentum and impulse are vectors.

In this particular case, equation 5.4 does not provide any extra information. When the force and the time for which the force acts are known, the problem can be solved without reference to impulse and linear momentum. Knowing the impulse is equivalent to knowing the force.

However, this is not always the case. We now try to generalise the idea of impulse to apply to a variable force so that equation 5.4 still holds. Replacing $\mathbf{a}$ by $\frac{d\mathbf{v}}{dt}$ in $\mathbf{F} = m\mathbf{a}$ we have for any force

$$m\frac{dv}{dt} = \mathbf{F}.$$

Integrating this equation from $t = t_1$ to $t = t_2$ gives

$$m(\mathbf{v} \text{ at } t = t_2) - m(\mathbf{v} \text{ at } t = t_1) = \int_{t_1}^{t_2} \mathbf{F}\, dt,$$

$$\text{change in momentum} = \int_{t_1}^{t_2} \mathbf{F}\, dt.$$

Definition We define the impulse of a general force $\mathbf{F}$ acting from $t = t_1$ to $t = t_2$ as

$$\int_{t_1}^{t_2} \mathbf{F}\, dt.$$

For a constant force $\mathbf{F}$ acting for a time t this reduces to $\mathbf{F}t$. The units of impulse are newton seconds (Ns).

The change in momentum produced by a force can then be measured without knowing the complete detailed form of the force provided that the impulse, which is the integral of the force with respect to the time, is known. In many practical situations the concept of impulse is not particularly useful because the change of momentum can be found simply by integrating the equation of motion. (This is how we found the impulse.) The concept, however, is important in quantifying the effects of large forces acting for very short intervals, such as those involved in collisions and the hitting of balls with bats or racquets. In these circumstances, the general structure of the force acting is not known, but the change of velocity is observed and from this the impulse due to the force can be calculated.

Problems are often set where an impulse is instantaneously applied and equation 5.4 used to find the change in velocity. In practice, the converse is more useful – that is, from a sharp change in the momentum we may deduce the impulse of the force acting on the body and thereby obtain some information about the force.

EXAMPLE 1 *A ball of mass* 0·05 kg *hits a vertical wall with a speed of* 10 ms^{-1}, *the direction of motion being perpendicular to the wall. The ball rebounds back with a speed of* 5 ms^{-1}. *Find the impulse exerted on the wall by the ball.*

This example illustrates clearly the vector nature of momentum, and it has to be remembered that all quantities in equation 5.4 are referred to some reference direction. We take this reference direction to be towards the wall so that the momentum in this direction before and after the collision is 0·5 Ns and −0·25 Ns respectively. Hence the impulse exerted on the ball in the reference direction is **0·75 Ns.**

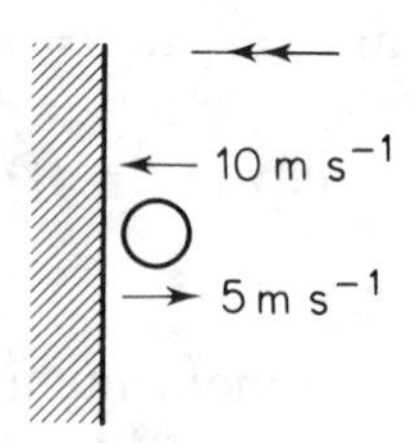

Fig. 5.6

EXERCISE 5.6

1 A cricket ball of mass 0·15 kg reaches a batsman with speed 12 ms^{-1} moving horizontally, and is hit straight back horizontally with a speed of 28 ms^{-1}. Calculate, in Ns, the magnitude of the impulse of the force of the bat on the ball. Given that the bat and the ball are in contact for 0·05 s, find the average force, in N, exerted by the bat on the ball. (*L*)

2 (i) A roller-skater, whose total mass with equipment is 80 kg, receives a horizontal impulse of magnitude 200 Ns when standing at rest. Calculate the initial speed of the skater and, if the resistance is 20 N, calculate the total distance moved across the rink. (*L*)

3 An impulse of magnitude 1·2 Ns is applied to a particle of mass 0·3 kg. The particle is free to move along the x-axis and the impulse is applied in this direction. Find the resulting speed of the particle.

4 A truck of mass 1200 kg, moving along a straight horizontal track at 4 ms^{-1}, runs into fixed buffers and rebounds with a speed of 3 ms^{-1}. The truck and the buffers are in contact for 0·2 s. Calculate
(a) the impulse of the force between the buffers and the truck in Ns,
(b) the average force, in N, exerted by the buffers on the truck. (*L*)

5 A tennis ball of mass 0·07 kg is moving horizontally with speed 5 ms^{-1} when it is hit by a racquet and returns straight back horizontally with a speed of 12 ms^{-1}. Calculate, in Ns, the magnitude of the impulse of the force of the racquet on the ball.

Given that the racquet and the ball are in contact for 0·06 s, find the average force, in N, exerted by the racquet on the ball. (*L*)

6 A concrete slab of mass 2 kg falls from rest at a vertical height of 10 m above firm horizontal ground from which it does not rebound. Calculate the impulse of the force exerted on the ground by the slab and state the units in which your answer is measured. (*L*)

7 A particle of mass m is moving with speed u when it receives an impulse which is of magnitude I and acting in the same direction as u. Find the speed of the particle immediately after the impulse. (*L*)

8 A ball of mass 0·15 kg falls vertically on to a horizontal concrete floor. The ball strikes the floor with a speed 9 ms^{-1} and rebounds with speed u ms^{-1} to reach a height of 2·5 m above the floor. Calculate the value of u and show that the impulse of the force exerted by the ball on the floor is 2·4 Ns. (Take g as 9·8 ms^{-2}.)

MISCELLANEOUS EXERCISE 5

1 A tug, of mass 9000 kg, is pulling a boat, of mass 5000 kg, by means of an inextensible horizontal tow rope along a straight canal. The resistive forces opposing the motions of the tug and the boat are 1500 N and 800 N respectively. Calculate, in N, the tension in the rope when the tug is accelerating at $\frac{1}{2}$ ms^{-2} and the driving force exerted by the tug. (*L*)

2 A car of mass 2500 kg is travelling along a straight level road at 72 kmh^{-1} when the engine is shut off. The car is brought to rest in a distance of 500 m by a constant retarding force F newtons. Calculate F and the time taken for the car to come to rest. (*L*)

3 A car of mass 480 kg, moving along a level road, has its speed reduced from 20 ms^{-1} to 10 ms^{-1} by a constant retarding force of 2000 N. Calculate, for this interval during which the car reduces speed,
(a) the time taken,
(b) the distance covered. (*L*)

4 A lift, starting from rest, moves with uniform acceleration for 6 s and with constant velocity for the next 6 s. It is then uniformly retarded and comes to rest 3 s later at a height of 4·2 m above its starting-point. Find the acceleration during the first stage of the motion.

A trunk weighing 36 kg rests on the floor of the lift. Find the reaction between the trunk and the lift during the three stages of the motion. (*O&C*)

5 A man of mass 50 kg stands on the floor of a lift which descends with acceleration 1 ms^{-2}. Find, in newtons, the total force exerted between the floor of the lift and the man's feet. Take $g = 9{\cdot}8$ ms^{-2}.) (*L*)

6 A man stands on a weighing machine which rests on the floor of a stationary lift, and the dial of the machine shows 80 kilograms. The lift then ascends for a period of 6 seconds, during which time the dial shows a constant value of 84 kilograms. Taking g as 10 ms^{-2}, find the upward acceleration of the lift, and show that the speed of the lift at the end of the 6 seconds in 3 ms^{-1}.

At this time the lift begins to slow down and the dial reading changes to a constant value of 74 kilograms. Find the retardation of the lift and the distance travelled, during the retardation, before it comes to rest. (*JMB*)

7 A heavy particle is suspended by a spring balance from the ceiling of a lift. When the lift moves up with constant acceleration $f\,\text{ms}^{-2}$ the balance shows a reading 1·8 kg. When the lift descends with constant acceleration $\frac{1}{3}f\,\text{ms}^{-2}$ the balance shows a reading 1 kg. Find the mass of the particle and the value of f. (*L*)

8 A particle of mass m hangs freely from a light inextensible string, the other end of which is attached to a lift which moves upwards with a constant acceleration f, and the tension in the string is T. Express T in terms of f, g and m.

In a lift at rest at the bottom of a vertical shaft, a particle of mass m is suspended from a light inextensible string. The lift then rises with constant acceleration for a time t_1 and then slows down with constant retardation (less than g) to rest after a further time t_2. During the period of acceleration the tension in the string is T_1 and during the period of retardation it is T_2. Show that

$$mg(t_1 + t_2) = T_1 t_1 + T_2 t_2.$$ (*JMB*)

9 A balloon is descending at a constant speed of $4\,\text{ms}^{-1}$. The crew releases a bag of ballast and as a result the balloon experiences an immediate upward acceleration of $2\,\text{ms}^{-2}$. Calculate

(i) the further distance which the balloon descends before it begins to ascend,

(ii) the vertical distance between balloon and ballast bag at the instant the balloon begins to ascend. (Take g to be $9{\cdot}8\,\text{ms}^{-2}$.) (*C*)

10 A particle of mass m moving vertically upwards with speed u enters a fixed horizontal layer of material of thickness a which resists its motion with a constant force of magnitude R. Show that it will pass through the layer if $u > u_0$, where

$$u_0^2 = 2a\left(\frac{R}{m} + g\right).$$

The particle emerges from the layer and moves freely under gravity until it re-enters the layer, which resists its motion as before. Show that the particle will pass completely through the layer in its downward motion only if $u^2 \geqslant 4aR/m$. (*JMB*)

11 A rough plane is inclined at an angle α to the horizontal, where $\tan\alpha = \frac{3}{4}$. A particle slides with acceleration $3{\cdot}5\,\text{ms}^{-2}$ down a line of greatest slope of this inclined plane. Calculate the coefficient of friction between the particle and the inclined plane. (*L*)

12 A particle moves along a line of greatest slope of a rough plane inclined at an angle β to the horizontal, where $\tan\beta = \frac{3}{4}$. The particle passes through a point A when moving upwards with speed u, comes momentarily to rest at a point C and subsequently passes through a point B when moving downwards at the same speed u. Given that the coefficient of friction between the particle and the plane is $\frac{1}{4}$, find AC and the speed of the particle on its return to A.

Show that $AB = AC$. (*L*)

13 A is the upper and B the lower of two points distant d apart on a line of greatest slope on a plane inclined to the horizontal at an angle $\tan^{-1}\frac{3}{4}$. A particle rests on the plane at A, its coefficient of friction with the plane being

4/5. Show that the speed with which it must be projected down the plane in order just to reach B is $\frac{1}{5}\sqrt{(2gd)}$, and that if it is then projected up the plane from B with the same initial speed it will only travel a distance $d/31$ before coming to rest.

The particle is removed and a second particle, whose coefficient of friction with the plane is 3/4, is projected down the plane from A with the same initial speed. Show that it reaches B in half the time taken by the first particle. (*L*)

14 A particle is projected from a point O with speed V up a line of greatest slope of a plane inclined at an angle α. The plane is rough with coefficient of friction μ for a distance a beyond O and is smooth beyond that. Given that $\mu > \tan\alpha$, prove that the particle cannot reach the smooth part of the plane and later return to O if $V^2 < 4\mu ag\cos\alpha$.

Prove also that the particle finally stops at a point on its way up if

$$V^2 \leqslant 2ag(\sin\alpha + \mu\cos\alpha)$$

and at a point on its way down if

$$2ag(\sin\alpha + \mu\cos\alpha) < V^2 < 4\mu ag\cos\alpha.$$ (*O&C*)

15 A particle of mass 2 kg is moving in a straight, smooth, horizontal groove with a speed of 5 $\mathrm{ms^{-1}}$. Calculate the possible values of the magnitude of the impulse which, applied to the particle along the groove, would result in the particle having a speed of 10 $\mathrm{ms^{-1}}$. (*L*)

16 An inelastic pile driver of mass 4000 kg falls freely from a height of 5 m on to a pile of mass 1000 kg driving the pile 20 cm into the ground. Find the speed with which the pile starts to move into the ground and also the average resistance to penetration of the ground in newtons. (Take g as 10 $\mathrm{ms^{-2}}$.) (*L*)

17 A particle P of mass 0·1 kg is acted on by a variable force so that it moves in a straight line with velocity v metres per second at time t seconds, where

$$v = 12 - t^2.$$

Calculate the distance covered by P from time $t = 0$ until it comes to rest instantaneously.

Find the acceleration of P at time t seconds and hence show that the force acting on P at that instant is $-\frac{1}{5}t$ newtons. (*L*)

18 Two bodies of masses 2 kg and 3 kg, initially at A and B respectively, start simultaneously from rest and move along the horizontal straight line AB. The first is acted on by a force of $2(6t + 2)$ newtons towards B and the second by a force of $3(12t^2 + 16)$ newtons towards A, where t seconds is the time since motion began. Calculate

(a) the speed of each body after 1 second,

(b) the distances covered by each body during the first second.

If the bodies collide after 1 second, find the distance AB. (*L*)

6 Work and Energy Concepts

Following the study in the last chapter of the effect of a force acting on a particle for a given time, we now consider the effect of a force when it moves a particle through a given distance. From this we derive, in §6.1, definitions of work done and of energy. In §6.3 we consider the energy stored in a stretched elastic spring. Finally, in §6.5, the time the operations take is introduced and the power developed is considered.

6.1 Work and energy for motion under gravity

If a particle of mass m is lifted from rest through a height h, then muscular effort is expended in doing this lifting and, in crude terms, the person who does the lifting experiences a loss of energy. If the particle is then released, from equation 4.8 it will fall back to its original position with speed $(2gh)^{\frac{1}{2}}$. This suggests that, in some way, the 'energy' expended in lifting the particle has been changed into 'energy of motion'. Newton's laws can be used to change this naive concept into a mathematical one.

The most effortless way of lifting anything is to do so extremely slowly so that the lifting force exactly balances the force of gravity. In ideal circumstances, therefore, the magnitude of the lifting force is mg. The higher the particle is lifted, the greater the effort exerted. Also, the heavier the weight, the greater the effort exerted. Therefore, a measure of the effort exerted in raising a particle of mass m vertically upwards through a distance h is mgh.

Definition The quantity mgh is known as the *work done* by the lifting force and, in general, *the work done by a force is the product of the force and the displacement, in the direction of the force, of the particle on which the force acts.* The work done by a force of 1 newton in moving a particle through 1 metre is the unit of work and this is called 1 joule (1 J). It is measured in $m^2\,kg\,s^{-2}$.

In moving upwards, the particle is moved a distance $-h$ in the direction of the force of gravity and, therefore, the work done by the force of gravity is $-mgh$. The work done by the lifting force is taken as a measure of the energy imparted to the particle and is called the *potential energy* (P.E.) imparted by the action of the lifting force. In general, the potential energy associated with a force acting on a body is the work that would have to be done against that force in moving the body from a standard position to the present one. The standard position is referred to as the position of zero potential energy.

If the ground is taken as the level of zero potential energy, a particle at a height h above the ground has a potential energy of mgh and a particle on the ground has zero potential energy. When the particle is dropped from a height h, it will reach the ground with speed v where $v = (2gh)^{\frac{1}{2}}$, so that $\frac{1}{2}mv^2$ is equal to its potential energy at height h. This quantity, $\frac{1}{2}mv^2$, is certainly associated with the motion of the particle, and so it is chosen as the energy associated with the motion. It is known as the *kinetic energy* (K.E.). Thus, the naive view that mechanical energy is converted to kinetic energy has been given a mathematical basis.

If the particle has dropped to a height x, it has fallen a distance $h-x$ from the highest point, and hence its speed is $[2g(h-x)]^{\frac{1}{2}}$. We have

at height h,	P.E. $= mgh$,	K.E. $= 0$,
at height x,	P.E. $= mgx$,	K.E. $= mg(h-x)$
at the ground,	P.E. $= 0$,	K.E. $= mgh$.

Therefore, throughout the fall,

$$\text{K.E.} + \text{P.E.} = mgh, \qquad 6.1$$

that is, K.E. plus P.E. is constant.

This result embodies an important principle.

The principle of energy states that the sum of the kinetic and potential energies remains constant.

So far the principle has only been proved for motion under gravity. An investigation will now be made to determine to which other types of forces the principle can be applied.

6.2 Validity of the principle of energy

Consider the work done by a force, the resolved part of which in the direction of x increasing is of magnitude F, where F is dependent on x only. For a very small displacement, δx, the force is approximately constant so that the work done is approximately $F\,\delta x$.

An approximation to the total work done in moving a particle from $x = x_1$ to $x = x_2$ under these conditions is $\sum_{x=x_1}^{x=x_2} F\,\delta x$. Taking the limit of this sum as $\delta x \to 0$, we deduce that

$$\text{total work done} = \int_{x_1}^{x_2} F\,\mathrm{d}x. \qquad 6.2$$

If the potential energy is measured from the position when $x = a$, then the potential energies at $x = x_1$ and $x = x_2$ are $\int_a^{x_1} -F\mathrm{d}x$ and $\int_a^{x_2} -F\mathrm{d}x$ respectively. Hence,

$$(\text{P.E. at } x = x_2) - (\text{P.E. at } x = x_1) = \int_{x_1}^{x_2} -F\mathrm{d}x = -\int_{x_1}^{x_2} F\mathrm{d}x.$$

Now the equation of motion of the particle, of mass m, is

$$ma = mv\frac{\mathrm{d}v}{\mathrm{d}x} = F.$$

Integrating this equation with respect to x from $x = x_1$ to $x = x_2$ gives

$$\begin{aligned}(\tfrac{1}{2}mv^2)_{x=x_2} - (\tfrac{1}{2}mv^2)_{x=x_1} &= \int_{x_1}^{x_2} F\mathrm{d}x \\ &= -[(\text{P.E. at } x=x_2) - (\text{P.E. at } x=x_1)] \\ (\tfrac{1}{2}mv^2)_{x=x_2} + (\text{P.E.})_{x=x_2} &= (\tfrac{1}{2}mv^2)_{x=x_1} + (\text{P.E.})_{x=x_1}.\end{aligned}$$

Thus, the principle of energy holds in this case.

It is important to realise that the principle has been obtained by integrating Newton's second law. Thus, the principle of energy contains no additional assumptions other than those involved in this law. The principle of energy is of value because its use can avoid setting up the equation of motion. It is most useful in problems where relations are required between speed and position – for example, in finding the speed acquired when a system is released and a particle moves a given distance. Also, as the energy equation is obtained by integrating the equation of motion, the latter can be found by differentiating the energy equation.

When considering the use of the principle of energy in a problem, it is necessary to decide first whether the principle is applicable. The proof given holds for forces which are constant or functions of position only. In general, the principle of energy is only valid in such cases; otherwise the potential energy is not a function of position alone.

There is one particular class of forces which has to be treated carefully – the frictional forces. At first sight they appear to be constant and yet they are not. If a particle slides on a rough horizontal plane, then there is a frictional force of magnitude μmg acting on the particle, where μ is the coefficient of friction. This appears to be constant but, as it is a force which always opposes relative motion, it changes direction with the direction of motion of the particle and hence the force is not constant. It is not possible to define uniquely the work done against friction as the particle moves from $x = 0$ to $x = a$. In moving directly, the work done is μmga. However, if the particle is moved from $x = 0$ to $x = 2a$ and then back to $x = a$, the work done against friction is $3\mu mga$. Different paths give different values. Thus, the principle of energy cannot be used with such forces.

If there is any doubt as to whether the principle holds and a potential energy is defined, a simple test is to find if a force does any work when a particle is moved from a reference point along a path and brought back to the reference point. If there is a net total of work then the principle is *not* applicable.

6.3 Elastic energy

The force of tension or thrust in a stretched or compressed spring is a function of position, as given by Hooke's Law,

$$T = \lambda \frac{x}{L},$$

where x is the extension or compression, L is the natural length of the spring and λ is the modulus of elasticity. To find the work done in stretching an elastic spring from an unstretched and uncompressed position to a total extension of x_1 we consider the work done in stretching the spring from an extension x to one of $x + \delta x$, where δx is small. During this small displacement, the force is approximately constant and equal to $\lambda \frac{x}{L}$. Hence, the work done is approximately $\lambda \frac{x}{L}\,\delta x$. An approximation to the total work done in stretching the spring from $x = 0$ to $x = x_1$ is, therefore,

$$\sum_{x=0}^{x_1} \lambda \frac{x}{L}\,\delta x.$$

Taking the limit of this sum as $\delta x \to 0$, we obtain that the total work done W is given by

$$W = \int_0^{x_1} \lambda \frac{x}{L}\,dx = \lambda \frac{x_1^2}{2L}. \qquad 6.3$$

If we allow the extension of the spring to change from x_1 to x_2 and then to return again to extension x_1, the work done in stretching is

$$\int_{x_1}^{x_2} \lambda \frac{x}{L}\,dx.$$

The work done in allowing it to return is

$$\int_{x_2}^{x_1} \lambda \frac{x}{L}\,dx = -\int_{x_1}^{x_2} \lambda \frac{x}{L}\,dx.$$

Hence, the total work done is zero and so this work done is a form of potential energy. To distinguish the two forms of potential energy so far discussed, we will refer to the one as gravitational potential energy and the other as elastic potential energy.

Similarly, the above theory would hold for an elastic string provided only extensions of the string are considered and not compressions. The elastic potential energy in a stretched or compressed spring or in a stretched string is

$$\lambda \frac{x^2}{2L}. \qquad 6.4$$

Validity of the principle of energy for a particle attached to an elastic spring and moving on a smooth horizontal table.

A particle of mass m is attached to one end of an elastic spring, the other end of which is fixed at a point O. The particle is projected away from O with speed u and the motion takes place on a smooth horizontal table. If T is the tension in the spring when the speed of the particle is v and the extension of the spring is x, then

$$T = -mv\frac{dv}{dx}.$$

If the spring is extended from O to x_1 while the speed of the particle changes from u to v_1, then

$$\int_o^{x_1} T\mathrm{d}x = \left[-\frac{1}{2}mv^2\right]_u^{v_1}$$

$$\lambda\frac{x_1^2}{2L} + \frac{1}{2}mv_1^2 = \frac{1}{2}mu^2,$$

that is, elastic energy at extension x_1 + kinetic energy at extension x_1 = the original kinetic energy.
Hence, the principle of energy holds in this case.

EXAMPLE 1 *A particle is thrown vertically upwards with an initial speed u. Find* (i) *the greatest height reached,* (ii) *the speed when the particle is at a height h which is less than the greatest height.*

We take the starting point as the level of zero gravitational energy. Note that this level may be chosen where we wish. We are only concerned in the energy equation with the difference in gravitational energy.

Initially, K.E. = $\frac{1}{2}mu^2$, P.E. = 0.

At the highest point, height H, K.E. = 0, P.E. = mgH.

Hence, by the principle of energy,

$$mgH = \tfrac{1}{2}mu^2$$

$$H = \frac{u^2}{2g}.$$

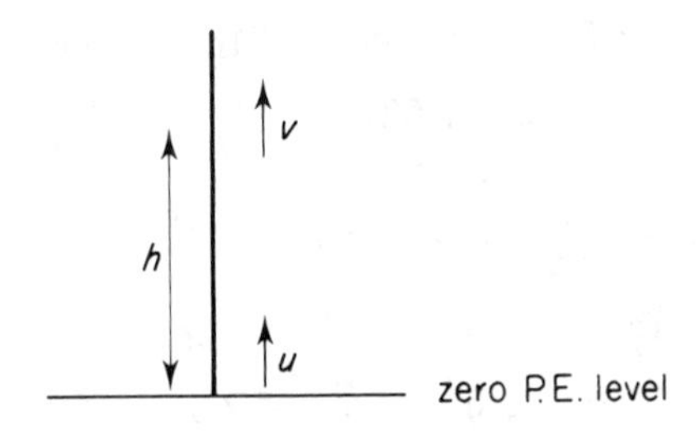

Fig. 6.1

Let v be the speed at height h (see Fig. 6.1). At this point, K.E. $= \frac{1}{2}mv^2$, P.E. $= mgh$.
By the principle of energy

$$\frac{1}{2}mv^2 + mgh = \frac{1}{2}mu^2$$
$$v^2 = u^2 - 2gh$$
$$v = \sqrt{(u^2 - 2gh)}.$$

Note There is no distinction between the cases when the particle is rising or falling. In both cases, the energy is the same. *Energy is a scalar quantity.* Also, the above problem can be solved just as simply by using the equations of motion under uniform acceleration.

EXAMPLE 2 *A particle of mass* 40 g *is attached to one end of an elastic spring of modulus* 1·2 N *and of natural length* 25 cm. *The other end of the spring is attached to a fixed point O on a smooth horizontal table. The spring is stretched to an extension of* 5 cm *and the particle is released from rest. Find*
(*i*) *the speed of the particle when the string returns to its unstretched and uncompressed position,*
(*ii*) *the speed when the extension is* 2 cm.

Initially, the elastic potential energy is

$$\lambda\frac{x^2}{2L} = \frac{1{\cdot}2 \times (0{\cdot}05)^2}{2 \times 0{\cdot}25}\text{ J} = 6 \times 10^{-3}\text{ J},$$

$$\text{K.E.} = 0.$$

When the spring is at its natural length,

$$\text{P.E.} = 0,$$
$$\text{K.E.} = \tfrac{1}{2} \times 0{\cdot}04v_1^2\text{ J},$$

where v_1 ms^{-1} is the speed in this position.
By the principle of energy,

$$0{\cdot}02\,v_1^2 = 6 \times 10^{-3}$$
$$20v_1^2 = 6$$
$$v_1 \approx \mathbf{0{\cdot}55}.$$

When the extension is 2 cm and the speed of the particle is v_2 ms^{-1},

$$\text{P.E.} = \frac{1{\cdot}2 \times (0{\cdot}02)^2}{2 \times 0{\cdot}25}\ \text{J} = 9{\cdot}6 \times 10^{-4}\ \text{J},$$

$$\text{K.E.} = \tfrac{1}{2} \times 0{\cdot}04 \times v_2^2\ \text{J}.$$

By the principle of energy,

$$0{\cdot}02\,v_2^2 + 9{\cdot}6 \times 10^{-4} = 6 \times 10^{-3}$$
$$0{\cdot}02\,v_2^2 = 5{\cdot}04 \times 10^{-3}$$
$$v_2^2 = 0{\cdot}252$$
$$v_2 \approx \mathbf{0{\cdot}502}.$$

EXAMPLE 3 *A particle of mass* 20 g *is attached to one end of an elastic string of modulus* 0·8 N *and of natural length* 50 cm. *The other end of the string is attached to a fixed point O and the particle is pulled vertically below O so that the extension of the string is* 30 cm. *The particle is released from rest in this position. Find*
(i) *the total height through which the particle rises,*
(ii) *the speed of the particle when the extension of the string is* 10 cm.

(i) Let the particle rise to the point B in Fig. 6.2, at a height of x cm above its point of release, and come to rest instantaneously. We choose the level of zero gravitational P.E. to be that of A (the start). It is important in any example to make the level chosen clear.)

At A, gravitational P.E. $= 0$,

K.E. $= 0$,

$$\text{elastic P.E.} = \frac{0{\cdot}8 \times (0{\cdot}3)^2}{2 \times 0{\cdot}5}\ \text{J} = 0{\cdot}072\ \text{J using equation 6.4.}$$

K.E. $= 0$.

At B, gravitational P.E. $= \dfrac{mgx}{100}$ (as x is in cm)

$= 0{\cdot}002x$ J (taking $g = 10$ ms^{-2}),

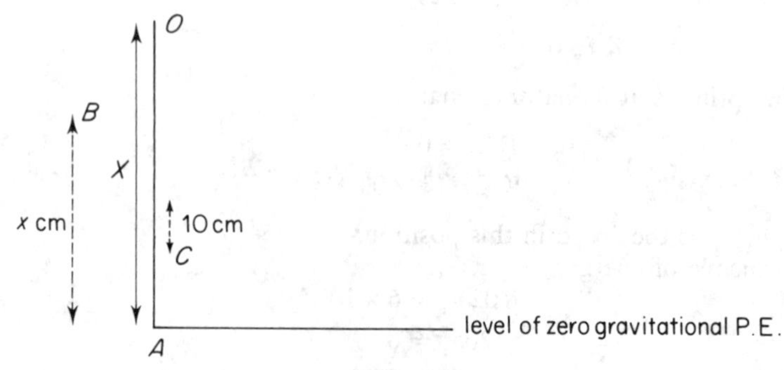

Fig. 6.2

Elastic P.E. = zero if $x \geqslant 30$ (assuming that the particle does not rise so high that the string becomes stretched above O). If, however, $x < 30$ the elastic energy is

$$\frac{0{\cdot}8 \times (30-x)^2}{2 \times 0{\cdot}5}\ \text{J}.$$

To decide which form for the elastic energy to use, we have to see whether the initial store of elastic energy is sufficient to raise the particle to the position when the string is unstretched. This would require an amount of energy $= 0{\cdot}002 \times 30$ J $= 0{\cdot}06$ J. Clearly, the initial elastic energy of $0{\cdot}072$ J is sufficient to raise the particle above the position in which the string is just unstretched. Thus, the elastic energy is zero when the particle is at instantaneous rest.

By the principle of energy, $\quad 0{\cdot}002x = 0{\cdot}072$

$$x = 36$$

So, the particle rises **36 cm**.

(ii) Let C be the position of the particle when the extension is 10 cm (see Fig. 6.2).

At C, gravitational P.E. $= 0{\cdot}04$ J,

K.E. $= \frac{1}{2} \times 0{\cdot}02 \times v^2$ J (where v is the speed of the particle at C),

$$\text{elastic P.E.} = \frac{0{\cdot}8 \times (0{\cdot}1)^2}{2 \times 0{\cdot}5}\ \text{J},$$

By the principle of energy,

$$0{\cdot}04 + 0{\cdot}01v^2 + 0{\cdot}008 = 0{\cdot}072$$

$$0{\cdot}01v^2 = 0{\cdot}024$$

$$v \approx 1{\cdot}55.$$

When the extension of the string is 10 cm, the speed is approximately **1·55 ms^{-1}**.

EXERCISE 6.4

Take $g = 10\ \text{ms}^{-2}$ unless otherwise stated.

1 A particle of mass m is thrown vertically upwards with a speed of 5 ms^{-1} from a point 2 m above the ground. Find
(a) the greatest height above the ground reached,
(b) the speed of the particle when it hits the ground.

2 A pebble of mass 20 g is given a speed of 20 ms^{-1} sliding on ice. The pebble travels 100 m on the ice before coming to rest under the force of friction alone. Calculate
(i) the kinetic energy of the pebble at the start,
(ii) the coefficient of friction, assuming that the force of friction is constant throughout the motion, by considering the work done by the force of friction.

3 A rock-climber of mass 60 kg does 6468 J of work in slowly climbing up a fixed vertical rope. Calculate the height through which he climbs. *(L)*

4 One end of a light elastic spring of natural length 0·5 m and modulus 1·6 N is attached to a fixed point A of a smooth horizontal table, and a particle P of mass 20 g is attached to the other end of the spring. The particle is released from rest on the table when the spring is straight and extended by 0·25 m. Calculate the speed of the particle

(i) when the spring has no extension or compression,
(ii) when the spring is compressed by 0·10 m.

5 An elastic string, of natural length l and modulus of elasticity λ, is stretched to a length $l+x$. As a result, the tension in the string is mg and the energy stored in it is E. Find x and λ in terms of E, g, l and m. (*L*)

6 A light elastic string is of natural length l and modulus $8mg$. Show by integration that the work required to extend the string from length l to length $l+k$, where $k > 0$, is $4mgk^2/l$.

One end of the string is attached to a fixed point A of a smooth horizontal table and a particle P of mass m is attached to the other end of the string. The particle is released from rest on the table with the string straight and extended. At time t after release and before the string becomes slack, $AP = l+x$. If initially $AP = 9l/8$, find the speed with which P reaches A. (*O&C*)

7 Prove that the work done in producing an extension x in an elastic string, of natural length l and modulus of elasticity λ, is $\dfrac{\lambda x^2}{2l}$.

A ball of mass m is attached to one end of such a light elastic string and the other end of the string is attached at a fixed point A. The ball is released from rest at A. Prove that, after it has fallen a distance $x+l$ $(0 < x < l)$, its speed v is given by

$$v^2 = 2g(l+x) - \frac{\lambda x^2}{ml}.$$ (*O&C*)

8 One end O of a light elastic string, obeying Hooke's law and of natural length l, is attached to a fixed point. To the other end P of the string is attached a particle of mass m which hangs in equilibrium with $OP = 5l/4$. The particle is pulled down vertically a further distance a and released from rest.

(i) If $a \leqslant l/4$, show that P rises a distance $2a$ before first coming to instantaneous rest.
(ii) If $l/4 < a < 3l/4$, show that P rises a distance $(l+4a)^2/(8l)$ before first coming to instantaneous rest and find its greatest speed during this motion. (*L*)

9 The gravitational force per unit mass at a distance r $(> R)$ from the centre of the earth is $gR^2/(r)^2$ where R is the radius of the earth and g is a constant. Show that the work done in raising a particle of mass m from $r = 2R$ to $r = 3R$ is $mgR/6$.

When a rocket is travelling directly away from the centre of the earth, it leaves behind a fuel container travelling at $(gR/2)^{\frac{1}{2}}$ at a distance of $2R$ from the centre of the earth. Show that the speed of the container when it reaches a distance of $3R$ from the earth's centre is $(gR/3)^{\frac{1}{2}}$.

6.4 Power

In practice, a machine is required not only to do a certain amount of work but to do that work in a limited interval of time. A powerful car demonstrates its power by rapid acceleration – it produces kinetic energy more rapidly than a car of lesser power.

Definition Power is the rate at which work is done. If 1 joule is provided in 1 second, the rate of working is 1 watt. The watt (W) and the kilowatt ($kW = 10^3 W$) are the standard units of power.

If the driving force exerted by a car's engine is constant and of magnitude F N, and if the car is driven s m in t s at a uniform speed v ms^{-1}, Fs J of work are done by the engine at a rate of $\frac{Fs}{t}$ W. Hence,

$$\text{Power developed} = Fv \text{ W}.$$

This argument applies to the case when the car moves at a constant speed and the engine exerts a constant force, If the driving force varies, the work done W in travelling a distance s is given by

$$W = \int_0^s F \mathrm{d}s \text{ (by equation 6.2).}$$

$$\text{Power} = \frac{\mathrm{d}W}{\mathrm{d}t} = \frac{\mathrm{d}W}{\mathrm{d}s} \cdot \frac{\mathrm{d}s}{\mathrm{d}t} = Fv, \qquad 6.5$$

that is, the same result as for constant speed.

EXAMPLE 1 *The engine of a lorry of total mass* 2 tonnes *is working at* 50 kW. *The lorry is travelling at a constant speed of* 20 ms^{-1} *along a level road. Taking g as* 10 ms^{-2}, *find the total resistance to the motion.*

If the power is increased to 60 kW, *find the acceleration of the lorry assuming that the resistances to motion remain constant and that the lorry is travelling at* 20 ms^{-1}.

When the power is 50 kW, the driving force F N is given by

$$F . 20 = 50\,000$$
$$F = 2\,500 \text{ N}.$$

As the lorry is moving at constant velocity, the resultant force acting on it must be zero. Hence the total resistance is **2 500 N.**

When the power is 60 kW, the driving force F_1 N is given by

$$F_1 . 20 = 60\,000$$
$$F_1 = 3\,000.$$

The net accelerating force = (3 000 – 2 500) N = 500 N. The equation of motion is

$$500 = 2000\,a$$
$$a = \tfrac{1}{4}.$$

The acceleration is $\mathbf{\frac{1}{4}}$ **ms^{-2}**.

The power developed by a pump

Suppose that a pump is raising water through a hose to a height h (Fig. 6.3) and it is delivering the water at the top with speed v. Assume that a mass m of water is raised and ejected in 1 s. During the motion, if the force exerted by the pump on this mass is of magnitude F,

$$F - mg = mv\frac{dv}{dx},$$

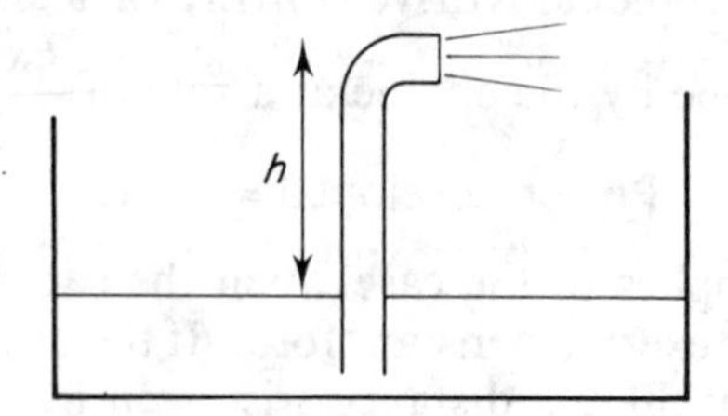

Fig. 6.3

where x is the distance the water is raised and

$$\int_0^h F\,dx = mgh + \left[\frac{1}{2}mv^2\right]_0^v,$$

that is, the work done = P.E. given to the water + K.E. given to the water.

As this is the work done in 1 second, this is the power developed, neglecting resistances and assuming that the rate of working is constant. So, the output of power equals the rate at which the total P.E. and K.E. is produced.

EXAMPLE 2 *A pump raises water from a lake* 7 m *below the nozzle of the pump. The water is discharged at* $15\ \text{ms}^{-1}$ *through the nozzle, the area of cross-section of which is* $2\ \text{cm}^2$. *Assuming that the water is discharged at full bore, find the output of power developed by the pump. (Take* g *as* $10\ \text{ms}^{-2}$ *and assume that the mass of* 1 l *of water is* 1 kg.)

In 1 s, $15 \times 2 \times 10^{-4}\ \text{m}^3$ of water is discharged, and $1\ \text{m}^3 = 10^6\ \text{cm}^3 = 10^3\ \text{l}$.

$$\text{Mass discharged per s} = 15 \times 2 \times 10^{-1}\ \text{kg} = 3\ \text{kg}.$$

$$\text{K.E. produced per s} = \tfrac{1}{2} \times 3 \times 225\ \text{J} = 337{\cdot}5\ \text{J}.$$

$$\text{P.E. produced per s} = 3 \times g \times 7 = 210\ \text{J}.$$

$$\text{Total energy produced per s} = 547{\cdot}5\ \text{J}.$$

$$\text{Power} = 547{\cdot}5\ \text{W} = \mathbf{0{\cdot}5475\ kW}.$$

EXERCISE 6.5

1 At a mine an engine raises coal of mass 1200 kg from a vertical depth of 400 m. The coal starts from rest and ends at rest and the time required to

complete this operation is 210 s. Calculate, in kW, the average rate of working by the engine which raises this coal. (*L*)

2 A train of mass 3×10^5 kg travels along a straight level track. The resistance to motion is $1{\cdot}5 \times 10^4$ N. Find the tractive force required to produce an acceleration of $0{\cdot}1\ \text{ms}^{-2}$, and the power in kW which is then developed by the engine when the speed of the train is $10\ \text{ms}^{-1}$.

Find also the maximum speed attainable on the same track when the engine is working at a rate of 360 kW. (*JMB*)

3 A car of mass 2000 kg travels up a slope of inclination α to the horizontal, where $\sin\alpha = \dfrac{1}{49}$, against constant frictional resistances of 3600 N. Calculate the maximum speed of the car when the engine works at a rate of 100 kW.

After reaching the top of the slope the power is shut off and the car descends a slope of inclination β to the horizontal against the same constant frictional resistances at constant speed. Calculate $\sin\beta$. ($g = 9{\cdot}8\ \text{ms}^{-2}$.) (*L*)

4 At the bottom of a coal mine of depth 400 m there is a cage of mass 250 kg containing 1200 kg of coal. The wire rope which raises the cage is of mass $2\frac{1}{2}\ \text{kg m}^{-1}$. Find, in joules, the work which must be done to raise the loaded cage to the surface and the average rate of working, in kilowatts, of the engine driving the hoist when the time taken to raise the cage from rest to rest is 49 seconds. (Take $g = 9{\cdot}8\ \text{ms}^{-2}$.) (*O&C*)

5 A car of mass 1000 kg has a maximum speed of $15\ \text{ms}^{-1}$ up a slope inclined at an angle θ to the horizontal where $\sin\theta = 0{\cdot}2$. There is a constant frictional resistance equal to one tenth of the weight of the car. Find the maximum speed of the car on a level road assuming that the engine works at the same rate.

If the car descends the same slope with its engine working at half this rate, find the acceleration of the car at the moment when its speed is $30\ \text{ms}^{-1}$. [Take g as $9{\cdot}8\ \text{ms}^{-2}$.] (*L*)

6 The resistive forces opposing the motion of a train, of total mass 50 000 kg, are constant and total 8 500 N.

(a) Find the power, in kW, which is required to keep the train moving along a straight level track at a constant speed of $12\ \text{ms}^{-1}$.

(b) If this power is suddenly increased by 18 kW as the train moves along the level track at $12\ \text{ms}^{-1}$, find the initial acceleration, in ms^{-2}, of the train.

(c) When the train climbs a hill, of inclination θ to the horizontal, at a constant speed of $8\ \text{ms}^{-1}$, the engine of the train is working at a rate of 180 kW. Find the value of $\sin\theta$. (*L*)

7 A lorry of mass M kg, with its engine working at K kilowatts, has a maximum speed on the level of u km per hour. If the maximum speed up a road inclined at an angle α to the horizontal is v km per hour, when the rate of work and the resistance are unchanged, prove that

$$18\,000\,K\,(u - v) = 49\,Muv\sin\alpha.$$

(Take $g = 9{\cdot}8\ \text{ms}^{-2}$.) (*O&C*)

8 A car of mass 1600 kg ascends a slope of inclination α to the horizontal, where $\sin\alpha = \frac{1}{20}$, at a steady speed of $15\ \text{ms}^{-1}$. Given that the constant frictional resistance is 400 N, calculate the power, in kW, developed by the car.

When the car reaches the top of the slope the power is cut off and the car descends a slope of inclination β to the horizontal, where $\sin\beta = \frac{1}{30}$. Assuming that the constant frictional resistance remains 400 N, calculate the acceleration with which the car descends this slope. (*L*)

9 Water is pumped through a pipe from a reservoir and discharged through a nozzle, whose cross-sectional area is $5\,\text{cm}^2$, at a steady rate of 10 litres per second. The nozzle is 20 m vertically above the water level of the reservoir. Taking 1 litre of water to be of mass 1 kg and volume $1\,000\,\text{cm}^3$, calculate
(a) the speed, in ms^{-1}, of the water at the instant it leaves the nozzle,
(b) the kinetic energy, in J, gained by the water delivered each second,
(c) the potential energy, in J, gained by the water delivered each second,
(d) the power, in W, required to give the water this kinetic and potential energy. (*L*)

10 A water pump raises 50 kg of water a second through a height of 20 m. The water emerges as a jet with speed $50\,\text{ms}^{-1}$. Find the kinetic energy and the potential energy given to the water each second and hence find the effective power developed by the pump. (*L*)

11 A pump raises 100 kg of water per second from a depth of 30 m. The water is delivered at a speed of $30\,\text{ms}^{-1}$. Find, in joules, (a) the potential energy, (b) the kinetic energy gained by the water delivered each second. Neglecting frictional losses calculate, in kW, the rate at which the pump is working. (Take g as $9{\cdot}8\,\text{ms}^{-2}$.) (*L*)

12 A pump delivers 280 kg of water per minute, the water being delivered in a horizontal jet at a speed of $30\,\text{ms}^{-1}$. Find, in joules, the kinetic energy of the water delivered each second. The efficiency of the engine driving the pump is 30 per cent. Find the rate, in kilowatts, at which this engine is working. (*O&C*)

13 A car of mass 1600 kg is moving along a horizontal road. The resistance to the motion of the car is 800 N. Calculate the accleration of the car at the instant when its speed is $7{\cdot}5\,\text{ms}^{-1}$ and its engine is working at 15 kW. (*L*)

14 State Newton's first law of motion.

A train, whose engine can develop a power of 150 kW, has a mass of 30 000 kg. Stating clearly where the above law is used, find the maximum speed at which the train can climb a plane inclined at an angle α to the horizontal, where $\sin\alpha = \frac{1}{50}$, assuming that the total frictional resistances are 1620 N.

Calculate also, at an instant when the speed is $15\,\text{ms}^{-1}$, the greatest acceleration up this incline. (*L*)

MISCELLANEOUS EXERCISE 6

1 A car is travelling at a steady speed of $30\,\text{ms}^{-1}$ on a horizontal road against a constant resistance of 1200 N. Calculate, in kW, the rate of working of the engine.

The car is then attached to a caravan by a towbar and the resistance to the motion of the caravan is 900 N. If the rate of working of the engine is now 42 kW, calculate the speed of the car and caravan on a horizontal road. State the tension in the towbar.

The total mass of the car and the caravan is 2 tonnes and the car is to pull the caravan directly up a hill which is inclined at an angle of arcsin (1/40) with the horizontal. If the car is to develop 44 kW, calculate the acceleration up the hill when the car and caravan are travelling at 16 ms^{-1}. You may assume that the non-gravitational resistances to the motion of the car and the caravan are constant.

2 The engine of a certain car can develop 16 kW. The maximum speed of this car on a level road is 144 kmh^{-1}. Calculate the total resistance to the motion of the car at this maximum speed.

Given that the non-gravitational resistance to the motion of the car varies as the square of the speed and that the mass of the car is 800 kg, calculate the power developed in kW by the engine when the car moves at a constant speed of 72 kmh^{-1} directly up a hill which is inclined at an angle of arcsin (1/20) with the horizontal.

3 A car of mass 500 kg is moving at a constant speed of 96 kmh^{-1} on a level road. The non-gravitational resistance to the motion of the car at all speeds and on all roads is 750 N. Calculate the rate at which the engine of the car is working.

When the car climbs directly up a certain hill, it has a maximum speed of 54 kmh^{-1} and the engine is working at a rate of 15 kW. Calculate the angle of inclination of the road to the horizontal.

Find the acceleration of this car when it is travelling up a hill inclined at an angle of arcsin (1/20) with the horizontal, the speed of the car being 48 kmh^{-1} with its engine working at the rate of 15 kW.

4 The maximum rate of working of a certain car is 30 kW. Its maximum speed on a horizontal road is 50 ms^{-1} and its maximum speed directly up a hill which is inclined at an angle of arcsin (1/10) with the horizontal is 30 ms^{-1}. Assuming that the non-gravitational resistance to the motion of the car is constant, calculate the mass of the car and the maximum speed it can travel up a hill inclined at an angle of arcsin (1/5) with the horizontal.

5 One end of an elastic string of natural length $4a$ is attached to a fixed point O and the other end has a particle P of mass m attached. The particle is released at a depth of $7a$ from a point vertically below O. P rises to a height of $2a$ above O before coming to rest. Calculate the modulus of the elastic string in terms of mg.

6 A locomotive of mass 60 tonnes pulls a set of coaches whose total mass is 240 tonnes. The engine of the locomotive works at a constant rate of 500 kW and non-gravitational resistances on both the locomotive and the coaches are x newtons per tonne. The track is level and the maximum speed of the train is 144 km/h. Find x. Find also the tension in the coupling between the locomotive and the first coach when the speed is

(a) 144 km/h,
(b) 72 km/h.

When the speed of the train is 72 km/h the coupling on the last coach breaks. Given that the mass of that coach is 20 tonnes, find how far this coach will travel before coming to rest. (*L*)

7 The engine of a car, of mass M kg, works at a constant rate of H kW. The non-gravitational resistance to the motion of the car is constant. The maximum speed on level ground is V m/s. Find, in terms of M, V, H, α and g, expressions for the accelerations of the car when it is travelling at speed $\frac{1}{2}V$ m/s

(a) directly up a road of inclination α,
(b) directly down this road.

Given that the acceleration in case (b) is twice that in case (a), find $\sin\alpha$ in terms of M, V, H and g. Find also, in terms of V alone, the greatest steady speed which the car can maintain when travelling directly up the road. *(L)*

8 A nylon climbing rope of length 40 metres is stretched by a force F. The relation between the force F and the corresponding extension x is given in the following table:

x (metres)	0	1	2	3	4	5	6	7	8	9	10
F (newtons)	0	1600	2400	3000	3600	4100	4600	5200	5700	6400	7900

Use Simpson's rule to estimate the energy stored in the rope when it is extended by 10 metres.

A climber of mass 64 kg is climbing on a vertical rock-face h metres above the point to which he is securely attached by the rope of length 40 metres described above. The climber slips and falls freely, being brought to rest for the first time by the rope when it has been extended by 10 metres. Calculate the value of h to the nearest metre. Determine also the velocity of the climber at the moment the rope becomes taut. (Take $g = 9{\cdot}8\text{ ms}^{-2}$.) *(JMB)*

9 A car of mass 1000 kg moves along a horizontal road with acceleration proportional to the cube root of the time t seconds after starting from rest. When $t = 8$, the speed of the car is 8 m/s. Neglecting frictional resistances, calculate, in kW, the rate at which the engine driving the car is working when $t = 27$. *(L)*

10 A particle of mass m is attached to one end of an elastic string, of modulus $4mg$ and natural length a, the other end of the string being attached to a fixed point O. If the particle is released from a point vertically below O at a distance $5a/3$, find
(i) the height to which the particle will rise,
(ii) the speed when the particle is at a depth of $3a/2$ below O.

11 A small ring of mass m is free to slide on a smooth straight wire which is fixed at an inclination $\pi/6$ to the horizontal. The ring is attached to one end of a light elastic string of natural length a and modulus $4mg$. The other end of the string is attached to a fixed point A of the wire and the particle rests in equilibrium at the point B. Find the distance AB.

If the particle is released from rest at A, show that the maximum speed of the particle in the ensuing motion is $\frac{1}{4}\sqrt{(17ag)}$. Find also the distance of the particle when it first comes to rest after leaving A.

12 A particle is suspended from a fixed point O by a light elastic string which is of natural length a and which obeys Hooke's law. When the particle hangs at rest, the length of the string is $5a/4$. If the particle is released from rest at O, find the distance it falls before it first comes to rest. *(L)*

7 Motion of Interacting Particles

In this chapter the work on the dynamics of a single particle is extended to cover problems involving particles which interact with each other in some way, for example, particles connected by a string. The basic principles involved are established in §7.1 and applied in §7.2 and §7.3 to problems involving impact between particles and impact between particles and rigid bodies. The solutions of some of the simpler basic problems of connected particle motion are considered in §7.4, whilst in §7.5 we examine the impulsive motion of connected particles.

7.1 Conservation of Linear Momentum

Two particles P_1 and P_2, of mass m_1 and m_2 respectively, are assumed to interact in some way. For example, they may be connected by a spring or string, or may just collide with each other and then either stick together (that is, coalesce) or separate.

At this stage we assume that the particles are constrained to move along a line and x_i $(i = 1, 2)$ is taken to denote their respective displacements from a fixed point O on that line. It will further be assumed that, apart from the interacting force between the particles, there are no other forces acting on them. Newton's third law states that if the force exerted by P_1 on P_2 is $\mathbf{F}$ then that exerted by P_2 on P_1 is $-\mathbf{F}$ (see Fig. 7.1). The force $\mathbf{F}$ can be written as $F\mathbf{i}$, where $\mathbf{i}$ denotes the unit vector in the sense in which the displacement is measured, so that the equations of motion of the particles are

$$m_1\ddot{x}_1 = -F, \qquad 7.1$$

$$m_2\ddot{x}_2 = F. \qquad 7.2$$

Adding these gives

$$m_1\ddot{x}_1 + m_2\ddot{x}_2 = 0$$

or, equivalently,

$$\frac{\mathrm{d}}{\mathrm{d}t}(m_1\dot{x}_1 + m_2\dot{x}_2) = 0, \qquad 7.3$$

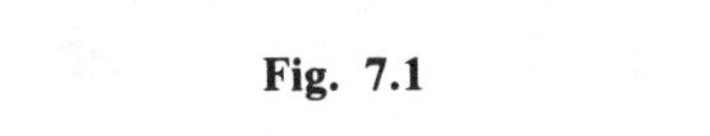

P_1 $-F$ F P_2

Fig. 7.1

so that $m_1\dot{x}_1 + m_2\dot{x}_2$ is constant. This latter quantity is the total linear momentum of the particles. (More precisely it is the component of the total linear momentum in the direction of the unit vector **i**). Therefore, the total linear momentum of the system has been shown to be constant, and this is a particular case of the following principle.

The principle of conservation of linear momentum which states that the total linear momentum of a system of particles is conserved whenever there are no external forces acting on the system.

This has only been proved above for a system of two particles but it is quite easy to generalise the proof so as to cover the case of any number of particles. The result is also true for any rectilinear motion in which there is no total force in the direction of motion. The principle of conservation of linear momentum stated above is also valid for general motion, but this will not be demonstrated in this book.

The principle of conservation of linear momentum is particularly useful for handling problems involving collisions where the nature of the interactive forces is not known but the momentum after a collision can be deduced immediately from that before collision. In order to apply the principle to collision problems, it is not necessary to exclude the action of external forces. In fact, it is sufficient to require that the total impulse, over the period of collision, of the external forces is zero. This can be shown as follows.

If it is assumed that there are external forces $F_1\mathbf{i}$ and $F_2\mathbf{i}$ acting on P_1 and P_2 respectively, then equations 7.1 and 7.2 become respectively

$$m_1\ddot{x}_1 = -F + F_1,$$
$$m_2\ddot{x}_2 = F + F_2.$$

Adding these equations and integrating with respect to t from $t = 0$ to $t = T$ gives

$$\text{(linear momentum at time } t = T) - \text{(linear momentum at time } t = 0)$$
$$= \int_0^T (F_1 + F_2)\,\mathrm{d}t.$$

The right-hand side is the total impulse (see §5.8) of the external forces, and momentum is conserved if this impulse vanishes. For impulses where the collision process is assumed to take place over a small interval of time, the impulse of normal external forces, such as the force of gravity, is very small and is normally neglected.

The application of the principle of conservation of linear momentum to collision problems is illustrated in the following examples. Most problems are relatively straightforward provided that you remember that the total linear momentum in a *particular reference direction* is conserved. It is, therefore, useful in any problem to draw a simple diagram and mark in the reference direction and the velocities (taking care over the signs) in the reference direction.

EXAMPLE 1 *A small particle A, of mass 2m moving with speed 10u, overtakes another particle B, of mass 3m and moving with speed 5u. Assuming that, after collision the particles have coalesced and move together, determine the speed of the composite particle after the collision.*

Fig. 7.2 shows the situation before and after collision, with the reference direction being to the right and v denoting the common speed after collision. Conservation of linear momentum gives

$$2m.10u + 3m.5u = 5m.v$$

so that $v = \mathbf{7u}$.

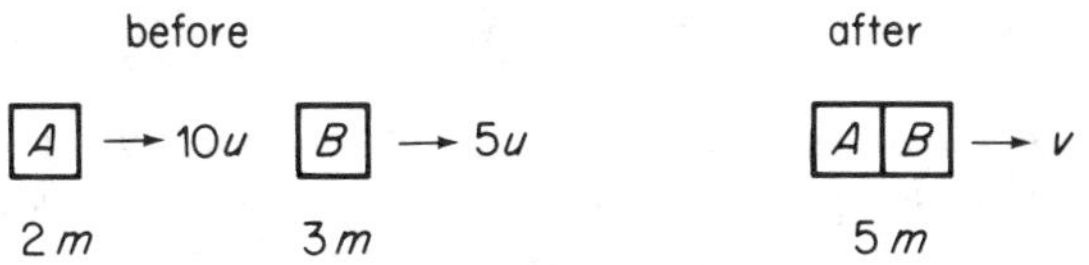

Fig. 7.2

EXAMPLE 2 *Determine the common speed in the previous example for the case when the particles are moving directly towards each other.*

The situation is now as in Fig. 7.3. The reference direction is again chosen in the direction of motion of A so that B has a velocity $-5u$ in that direction. Conservation of linear momentum now gives

$$2m.10u - 3m.5u = 5m.v$$

so that $v = \mathbf{u}$.

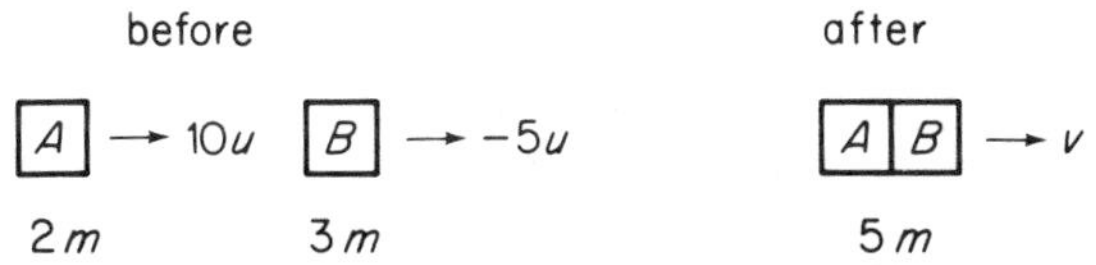

Fig. 7.3

EXAMPLE 3 *Two particles P and Q, of mass 2m and 4m respectively and free to move on a smooth straight line, are joined together by a light inextensible string. Initially, the particles are at rest with the string slack and Q is projected away from P with speed u. Determine the subsequent motion.*

Until the string becomes taut, Q will move with speed u. Immediately after the string becomes taut, the two particles will, since the string is inextensible, move together with common speed v. Fig. 7.4 shows the situation just before and just after the string tautens. Conservation of linear momentum gives

$$4mu = 6mv$$

so that $v = \frac{2}{3}\mathbf{u}$.

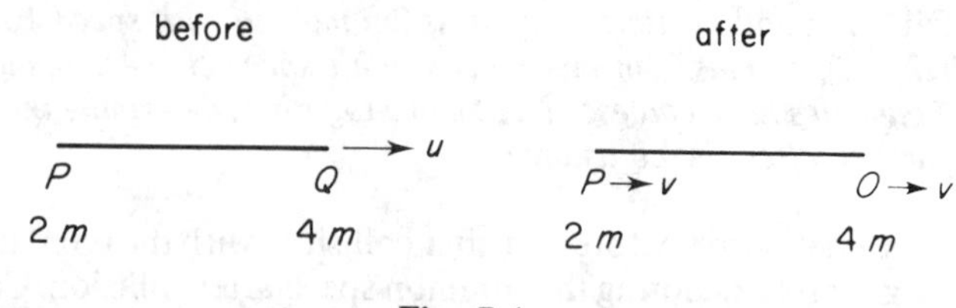

Fig. 7.4

The velocity of P is changed instantaneously from zero to $2u/3$, so an impulse of magnitude $2m \,.\, 2u/3$ has been applied to it and this is the impulse generated in the string. (This is usually referred to as the impulse of the tension.)

EXAMPLE 4 *A gun of mass M is free to recoil horizontally and a shell of mass m is fired from it so that the speed of the shell just after it leaves the barrel is v and the shell is moving at an angle θ to the horizontal. Find the speed at which the gun is moving just after the shell has left the barrel.*

The situation immediately after the shell has left the barrel is shown in Fig. 7.5, where U denotes the speed of the gun to the right. Initially, the momentum of gun and shell were zero and there are no impulses horizontally on the system of shell and gun, so the conservation of linear momentum gives

$$MU + mv\cos\theta = 0.$$

Therefore, the gun moves to the left with speed $\boldsymbol{(mv\cos\theta)/M}$.

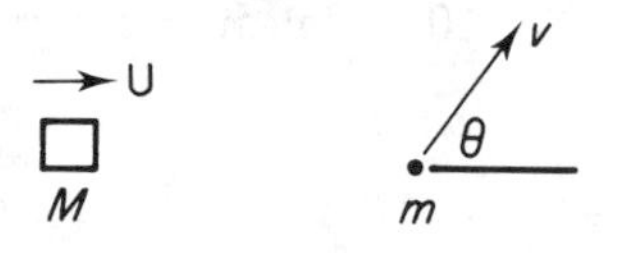

Fig. 7.5

If it had been stated that v was the speed of the shell relative to the barrel then the actual speed of the shell to the right would have been $U + v\cos\theta$, and the speed of recoil would then have been $(mv\cos\theta)/(M+m)$.

EXERCISE 7.1

1 A particle of mass $3m$ moving with speed $2u$ overtakes and coalesces with a particle of mass $4m$ moving with speed $9u$. Find their common speed after collision.

2 A particle of mass $11m$ moving with speed $3u$ collides directly and coalesces with a particle of mass $4m$ moving in the opposite direction with speed $0{\cdot}75u$. Find their common speed after collision.

3 A particle of mass m moving with speed $6u$ overtakes and collides with a particle of mass $24m$ moving with speed u in the same direction. The direction of motion of the lighter particle is reversed at impact and its speed is reduced to $2u$. Find the speed of the other particle.

4 A bullet of mass 0·01 kg is fired into a stationary block of wood of mass 1 kg and thereafter remains embedded in the wood. Given that the initial speed of the bullet was $505\,\mathrm{ms}^{-1}$, find the speed of the block immediately after impact.

5 Two particles of masses m and $4m$ move directly towards each other with speeds u and $4u$ respectively. After impact the direction of motion of the lighter particle is reversed and it moves away with speed $5u$. Find the speed of the heavier particle after collision.

6 A particle of mass $2m$ moving with speed $6u$ overtakes and collides with a particle of mass m moving with speed u. The collision is such that the second particle receives an impulse of magnitude $3\,mu$ in the direction of its motion. Find the speeds of the particles after collision.

7 A gun of mass 10 tonnes, free to recoil in the direction of the barrel, fires a shot of mass 100 kg with a speed of $400\,\mathrm{ms}^{-1}$. Find the speed of recoil of the gun.

8 Two particles of masses m and $3m$ are connected by a light inextensible string and are initially at rest on a smooth table, with their separation being much less than the length of the string. They are then both projected in the same direction with speeds $4u$ and $12u$ respectively. Find their common speed immediately after the string has tautened.

9 A train of trucks is being started from rest and, just before the last coupling becomes taut, the moving part has acquired a velocity of $3\,\mathrm{ms}^{-1}$. If the moving part of the train weighs 45 tonnes and the last truck weighs 5 tonnes, find the impulse in the last coupling.

10 Two particles of masses m and $2m$, connected by a light inextensible string, are at rest on a smooth table with the string just taut. An impulse of magnitude $6\,mu$ is applied to the heavier particle in the direction along the line of the string but away from the other particle. Find the speed with which the particles move and the impulse in the string.

7.2 Direct impact

We now consider in slightly more detail problems of collision between two small smooth spheres or between one such sphere and a smooth plane. The assumption of smoothness ensures that the interactive impulse at impact will be along the common normal, which will, therefore, be along the line of centres in the first case and along the radius to the point of contact in the second case. Our discussion will also be restricted to direct impact, that is, to the case when the motion is in the line of common normal.

When a smooth sphere, of mass m moving with speed u perpendicular to a smooth wall, collides with the wall it experiences an impulse perpendicular to the wall so that it rebounds with speed v (see Fig. 7.6). The momentum in the direction away from the wall, is $-mu$ before collision and mv after, so an impulse, J, where

$$J = m(v + u) \qquad 7.4$$

has acted on the sphere.

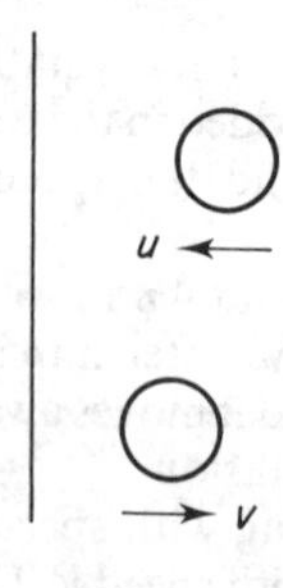

Fig. 7.6

Experimental observation has shown that for any particular sphere and wall, the ratio v/u is independent of the speed u and is a positive constant, e, which is known as the coefficient of restitution between the sphere and the wall. Therefore,

$$v = eu. \qquad 7.5$$

The kinetic energy loss due to the collision is $\frac{1}{2}mu^2(1-e^2)$ and, as the collision does not create energy, $0 \leqslant e \leqslant 1$. (When $e = 0$ the collision is said to be inelastic.)

Problems of direct collision with a smooth plane can normally be solved by using only equations 7.4 and 7.5.

EXAMPLE 1 *A smooth sphere is dropped vertically from a height* $2a$ *so as to impinge directly on a smooth horizontal plane. The coefficient of restitution for collision between the sphere and the plane is* $\frac{1}{2}$. *Find the total distance travelled by the sphere up to the instant when it hits the plane for the third time.*

The situation is shown in Fig. 7.7. The sphere will have a speed $2(ga)^{\frac{1}{2}}$ just before it first hits the plane. By equation 7.5 it rebounds off the plane with speed $(ga)^{\frac{1}{2}}$. It rises to a height $\frac{1}{2}a$ and then starts falling, returning to the plane with speed $(ga)^{\frac{1}{2}}$. It now rebounds off with a speed $(ga)^{\frac{1}{2}}/2$ and rises to a height $a/8$ before again starting to fall. Therefore, between the first and second rebounds the sphere travels a distance a, and between the second and third rebounds it travels a distance $a/4$. Adding to these the distance that the sphere dropped before reaching the plane, the total distance travelled by the sphere is $\mathbf{13a/4}$.

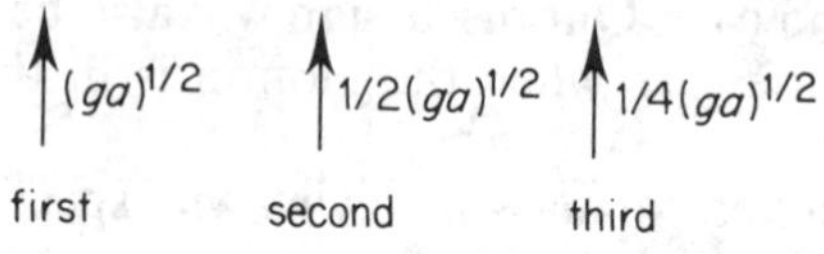

$(ga)^{1/2}$ $1/2(ga)^{1/2}$ $1/4(ga)^{1/2}$

first second third

Fig. 7.7

We now consider the collision of two small spheres. Fig. 7.8(a) shows two such spheres S_1 and S_2, of masses m_1 and m_2 respectively, and with speeds u_1 and u_2 respectively. When $u_1 > u_2$, S_1 overtakes S_2 and the spheres

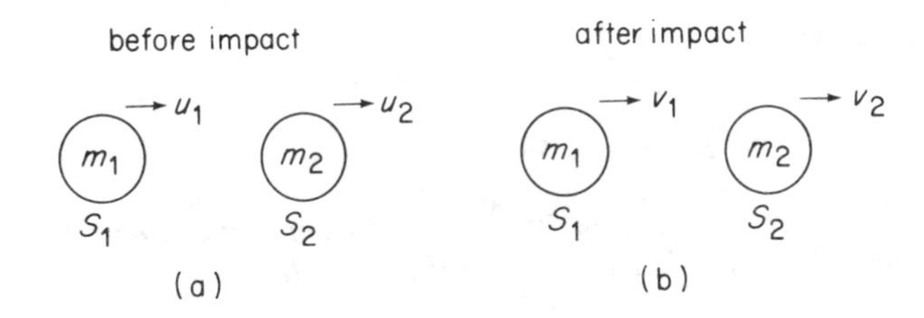

Fig. 7.8

collide. Assuming they do not coalesce, the spheres will then separate as shown in Fig. 7.8(b). The principle of conservation of linear momentum gives

$$m_1u_1 + m_2u_2 = m_1v_1 + m_2v_2 \tag{7.6}$$

But given u_1 and u_2, equation 7.6 is insufficient to determine v_1 and v_2, and further information is required. This is provided by Newton's experimental law which states that for any particular pair of spheres, the relative velocity of the spheres after separation is in the opposite direction to that before collision and that the ratio (relative speed after collision/relative speed before collision) is a constant. This constant is called the coefficient of restitution between the spheres and is denoted by e, where $0 \leqslant e \leqslant 1$. (The case $e = 0$ corresponds to coalescence.) Symbolically, Newton's law is

$$v_1 - v_2 = -e(u_1 - u_2). \tag{7.7}$$

Equations 7.6 and 7.7 are sufficient, given any four of m_1, m_2, v_1, v_2, u_1 and u_2, to determine the other two. Direct collision problems reduce, in essence, to solving equations 7.6 and 7.7 under various conditions. It is essential to draw a clear diagram, showing conditions before and after collision separately, and to show all velocities referred to the same reference direction.

EXAMPLE 2 *A sphere S_1, of mass 4m and moving with speed 3u, overtakes and collides directly with a second sphere S_2, of mass 2m and moving with speed u. The coefficient of restitution for collision between the spheres is $\frac{1}{4}$. Find their velocities after collision.*

After collision, the spheres are assumed to move in the direction of the original motion with speeds v and w, as shown in Fig. 7.9. Conservation of linear momentum gives

$$14mu = 4mv + 2mw.$$

Fig. 7.9

Newton's law gives

$$w - v = -\tfrac{1}{4}(u - 3u) = \tfrac{1}{2}u.$$

Solving these equations gives $w = \mathbf{8u/3}$, $v = \mathbf{13u/6}$.

EXAMPLE 3 *Determine the velocities after collision if, in example 2, the spheres are moving towards each other and not in the same direction.*

If the reference direction is chosen in the initial direction of motion of S_1, the situation before and after collision is as shown in Fig. 7.10. We now have

$$10mu = 4mv + 2mw$$
$$v - w = -\tfrac{1}{4}(3u + u) = -u.$$

Solving these equations gives $v = \mathbf{4u/3}$, $w = \mathbf{7u/3}$.

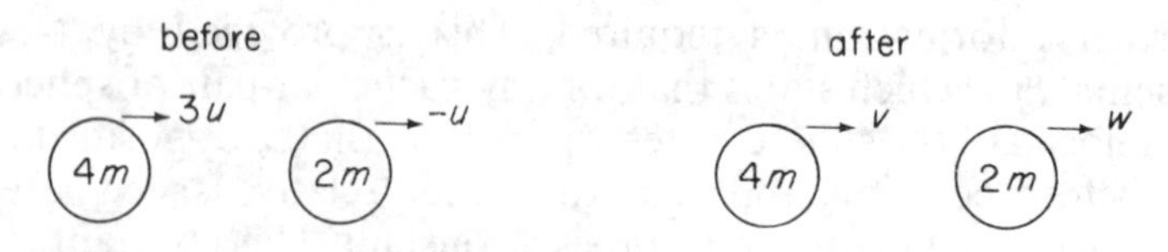

Fig. 7.10

EXERCISE 7.2

In numerical questions, take $g = 10\,\text{ms}^{-2}$.

Questions 1 to 8 refer to a small ball dropped from rest at a height h on to a smooth plane, the coefficient of restitution being e, and h_1 and h_2 being the heights reached after the first and second bounce respectively.

1 $h = 4{\cdot}05\,\text{m}$, $e = \tfrac{1}{3}$; find h_1 and the time taken to reach the plane again.
2 $h = 4\,\text{m}$, $h_1 = 1\,\text{m}$; find e.
3 $h = 2{\cdot}5\,\text{m}$, $h_2 = 0{\cdot}04\,\text{m}$; find e.
4 $h = 12{\cdot}8\,\text{m}$, $e = 0{\cdot}25$; find the speed of the ball after the first and second impacts.
5 $e = \tfrac{1}{5}$; find h given that speed after the second impact is $2\,\text{ms}^{-1}$.
6 $h = 2\,\text{m}$, $e = \tfrac{1}{3}$; find the total distance travelled by the ball before it comes to rest.
7 $h = 0{\cdot}45\,\text{m}$, $e = 0{\cdot}7$; find the total time taken by the ball before coming to rest.
8 $h = 7{\cdot}2\,\text{m}$, $e = 0{\cdot}25$; find, given that the ball is of mass $0{\cdot}2\,\text{kg}$, the magnitude of the impulse on the ball at the first and second bounces.

Questions 9 to 11 refer to a sphere of mass m_1 kg moving with speed $u_1\,\text{ms}^{-1}$ colliding directly with a sphere of mass m_2 moving with speed $u_2\,\text{ms}^{-1}$ in the same direction; $v_1\,\text{ms}^{-1}$ and $v_2\,\text{ms}^{-1}$ denote the speed after impact, the positive direction being that of the original motion of the spheres.

9 $m_1 = 4$, $m_2 = 1$, $u_1 = 4$, $u_2 = 1$, $e = 0{\cdot}5$; find v_1 and v_2.
10 $m_1 = 2$, $m_2 = 3$, $u_1 = 6$, $u_2 = 2$, $v_1 = 3$; find e and v_2.
11 $m_2 = 12$, $u_1 = 10$, $u_2 = 2$, $v_1 = 2$, $v_2 = 4$; find e and m_1.

Questions 12 to 14 refer to a sphere A of mass m_1 kg moving with speed u_1 ms^{-1} colliding directly with a sphere B of mass m_2 kg moving directly towards A with speed u_2 ms^{-1}. After collision, the speeds of the spheres A and B, in the original direction of motion of sphere A, are v_1 and v_2 respectively.

12 $m_1 = 4$, $m_2 = 1$, $u_1 = 4$, $u_2 = 1$, $e = 0{\cdot}5$; find v_1 and v_2.

13 $m_2 = 2$, $u_1 = 6$, $u_2 = 4$, $v_1 = 4$, $e = 0{\cdot}1$; find m_1 and v_2.

14 $m_1 = 2$, $u_1 = 2$, $u_2 = 3$, $v_1 = -1$, $e = 0{\cdot}2$; find m_2 and v_2.

15 A sphere of mass 3 kg moving with speed 8 ms^{-1} collides directly with a sphere of mass 12 kg, which is at rest. Find the range of e so that the spheres move in opposite directions after collision.

16 A sphere mass m moving with speed $4u$ overtakes and collides with a similar sphere of mass $3m$ moving with speed $6u$. The kinetic energy lost in the collision is $9mu^2/8$. Find the coefficient of restitution.

7.3 Continuous impact

A simple example of continuous impact is provided by a jet of water striking a smooth wall. The jet will be composed of a large number of particles and the effect of the collision of each particle can be treated as in §7.1. If it is assumed

(i) that the water has a speed v when it hits the wall,
(ii) that its speed is completely destroyed by the impact,
(iii) that a mass m hits the wall in unit time,

then the impulse exerted on the wall in time t is mv. If the magnitude of the force exerted by the jet on the wall is denoted by F, then

$$\int_0^t F\,\mathrm{d}w = mvt$$

and differentiating this gives

$$F = mv, \qquad 7.8$$

enabling the force to be found in terms of the speed of the water and the mass per unit time which hits the wall.

EXERCISE 7.3

In numerical questions, take $g = 10\,\text{ms}^{-2}$.

1 A jet discharges water at a rate of 10 kgs^{-1} with a speed of 20 ms^{-1}. The water strikes a vertical wall at right angles and does not rebound. Find the force exerted on the wall.

2 Water is discharged from a nozzle of area $0{\cdot}3\,\text{m}^2$ at a speed of 50 ms^{-1} and impinges at right angles on a vertical wall. Find the force exerted on the wall.

3 Rain falls steadily on level ground so that the total rainfall in 100 s is 1 cm. The speed of the rain on striking the ground is that acquired by falling from rest through 245 m. Find, assuming that the rain does not rebound, the total force per m^2 exerted on the ground.

4 A jet of water issues vertically at a speed of $32\,\text{ms}^{-1}$ from a nozzle of area $10^{-5}\,\text{m}^2$. A small ball weighing 1 kg is balanced in the air by the impact of the water on it. Find the distance between the jet and the lowest point of the ball.

7.4 Motion of connected bodies

The motion of two or more bodies which are connected together in some way – for example, a car pulling a caravan – is slightly more complicated than the motion of one body, because some account has to be taken of the interactive forces produced by the connection. The most important fact to remember about interactive forces is that the force on body A due to body B is equal in magnitude, but opposite in direction, to that exerted by body A on body B. The basic method to be used in solving 'connected-body' problems is a slight extension of that for a single body, in that the total force, including the interactive ones, on each body are found and Newton's law applied separately to each body. For bodies moving in the same straight line, the calculation can sometimes be simplified by considering the motion of all the bodies as that of one particle. The equation so obtained will not involve the interactive forces occurring, but it will be equivalent to that obtained by eliminating the interactive forces from the separate equations. The following examples illustrate some commonly occurring problems involving connected bodies.

EXAMPLE 1 *A car of mass* 950 kg *tows a caravan of mass* 550 kg *along a straight level road. The engine of the car produces a constant tractive force of* 1500 N *on both the car and the caravan. There is a constant resistance of magnitude* 0·25 N *per kg acting on both the car and the caravan. Find the acceleration of the car and caravan. Find also the tension in the tow bar.*

Fig. 7.11 shows all the horizontal forces acting on the system with T N denoting the tension in the tow bar. Note that the tension acts to the right on the caravan (that is, pulling it) and to the left on the car (that is, impeding its motion). Since the car and caravan are joined rigidly together they will have a common acceleration $a\,\text{ms}^{-2}$. Treating the system as one body gives

$$\begin{aligned} 1500a &= 1500 - 237{\cdot}5 - 137{\cdot}5 \\ &= 1125. \end{aligned}$$

giving $a = 0{\cdot}75$, so the acceleration is $\mathbf{0{\cdot}75\,ms^{-2}}$.

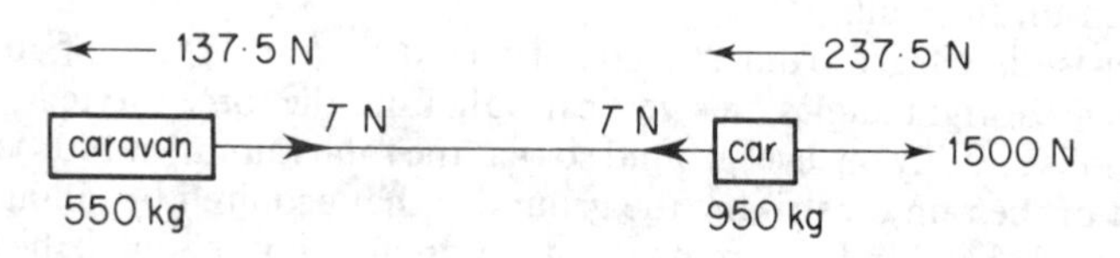

Fig. 7.11

The total force acting to the right on the car is $(1500 - 237{\cdot}5 - T)$ N, so applying Newton's law to the car gives

$$950a = 1500 - T - 237{\cdot}5.$$

The force to the right on the caravan is $(T - 137{\cdot}5)$ N so that

$$550a = T - 137{\cdot}5.$$

As stated earlier, adding these equations gives the equation of motion of the whole system. Substituting $a = 0{\cdot}75$ in either equation gives $T = 550$, and the tension in the bar is **550 N**.

EXAMPLE 2 *A light inextensible string passes over a smooth pulley and particles of masses m_1 and m_2, where $m_1 > m_2$, are attached to the ends of the string and can move, with the parts of the string not in contact with the pulley being taut and vertical. Find the acceleration of the particles and the tension in the string.*

Since the string is light the tension will be constant throughout any straight length, and the smoothness of the pulley ensures the tension is the same throughout the string. Fig. 7.12 shows the forces acting on the particles. Since the string is taut, the particles will both have accelerations of the same magnitude a but in opposite directions, as shown in Fig. 7.12. The downward force on the heavier particle is $m_1g - T$ so that

$$m_1a = m_1g - T. \qquad 7.9$$

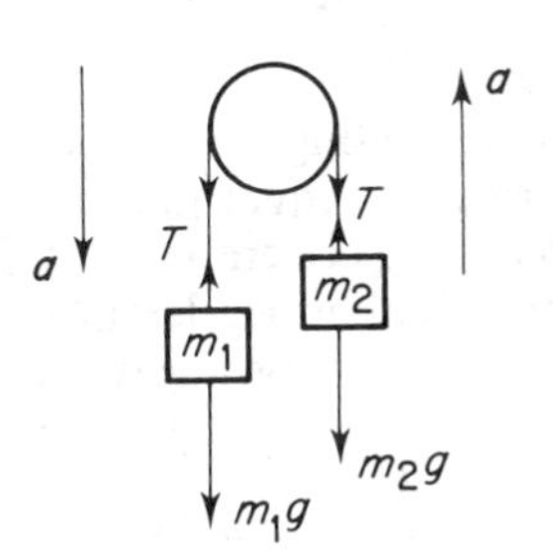

Fig. 7.12

The upward force on the other particle is $T - m_2g$ so that

$$m_2a = T - m_2g. \qquad 7.10$$

Adding the equations gives

$$a = \frac{(m_1 - m_2)}{m_1 + m_2}g$$

and substituting this in any one of the equations gives

$$T = \frac{2m_1m_2g}{m_1 + m_2}.$$

It should be noted that the equation obtained by adding equations 7.9 and 7.10 is

$$(m_1 + m_2)a = (m_1 - m_2)g.$$

This is precisely the equation that would have been obtained by considering the horizontal motion of the system formed by rotating clockwise that part of the string hanging to the left of the pulley and rotating the other part counter-clockwise, so that the string becomes straight and horizontal with there being no change, relative to the string, in the forces acting. It is certainly not a correct method to use for this problem and others involving strings over pulleys. However, it serves as a useful check on the algebra in eliminating the tension.

EXAMPLE 3 *Fig. 7.13 shows a particle A, which is of mass 7m and lies on a smooth rectangular table. Particle A is connected by light inextensible strings to two particles B and C of masses 5m and 4m respectively, passing over smooth pulleys at the edges of the table. The strings are both perpendicular to the edges of the table. Find the acceleration of A, B and C and the tension in the string attached to C.*

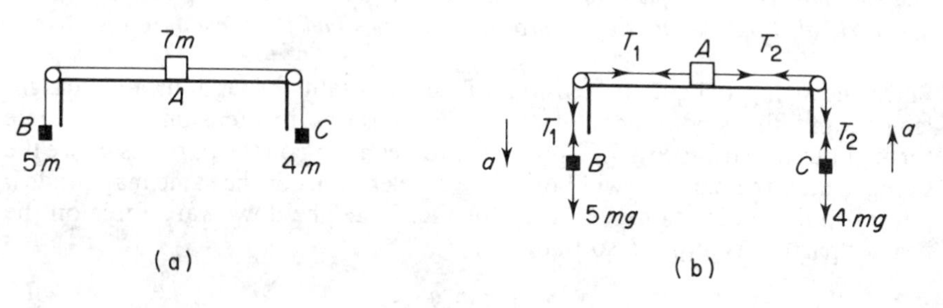

Fig. 7.13

Figure 7.13(b) shows the forces acting on the particles; there is no reason to assume the tensions in the strings are equal and so they are denoted by T_1 and T_2. If it is assumed that B has an acceleration a downwards, then A has an acceleration of a to the left and C has an acceleration a upwards. The equations of motion of the particles A, B, C are respectively

$$5mg - T_1 = 5ma,$$
$$T_1 - T_2 = 7ma,$$
$$T_2 - 4mg = 4ma.$$

Adding these gives $a = \boldsymbol{g/16}$ and substituting in the last equation gives $T_2 = \boldsymbol{17mg/4}$.

EXERCISE 7.4

Take $g = 10\,\text{ms}^{-2}$.

Questions 1 to 4 refer to a car of mass M kg towing a caravan of mass m kg up a slope inclined at an angle α to the horizontal. The resistances to the motion of the car and the caravan are denoted respectively by R_1 N and R_2 N and the tractive force is denoted by P N.

1 $M = 1\,200$, $m = 400$, $P = 2\,000$, $R_1 = 300$, $R_2 = 100$, $\alpha = 0°$; find the acceleration of the car and the tension in the tow bar.

2 $M = 1\,000$, $m = 200$, $P = 1\,900$, $R_1 = 250$, $R_2 = 150$, $\alpha = 0°$; find the acceleration of the car and the tension in the tow bar.

3 $M = 1\,200, m = 300, P = 1\,800, R_1 = 275, R_2 = 100, \sin\alpha = 1/200$; find the acceleration of the car and the tension in the tow bar.

4 $M = 1\,300, m = 200, R_1 = 350, R_2 = 150, \sin\alpha = 1/10$, and the car engine works at a steady rate of 20 kw. Find the steady speed at which the car can move up the slope and the tension in the tow bar at this speed.

Questions 5 to 7 refer to a light inextensible string passing over a small smooth pulley, with particles of masses m_1 and m_2 attached one at each end of the string. Find, in each case, the magnitude of the acceleration of the particles and the tensions in the string.

5 $m_1 = 5$ kg, $m_2 = 3$ kg.

6 $m_1 = 7$ kg, $m_2 = 3$ kg.

7 $m_1 = 4M$, $m_2 = 6M$.

Questions 8 to 10 refer to a particle A of mass m_1 on a horizontal plane and attached by a light inextensible string, which passes over a small smooth pulley at the edge of the plane, to a particle B of mass m_2 which hangs freely with the string vertical. The vertical plane through B and the pulley contains that part of the string between A and the pulley.

8 The plane is smooth, $m_1 = 2$ kg, $m_2 = 3$ kg. Find the magnitude of the acceleration of the particles and the tension in the string.

9 The plane is smooth, $m_1 = 2$ kg, $m_2 = 6$ kg. Find the magnitude of the force exerted on the pulley.

10 The plane is rough with coefficient of friction 0·5, $m_1 = 3$ kg, $m_2 = 7$ kg. Find the tension in the string, given that the system is released from rest.

Questions 11 to 14 refer to a particle A of mass m_1 on a plane inclined at an angle α to the horizontal. It is connected by a light inextensible string, which passes over a small smooth pulley at B, to a particle C of mass m_2 and which hangs freely with the string BC vertical. The portion of the string AB is parallel to a line of greatest slope of the plane.

11 The plane is smooth with $\alpha = 30°$, $m_1 = 1$ kg, $m_2 = 3$ kg. Find the magnitude of the acceleration of the particles.

12 The plane is smooth with $\sin\alpha = \frac{4}{5}$, $m_1 = 5$ kg, $m_2 = 3$ kg. Find the tension in the string.

13 The plane is rough with coefficient of friction 0·25, $\sin\alpha = \frac{3}{5}$, $m_1 = 3$ kg, $m_2 = 7$ kg. Find the magnitude of the acceleration of the particles given that the system is released from rest.

14 The plane is rough with coefficient of friction 0·25, $\sin\alpha = 5/13$, $m_1 = 13$ kg, $m_2 = 1$ kg. Find the magnitude of the acceleration of the particles given that the system is released from rest.

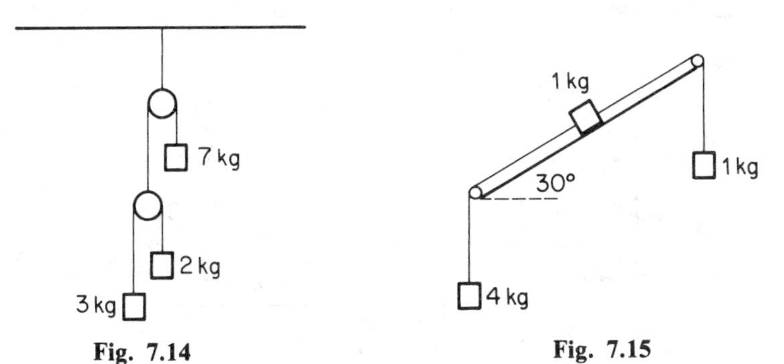

Fig. 7.14 **Fig. 7.15**

15 Find the acceleration of the particle of mass 7 kg in the configuration shown in Fig. 7.14. The pulleys are light, and the strings light and inextensible.

16 Find the tensions in the strings for the configuration in Fig. 7.15 when the plane is smooth.

7.5 Impulsive motion of connected bodies

Problems involving the impulsive motion of connected bodies free to move in the same straight line can be solved by straightforward application of the principle of conservation of momentum as in §7.1, example 3. However, similar problems for particles connected by a string passing over a pulley, as in §7.4, example 2, are slightly more complicated and have to be handled rather more carefully. Some of the difficulties which may be encountered are illustrated in the following examples.

EXAMPLE 1 *In the motion described in §7.4, example 2, the particle of mass m_1 hits an inelastic plane so that it comes to rest and, at this instant, the other particle is moving upwards with speed u. Find the common speed of the particles after the string has just become taut again.*

Once one of the particles has been reduced to rest, the other will move freely under gravity with the string slack. It will, therefore, return to its original position moving downwards with speed u, until the string tautens and the two particles move together with a common speed v, say. The situation just before and just after the string tautens is shown in Fig. 7.16.

There will be an impulse J in the string immediately after it tautens, and equating change of momentum to impulse for each particle gives

$$m_1 v = J,$$

$$m_2 v - m_2 u = -J.$$

Adding these

$$(m_1 + m_2)v = m_2 u.$$

This is precisely the equation which would have been obtained by imagining the string to be horizontal, as mentioned in §7.4, example 2. It is often regarded as permissible to use this approach for impulsive motion, but the approach of considering the impulse on each particle separately is more satisfactory.

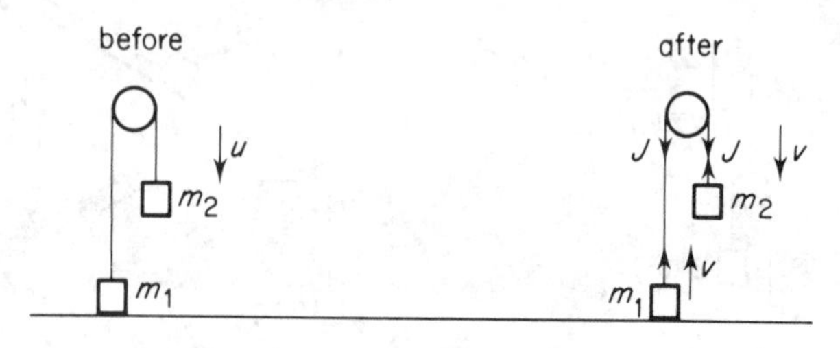

Fig. 7.16

EXAMPLE 2 *When both particles in §7.4, example 2 are moving with speed u, with the heavier particle of mass m_1 moving downwards, a particle of mass m_3 moving with speed v downwards collides with the lighter particle and coalesces with it. Find the common speed of the particles after the collision.*

Fig. 7.17 shows the situation immediately before and immediately after collision, the impulse in the string again being denoted by J. If we consider the system formed by the particles of masses m_2 and m_3 then, as shown in §7.1, the change in momentum is the total external impulse on the system and this is the impulse in the string. Therefore,

$$J = (m_2 + m_3)w - m_2 u + m_3 v.$$

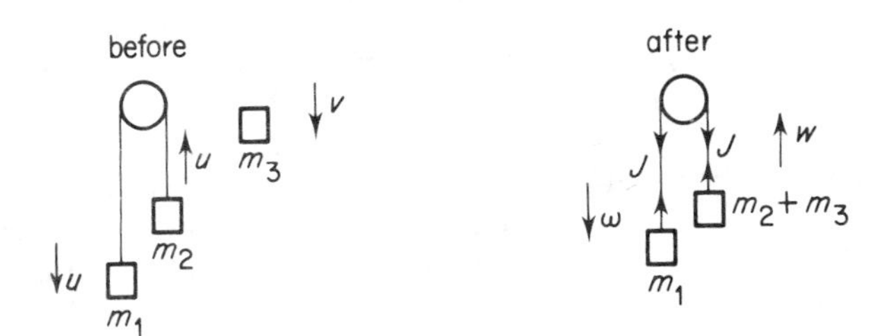

Fig. 7.17

Also $$J = -m_1 w + m_1 u.$$

Subtracting these equations gives

$$(m_1 + m_2 + m_3)w = (m_1 + m_2)u - m_3 v,$$

which is the same result as would have been obtained by applying conservation of momentum in a straight line.

A kind of motion, which is apparently impulsive, can occur in problems involving strings over pulleys where, during the motion, part of a particle is removed. In this case, there is no impulse applied to the remaining particle so there is no change in speed. There will, however, be a change in acceleration, because the masses will effectively be different.

EXERCISES 7.5

The following questions refer to two particles A and B of masses m_1 and m_2 respectively, connected by a light inextensible string passing over a small smooth pulley. Take $g = 10\,\text{m s}^{-2}$ in numerical questions.

1 $m_1 = 3\,\text{kg}$, $m_2 = 2\,\text{kg}$, the system is set off from rest and after 1 s the particle B picks up, from rest, an additional particle of mass 3 kg. Find the further distance moved before the system first comes to instantaneous rest.

2 $m_1 = 5\,\text{kg}$, $m_2 = 3\,\text{kg}$, the system is set off from rest, and after descending 5 m the particle A strikes an inelastic plane and comes to rest. Determine the time that it first remains on the plane and the speed with which it is jerked off.

3 $m_1 = 2{\cdot}6\,\text{kg}$, $m_2 = 2{\cdot}4\,\text{kg}$, both masses are set off initially to move with speed

$1\,\mathrm{ms}^{-1}$ with A moving downwards. After 5 s, A passes through a small ring and a mass 0·5 kg is removed from it. Find the time taken before A next reaches the ring.

4 $m_1 = 2\,\mathrm{kg}$, $m_2 = 1{\cdot}8\,\mathrm{kg}$, the system is at rest with A resting on a horizontal plane. A third particle C of mass 0·7 kg falls a distance 4·5 m from rest and coalesces with B. Find the height to which A rises.

5 $m_1 = m_2 = M$. When the system is at rest, an additional particle of mass m is placed on A, and when A has descended a distance h, this additional particle is removed at the same instant as B picks up an identical particle of mass m from rest. Find the further distance that A falls before coming to instantaneous rest.

MISCELLANEOUS EXERCISE 7

In numerical questions, g should, unless otherwise stated, be taken as $10\,\mathrm{ms}^{-2}$.

1 A particle of mass m is dropped from rest from a point A which is at a height h above a horizontal surface. After hitting the surface the particle rebounds vertically and reaches a maximum height of $\frac{1}{2}h$ above the surface at the point B. Find expressions for the loss of energy and the impulse on the particle at the impact.

At the instant of the particle's impact with the surface, a second particle, of mass $4m$, is dropped from rest at A. Show that the two particles collide at B.

The two particles coalesce when they collide at B. When the combined particle hits the surface, one-half of its kinetic energy is lost in the impact. Show that, after the impact, the combined particle reaches a greatest height of $\frac{41}{100}h$ above the surface. *(C)*

2 A ball bearing falling from rest at a height h above a horizontal surface rebounds to a height e^2h where e $(0 < e < 1)$ is a constant. If the ball bearing is released from rest at a height H above the surface, prove that the *total* distance travelled by the ball bearing before coming to rest is

$$\frac{1+e^2}{1-e^2}H.$$

If $H = 10\,\mathrm{cm}$ and $e = \frac{3}{5}$, find how many times the ball bearing hits the surface before travelling a total distance of 20 cm from the instant of its release. *(C)*

3 Three trucks A, B, C have masses 5, 10 and 20 tonnes respectively. They are close together on a straight, level track with B and C at rest. Frictional forces are too small for consideration.

(i) A is given a velocity of 5·6 m/s towards B so that A hits B and then B hits C. As a result, both A and B are reduced to rest. Find the velocity of C after the impacts and the kinetic energy lost in each collision.

(ii) On another occasion A is given a velocity of 5·6 m/s towards B but on hitting B the two trucks become coupled together and on hitting C all three become coupled together. Find the total loss of kinetic energy.

(iii) On a third occasion A has a velocity of 10 m/s towards B and both B and C have velocities of 4·4 m/s away from A. As in (ii) all become coupled together. Explain why the loss of kinetic energy is the same as in (ii).

(AEB 1979)

4 A straight railway line is inclined at an angle θ to the horizontal, where $\sin\theta = 0{\cdot}02$. A truck of mass 5 tonnes ($= 5000$ kg) is released from rest on this track. Neglecting any frictional resistances, find the speed of the truck after it has travelled a distance of 250 m.

A second truck, also of mass 5 tonnes, is released from rest, and after travelling a distance of 250 m down the same slope its speed is found to be $5\,\text{ms}^{-1}$. Calculate the (constant) frictional resistance to the motion of this truck.

A truck of mass 5 tonnes moving with a speed of $5\,\text{ms}^{-1}$ along a horizontal track collides with a truck of mass 10 tonnes which is at rest on the track. After impact both trucks move on the track in the same direction with the front truck moving twice as fast as the rear one. Calculate the speeds of the trucks, and the amount of energy lost in the impact. *(C)*

5 A small sphere A, of mass m moving with velocity $2u$ on a smooth horizontal plane, impinges directly on a small sphere B, of equal radius and mass $2m$ moving with velocity u in the same direction on the plane. Given that after impact B moves with velocity $3u/2$, calculate

(i) the coefficient of restitution between A and B,

(ii) the loss in kinetic energy due to the impact.

The sphere B continues to move with velocity $3u/2$ until it hits a vertical wall from which it rebounds and is then brought to rest by a second impact with the approaching sphere A. Calculate

(iii) the coefficient of restitution between B and the wall,

(iv) the final speed of the sphere A. *(AEB 1980)*

6 Two particles A and B of equal mass are sliding towards each other on a smooth horizontal table with velocities u and v respectively where $u > v$. The coefficient of restitution between the two particles is e. After the collision the increase in kinetic energy of B is equal to e times the loss in kinetic energy of A. Show that

$$\frac{u}{v} = \frac{3+e}{1-e}.$$ *(C)*

7 A small smooth sphere moves on a horizontal table and strikes an identical sphere lying at rest on the table at a distance d from a vertical wall, the impact being along the line of centres and perpendicular to the wall. Prove that the next impact between the spheres will take place at a distance

$$2de^2/(1+e^2)$$

from the wall, where e is the coefficient of restitution for all impacts involved. *(L)*

8 Two beads A and B, of mass $2m$ and $3m$ respectively, are threaded on to a smooth thin circular wire hoop of radius a. The wire is fixed horizontally and the beads placed at opposite ends of a diameter. At time $t = 0$, bead A is projected along the wire with speed $\sqrt{(ag)}$. Find the force exerted by the wire on bead A before A strikes B. Find t when A first strikes B.

At the collision, the coefficient of restitution is e. Show that the beads collide for the second time when

$$t = \frac{\pi(e+2)}{e}\sqrt{\left(\frac{a}{g}\right)}.$$

Find the speeds of the particles just before this second collision. (*L*)

9 Three smooth spheres A, B, C of equal radii and masses m, $2m$, $2m$ respectively, lie in that order on a smooth horizontal table with their centres in a straight line. The coefficient of restitution between A and B is $\frac{1}{3}$ and that between B and C is e. Sphere A is projected with speed V to strike B directly. Show that after two impacts the total loss of kinetic energy is

$$8mV^2(4-e^2)/81.$$

Given that $e < \frac{1}{2}$, show also that there will be no more impacts. (*L*)

10 Particles of masses $m, me, \ldots, me^{n-1}$ are placed in this order in a straight line on a smooth horizontal table. The first particle is projected with velocity u towards the second. Show that when the last particle begins to move the loss in kinetic energy of the system is $\frac{1}{2}mu^2(e-e^n)$, e being the coefficient of restitution between any two particles. (*O*)

11 A particle is at rest inside a smooth straight tube, which is closed at both ends, and lies at rest on a smooth horizontal table. The mass of the tube is k times the mass of the particle, and the coefficient of restitution is e. If the tube is given a velocity u along its own length, prove that after n impacts the velocities of the tube and the particle are respectively

$$\frac{u[k+(-e)^n]}{k+1} \text{ and } \frac{ku[1-(-e)^n]}{k+1}. \qquad (O)$$

12 Suppose that, when it rains, water drops fall from a height of 500 m above ground level. Assuming that gravity is the only force acting, calculate the speed with which the rain hits the ground; take g as $10\,\text{ms}^{-2}$.

In fact, due to air resistance, the speed near the ground is found to be $15\,\text{ms}^{-1}$. During a storm 1 cm of rain falls in half an hour. It may be assumed that, when the rain hits the horizontal flat roof of a house, it loses all its vertical speed instantaneously. Calculate the force, assumed constant, on one square metre of the roof during the storm due to rain hitting the roof. (The density of water is $1000\,\text{kg m}^{-3}$.) Show that this force is very small compared with the weight of water that would collect on this area of the roof during the storm if it were prevented from draining away.

Find the rate at which the kinetic energy of the rain falling on to one square metre of the roof is being lost during the storm, and show that if all the kinetic energy of the falling rain could be converted into useful power, an area of $1600\,\text{m}^2$ would be needed to generate each kilowatt. (*C*)

13 A pump raises $3\,\text{m}^3$ of water per hour through a height of 3·7 m and delivers it in a horizontal jet through a circular nozzle of diameter 12·5 mm. Show that the speed at which the water leaves the nozzle is about $6{\cdot}8\,\text{ms}^{-1}$. If the pump works at 60% efficiency, calculate the power necessary to drive it.

The jet plays on a wall close to the nozzle in a direction normal to the wall. Assuming that the water does not splash back, calculate the force exerted on the wall. (*C*)

14 Sugar is poured steadily on to a scale pan so that 0·05 kg per second strikes pan, or the sugar in the pan, at a speed of $2\,\text{ms}^{-1}$. The sugar is assumed not to bounce after impact. Given that there is a mass of 1 kg in the other scale pan, find the mass of sugar in the pan when it is about to tip.

15 A car of mass 1500 kg is pulling a caravan of mass 900 kg along a straight horizontal road by means of a tow bar. At all speeds there is a force of magnitude 300 N resisting the motion of the car and a force of magnitude 200 N resisting the motion of the caravan. At the instant when the speed of the car is 15 m/s and the engine of the car is working at a rate of 12 kW, calculate

(i) the acceleration of the car,
(ii) the tension in the tow bar.

The car and caravan move up a road which is inclined at an angle $\sin^{-1}(1/12)$ to the horizontal at a steady speed of 20 m/s. Calculate

(iii) the rate at which the engine of the car is working,
(iv) the tension in the tow bar. (*AEB 1979*)

16 A locomotive of mass 60 tonnes pulls a set of coaches whose total mass is 240 tonnes. The engine of the locomotive works at a constant rate of 500 kW and non-gravitational resistances on both the locomotive and the coaches are x newtons per tonne. The track is level and the maximum speed of the train is 144 km/h. Find x. Find also the tension in the coupling between the locomotive and the first coach when the speed is

(a) 144 km/h,
(b) 72 km/h.

When the speed of the train is 72 km/h the coupling on the last coach breaks. Given that the mass of that coach is 20 tonnes, find how far this coach will travel before coming to rest. (*L*)

17 The maximum rate of working of the engine of a car is S kW. Against a constant resistance, the car can attain a maximum speed of u m/s on level ground and a maximum speed of $\frac{1}{2}u$ m/s directly up a slope of inclination α where $\sin\alpha = \frac{1}{16}$. Calculate the maximum speed of the car up a slope of inclination β, where $\sin\beta = \frac{1}{8}$, assuming that the resistance remains unchanged.

Given that $S = 15$ and that $u = 20$, find the maximum acceleration that can be attained when the car is towing a trailer of mass 300 kg at 10 m/s on level ground. It may be assumed that the resistance to the motion of the car remains unchanged and that the resistance to the motion of the trailer can be neglected. (*L*)

18 Two particles A and B, of masses $3m$ and m respectively, are connected by a light inextensible string which passes over a fixed light smooth pulley. The system is released from rest with the string taut and the straight parts of the string vertical. Find the acceleration of either particle and the tension in the string.

The particles A and B start at the same level. The string breaks after time t_0 and the string is sufficiently long so that B does not hit the pulley. Show that the vertical distance between A and B when B reaches its highest point is gt_0^2. (*L*)

19 Two particles, of masses m and $2m$, are attached to the ends A and B respectively of a light inextensible string which passes over a smooth peg C fixed at the top of a smooth plane inclined at an angle of 30° to the horizontal. Initially the particles are held at rest with the string taut, the portion AC of the string lying along a line of greatest slope of the plane and the portion CB hanging vertically. When the system is released, B moves a vertical distance of 6 metres before colliding with an inelastic horizontal table. Find the total distance A moves up the plane, assuming that it does not reach C. (*L*)

20

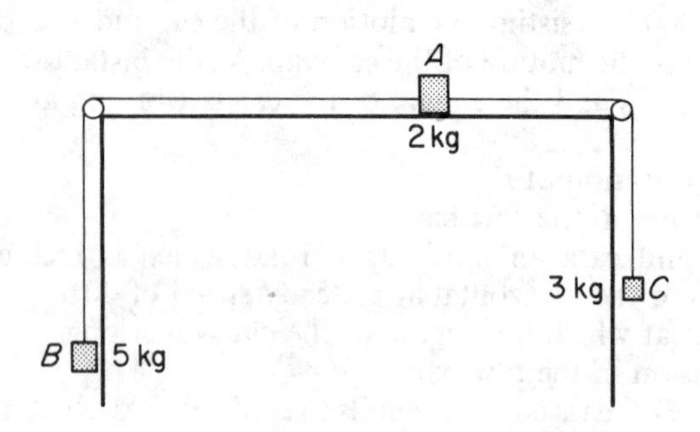

The diagram shows a particle A of mass 2 kg resting on a horizontal table. It is attached to particles B of mass 5 kg and C of mass 3 kg by light inextensible strings hanging over light smooth pulleys. Initially the plane ABC is vertical, the strings are taut and pass over the pulleys at right angles to the edges of the table. In the ensuing motion from rest find the common acceleration of the particles and the tension in each string before A reaches an edge

(a) when the table is smooth,

(b) when the table is rough and the coefficient of friction between the particle A and the table is $\frac{1}{2}$. *(L)*

21 Two particles, each of mass m, are connected by a light inextensible string which passes over a smooth pulley at the top of a fixed plane inclined at an angle $\tan^{-1}(\frac{5}{12})$ to the horizontal. The particle A is on the plane and the particle B hangs freely (see diagram). The system is released from rest with the string in a vertical plane through a line of greatest slope of the plane. The coefficient of friction between A and the plane is $\frac{1}{3}$. When B has fallen a distance h the string breaks. A comes to rest after travelling a further distance s up the plane. B falls a further distance h to strike a horizontal plane and rise to a height h above that plane. Find

(i) the speed of the particles when the string breaks,

(ii) the value of s,

(iii) the coefficient of restitution between B and the horizontal plane and the impulse of the blow the particle B strikes this plane. *(AEB 1970)*

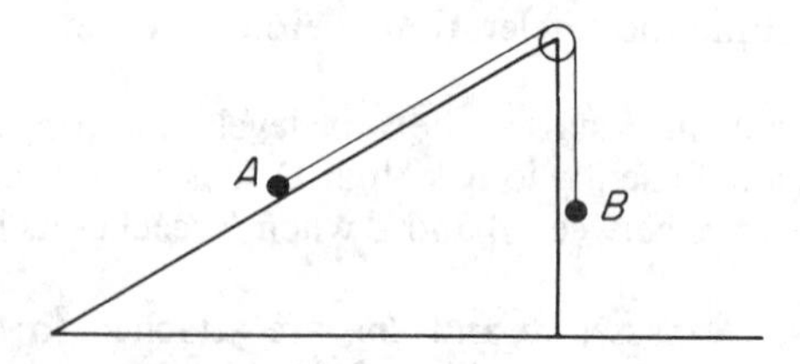

8 Vectors

Mechanics consists of dynamics—the study of motion—and statics—the study of systems of forces in equilibrium. Quantities include force, velocity, and acceleration and all of these are vector quantities. We first give a brief summary of those parts of the theory of vectors required for our work in mechanics.

We introduce differentiation of vector functions of the time and use vectors to study velocity, acceleration, relative motion and the magnitude, and direction and line of action of the resultant of a given set of forces.

Units

The units used are the metre for length, the second for time, the kilogram for mass and the newton for force. So, for example, position vector $\mathbf{r} = (\mathbf{i}+2\mathbf{j}+3\mathbf{k})$ is measured in m, velocity $\mathbf{v} = (\mathbf{i}+2\mathbf{j}+3\mathbf{k})$ in ms^{-1}, acceleration $\mathbf{f} = (\mathbf{i}+2\mathbf{j}+3\mathbf{k})$ in ms^{-2} and force $\mathbf{F} = (\mathbf{i}+2\mathbf{j}+3\mathbf{k})$ in N.

8.1 Velocity

Differentiation of a vector function of time

The velocity of a particle moving in a straight line Ox has been defined in §4.1 by

$$v = \frac{\mathrm{d}x}{\mathrm{d}t} = \dot{x}.$$

It is implicit in this definition that v is associated with the direction Ox. If $\mathbf{i}$ is a unit vector in the direction of Ox, the velocity $\mathbf{v}$ is given by

$$\mathbf{v} = \frac{\mathrm{d}x}{\mathrm{d}t}\mathbf{i}.$$

If the particle is moving so that it has components of velocity in both the Ox and Oy directions, then its velocity $\mathbf{v}$ is given by

$$\mathbf{v} = \frac{\mathrm{d}x}{\mathrm{d}t}\mathbf{i}+\frac{\mathrm{d}y}{\mathrm{d}t}\mathbf{j}. \qquad 8.1$$

Equation 8.1 is the definition of the velocity of a particle moving in two dimensions. If the particle is moving in three dimensions, then its velocity is defined by

$$\mathbf{v} = \frac{dx}{dt}\mathbf{i} + \frac{dy}{dt}\mathbf{j} + \frac{dz}{dt}\mathbf{k}. \qquad 8.2$$

Here, $\frac{dx}{dt}, \frac{dy}{dt}, \frac{dz}{dt}$ are the components of the velocity in the directions Ox, Oy, Oz, respectively.

When the velocity is a constant vector $\mathbf{u}$, where $\mathbf{u} = u_1\mathbf{i} + u_2\mathbf{j}$, then

$$\frac{dx}{dt} = u_1 \quad \text{and} \quad \frac{dy}{dt} = u_2,$$

where u_1 and u_2 are constants. Thus,

$$x = u_1 t + a_1 \quad \text{and} \quad y = u_2 t + a_2,$$

where (a_1, a_2) are the coordinates of the position of the particle at $t = 0$. Thus, the position vector of the particle at time t, $\mathbf{r}(t)$ is given by

$$\mathbf{r}(t) = x\mathbf{i} + y\mathbf{j} = a_1\mathbf{i} + a_2\mathbf{j} + t(u_1\mathbf{i} + u_2\mathbf{j}) = \mathbf{a} + \mathbf{u}t, \qquad 8.3$$

where $\mathbf{a}$ is the position vector of the particle at $t = 0$.

The notation $\mathbf{r}(t)$ is analogous to the notation $f(x)$ for a scalar function of x. Thus

$f(x)$ denotes the value of the function f for a value of x,

$\mathbf{r}(t)$ denotes the value of $\mathbf{r}$ at time t.

To differentiate a vector function of time with respect to time, we use an analogous procedure to the differentiation of a function of x from first principles. Consider the case when

$$\mathbf{r} = \mathbf{a} + \mathbf{u}t,$$

where $\mathbf{u}$ is a constant. If t is increased by δt,

$$\mathbf{r}(t + \delta t) = \mathbf{a} + (t + \delta t)\mathbf{u}.$$

Hence $\delta\mathbf{r}$, the increase in $\mathbf{r}$ is given by $\delta\mathbf{r} = \mathbf{u}\,\delta t$. Thus,

$$\frac{\delta\mathbf{r}}{\delta t} = \mathbf{u}$$

and proceeding to the limit as $\delta t \to 0$, we write $\frac{d\mathbf{r}}{dt} = \mathbf{u}$. Thus, when $\mathbf{u}$ is a constant,

$$\frac{d\mathbf{r}}{dt} = \mathbf{u}. \qquad 8.4$$

If the velocity is not constant, so that as t increases by δt, x increases by δx and y by δy, then $\delta\mathbf{r}$, the corresponding increase in $\mathbf{r}$ is given by

$$\mathbf{r} = x\mathbf{i} + y\mathbf{j}$$

and so
$$\frac{\delta \mathbf{r}}{\delta t} = \frac{\delta x}{\delta t}\mathbf{i} + \frac{\delta y}{\delta t}\mathbf{j}$$

leading to
$$\frac{d\mathbf{r}}{dt} = \frac{dx}{dt}\mathbf{i} + \frac{dy}{dt}\mathbf{j} = \mathbf{v}$$

by equation 8.1. That is,
$$\frac{d\mathbf{r}}{dt} = \mathbf{v} \qquad 8.5$$

The notation $\dot{\mathbf{r}}$ is often used for $\frac{d\mathbf{r}}{dt}$ just as $\dot{x}$ is used for $\frac{dx}{dt}$. Thus, we may write $\dot{\mathbf{r}} = \dot{x}\mathbf{i} + \dot{y}\mathbf{j}$.

If we considered the motion of a particle in three dimensions, by a similar argument we would find that
$$\frac{d\mathbf{r}}{dt} = \frac{dx}{dt}\mathbf{i} + \frac{dy}{dt}\mathbf{j} + \frac{dz}{dt}\mathbf{k} = \mathbf{v}.$$

Differentiation of $c\mathbf{r}$, where c is a scalar constant, and of a scalar product $\mathbf{r}_1 . \mathbf{r}_2$, where $\mathbf{r}_1$ and $\mathbf{r}_2$ are vectors which are functions of time.

If $\mathbf{R} = c\mathbf{r}$, where c is a scalar constant, we note that $\delta\mathbf{R} = c\,\delta\mathbf{r}$ and hence
$$\frac{\delta \mathbf{R}}{\delta t} = c\frac{\delta \mathbf{r}}{\delta t} \Rightarrow \frac{d\mathbf{R}}{dt} = c\frac{d\mathbf{r}}{dt}.$$

For a scalar product $\mathrm{R} = \mathbf{r}_1 . \mathbf{r}_2$, if
$$\mathbf{r}_1 = x_1\mathbf{i} + y_1\mathbf{j} \quad \text{and} \quad \mathbf{r}_2 = x_2\mathbf{i} + y_2\mathbf{j},$$

then
$$\mathrm{R} = x_1x_2 + y_1y_2$$
$$\begin{aligned}\frac{d\mathrm{R}}{dt} &= \frac{dx_1}{dt}x_2 + \frac{dy_1}{dt}y_2 + x_1\frac{dx_2}{dt} + y_1\frac{dy_2}{dt} \\ &= \left(\frac{dx_1}{dt}\mathbf{i} + \frac{dy_1}{dt}\mathbf{j}\right)\cdot(x_2\mathbf{i} + y_2\mathbf{j}) \\ &\quad + (x_1\mathbf{i} + y_1\mathbf{j})\cdot\left(\frac{dx_2}{dt}\mathbf{i} + \frac{dy_2}{dt}\mathbf{j}\right) \\ &= \frac{d\mathbf{r}_1}{dt}.\mathbf{r}_2 + \mathbf{r}_1.\frac{d\mathbf{r}_2}{dt} = \dot{\mathbf{r}}_1.\mathbf{r}_2 + \mathbf{r}_1.\dot{\mathbf{r}}_2.\end{aligned}$$

A similar proof may be obtained when $\mathbf{r}_1$ and $\mathbf{r}_2$ are given in terms of three dimensional components. Thus, the derivative of a scalar product is given by a similar rule to that for the derivative of the product of two scalar functions.

Frame of reference

When an observer on the platform of a station notes that a train is moving out of the station, he or she is observing the change of position of the train with respect to the platform. There is relative motion between the train and the platform. The platform, the station and all the other localities among which there is no relative motion are *the frame of reference* with respect to which the changes in position of the train and other moving objects are observed.

It is possible to choose different frames of reference. If the frame of reference is chosen to be the moving train, for example, then the platform would be observed to move in the opposite direction (see Fig. 8.1(b)). Some people might say that this would be an unrealistic choice because it is the train that is in motion and the platform that is at rest. However, it is well known that the earth moves as a planet round the sun in an ellipse with the sun as a focus, and the earth also rotates on its axis, so the platform is far from at rest.

Nevertheless, we will commence by choosing frames of reference, such as the platform, which are at rest relative to the local motion of the earth. We will use the term *fixed point* for a point at rest relative to such frames of reference.

Suppose that a particle P moves from a point A to a point B in time t. Taking a fixed point O as the origin of reference (a Newtonian origin), the

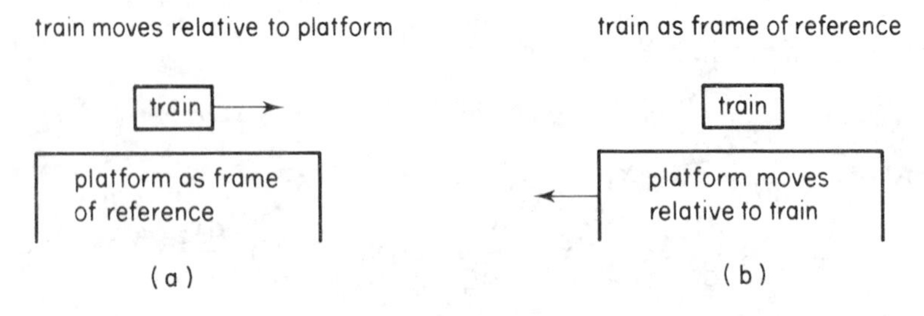

Fig. 8.1

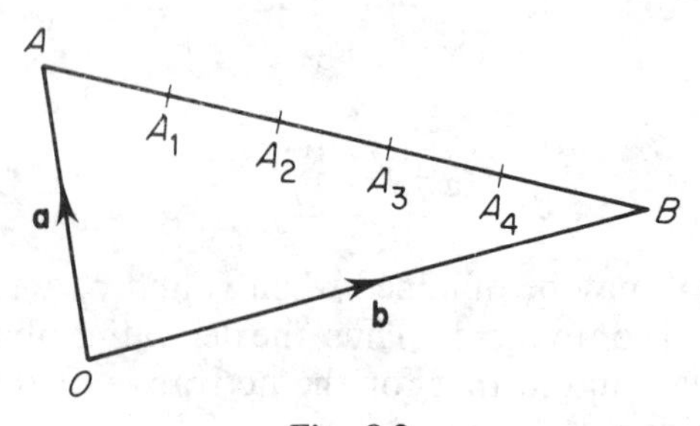

Fig. 8.2

position vectors of A and B with respect to O are denoted by $\mathbf{a}$ and $\mathbf{b}$ respectively, and

$$\text{vector } \overrightarrow{AB} = \mathbf{b} - \mathbf{a}.$$

If P describes AB with uniform velocity in time t, at equal intervals of time it will describe segments of AB of equal length – AA_1, A_1A_2, A_2A_3, and so on. The speed with which P moves will be $|(\mathbf{b}-\mathbf{a})/t|$ and the direction of this velocity is along $\overrightarrow{AB}$, so that the uniform velocity is $(\mathbf{b}-\mathbf{a})/t$. If we denote this uniform velocity by $\mathbf{u}$,

$$\mathbf{u} = \frac{\mathbf{b}-\mathbf{a}}{t}$$

$$\mathbf{b}-\mathbf{a} = \mathbf{u}t$$

$$\mathbf{b} = \mathbf{u}+\mathbf{a}t \qquad 8.6$$

Equation 8.6 gives the position vector of the particle at any time t, provided the particle continues to move with constant velocity $\mathbf{u}$.

EXAMPLE 1 *In this example,* $\mathbf{i}$ *and* $\mathbf{j}$ *are unit vectors parallel to the axes* Ox, Oy, *respectively. At time* $t = 0$, *a particle* A *is at a point with cartesian coordinates* $(3, 4)$ *referred to an origin* O, *and* A *is moving with constant velocity* $(8\mathbf{i}-6\mathbf{j})\,\text{ms}^{-1}$. *At the same instant, a second particle* B *is at a point with coordinates* $(5, 6)$ *relative to the same frame of reference, and* B *is moving with constant velocity* $(7\mathbf{i}+5\mathbf{j})\,\text{ms}^{-1}$. *Show that* A *and* B *will meet at a point and determine the coordinates of this point.*

Relative to O, the position vector of A at $t = 0$ is $(3\mathbf{i}+4\mathbf{j})$ m. Its constant velocity is $(8\mathbf{i}+6\mathbf{j})\,\text{ms}^{-1}$. Hence, using equation 8.6, the position vector $\mathbf{r}_A$, at time t s is given by

$$\mathbf{r}_A = [(3\mathbf{i}+4\mathbf{j})+t(8\mathbf{i}+6\mathbf{j})]\,\text{m}.$$

Similarly, the position vector $\mathbf{r}_B$ of B, at time t s is given by

$$\mathbf{r}_B = [(5\mathbf{i}+6\mathbf{j}+t(7\mathbf{i}+5\mathbf{j})]\,\text{m}.$$

In order that the two particles may meet, there must be a value of t such that the components of $\mathbf{r}_A$ and $\mathbf{r}_B$ along Ox and along Oy are equal, so

$$3+8t = 4+6t,$$
$$5+7t = 6+5t.$$

Both these equations give $t = \frac{1}{2}$ and, thus, A and B will meet at the point with position vector $(7\mathbf{i}+8\frac{1}{2}\mathbf{j})$ m, that is, the point with coordinates **(7, 8½)**.

EXAMPLE 2 *Particles* A *and* B *are moving in three dimensions. Initially,* A *is at the point with coordinates* $(5, 6, 2)$ *referred to an origin* O *and axes* Ox, Oy *and* Oz. *Particle* A *moves with constant velocity at a speed of* $12\,\text{ms}^{-1}$ *in the direction of the vector* $(\mathbf{i}+2\mathbf{j}+2\mathbf{k})$. *Particle* B *is initially at the point with coordinates* $(-13, 0, 11)$ *and* B *moves with constant velocity in the direction* $(2\mathbf{i}+2\mathbf{j}+\mathbf{k})$. *Find the speed of* B, *given that the particles collide.*

A is moving at $12\,\text{ms}^{-1}$ in the direction $(\mathbf{i}+2\mathbf{j}+2\mathbf{k})$. Since

$$|\mathbf{i}+2\mathbf{j}+2\mathbf{k}| = 3,$$

$$\text{the velocity of } A = 4(\mathbf{i}+2\mathbf{j}+2\mathbf{k})\,\text{ms}^{-1}.$$

Let the speed of $B = u\,\text{ms}^{-1}$.

As

$$|2\mathbf{i}+2\mathbf{j}+\mathbf{k}| = 3,$$

the velocity of $B = \dfrac{u}{3}(2\mathbf{i}+2\mathbf{j}+\mathbf{k})\,\text{ms}^{-1}$.

At time t seconds after the start, the position vectors of A and B are respectively $\mathbf{r}_A$ and $\mathbf{r}_B$ where

$$\mathbf{r}_A = [(5\mathbf{i}+6\mathbf{j}+2\mathbf{k}+t(4\mathbf{i}+8\mathbf{j}+8\mathbf{k})]\,\text{m},$$

$$\mathbf{r}_B = [-13\mathbf{i}+11\mathbf{k}+t\frac{u}{3}(2\mathbf{i}+2\mathbf{j}+\mathbf{k})]\,\text{m}.$$

The particles meet if there is a value of t for which the three equations

$$5+4t = -13+\tfrac{2}{3}ut, \qquad \text{(i)}$$
$$6+8t = \tfrac{2}{3}ut, \qquad \text{(ii)}$$
$$2+8t = 11+\tfrac{1}{3}ut, \qquad \text{(iii)}$$

are satisfied simultaneously. From (i) and (ii),

$$18+4t = 6+8t = \tfrac{2}{3}ut, \text{ so } t = 3$$

Substituting $t = 3$ in (ii) we obtain $30 = 2u$, $u = 15$. Substituting the values for t and u in (iii), both sides of the equation equal 26. Hence, the particles meet at the point with coordinates (17, 30, 26) and the speed of B is $\mathbf{15\,ms^{-1}}$.

EXERCISE 8.2

1 Find the position vector of the point of intersection of the lines with vector equations

$$\mathbf{r} = 3\mathbf{i}+2\mathbf{j}+t(2\mathbf{i}-\mathbf{j}),$$
$$\mathbf{r} = 6\mathbf{i}-3\mathbf{j}+s(\mathbf{i}+3\mathbf{j}),$$

where t and s are parameters.

2 Determine whether the pairs of lines (a) and (b) meet and, if they do so, determine the position vector of their point of intersection.
(a) $\mathbf{r} = 2\mathbf{i}+3\mathbf{j}+4\mathbf{k}+t(\mathbf{i}+\mathbf{j}+\mathbf{k})$, $\mathbf{r} = 3\mathbf{i}+2\mathbf{j}+\mathbf{k}+s(\mathbf{i}-2\mathbf{j}+3\mathbf{k})$.
(b) $\mathbf{r} = 3\mathbf{i}+5\mathbf{j}+9\mathbf{k}+t(2\mathbf{i}-\mathbf{j}+\mathbf{k})$, $\mathbf{r} = 9\mathbf{i}+9\mathbf{j}+7\mathbf{k}+s(\mathbf{i}+3\mathbf{j}-2\mathbf{k})$.

3 At time $t = 0$, a sailing boat is at a point A with position vector $(\mathbf{i}+2\mathbf{j})\,\text{m}$ relative to a fixed origin. The constant velocity vector of the boat is $(5\mathbf{i}+7\mathbf{j})\,\text{ms}^{-1}$. At time $t = 0$, a second boat is at the point B with position vector $(5\mathbf{i}+6\mathbf{j})\,\text{m}$ relative to the same origin and moves with constant velocity $(4\mathbf{i}+6\mathbf{j})\,\text{ms}^{-1}$. Show that the two boats will collide and find the position vector of the point of collision.

4 At a certain instant a particle A is at a point with position vector

$(3\mathbf{i}+4\mathbf{j}+7\mathbf{k})$ m referred to a fixed origin, and A is moving with speed $15\,\text{ms}^{-1}$ in the direction of the vector $(2\mathbf{i}+\mathbf{j}+2\mathbf{k})$. At the same instant a second particle is at a point with position vector $(15\mathbf{i}+7\mathbf{j}+\mathbf{k})$ m referred to the same origin, and moves with speed $14\,\text{ms}^{-1}$ in the direction of the vector $(3\mathbf{i}+2\mathbf{j}+6\mathbf{k})$. Find the position vector of the point at which the particles meet.

5 A particle is in equilibrium under the action of three forces **P**, **Q** and **R**. Given that $\mathbf{P} = (3\mathbf{i}+5\mathbf{j})$ N, $\mathbf{Q} = (-2\mathbf{i}+6\mathbf{j})$ N, calculate
(a) the magnitude of **R**,
(b) the tangent of the acute angle made by the line of action of **R** with the positive x-axis. (*L*)

6 The forces $\mathbf{F}_1$ and $\mathbf{F}_2$, where $\mathbf{F}_1 = (\mathbf{i}+3\mathbf{j})$ N and $\mathbf{F}_2 = (2\mathbf{i}+\mathbf{j})$ N, both act through the point with position vector $(\mathbf{i}+\mathbf{j})$ m. Find the resultant of the two forces and show that its magnitude is 5 N. Obtain, in both vector and Cartesian form, equations for the line of action of this resultant. Determine the cosine of the angle between the directions of the forces $\mathbf{F}_1$ and $\mathbf{F}_2$. (*L*)

7 The forces **P** and **Q** act through the points with position vectors $(\mathbf{i}+3\mathbf{j})$ m and $-3\mathbf{j}$ m respectively. Given that $\mathbf{P} = 3\mathbf{i}$ N and $\mathbf{Q} = 4\mathbf{j}$ N, find the position vector of the point where the line of action of the resultant of **P** and **Q** meets the x-axis. (*L*)

8 The coplanar forces **i** N and **j** N act in the directions due east and due north, respectively.
(a) Find the magnitude of the resultant of the three forces $(2\mathbf{i}+\mathbf{j})$ N, $(4\mathbf{i}-9\mathbf{j})$ N and $(\mathbf{i}-16\mathbf{j})$ N.
(b) Find the direction of the resultant to the nearest degree. (*L*)

9 The position vectors of the points A and B are $2\mathbf{i}$ m and $(3\mathbf{i}+4\mathbf{j})$ m, respectively with O as origin. Forces of magnitude 6 N and 10 N act along $\overrightarrow{OA}$ and $\overrightarrow{OB}$ respectively. Calculate
(a) the magnitude of the resultant of these two forces,
(b) the tangent of the angle between the line of action of the resultant and the vector **i**,
(c) the position vector of the point in AB through which the line of action of the resultant passes. (*L*)

10 Three points A, B and C have position vectors $(\mathbf{i}+\mathbf{j}+\mathbf{k})$ m, $(\mathbf{i}+2\mathbf{k})$ m and $(3\mathbf{i}+2\mathbf{j}+3\mathbf{k})$ m respectively, relative to a fixed origin O. A particle P starts from B at time $t = 0$, and moves along $\overrightarrow{BC}$ towards C with constant speed $1\,\text{ms}^{-1}$. Find the position vector of P after t seconds
(a) relative to O,
(b) relative to A.

If angle PAB $= \theta$, find an expression for $\cos\theta$ in terms of t. (*L*)

11 A system of three forces, $\mathbf{F}_1 = (\mathbf{i}-5\mathbf{j}+\mathbf{k})$ N acting through a point with position vector $(5\mathbf{i}-10\mathbf{j}+10\mathbf{k})$ m, $\mathbf{F}_2 = (\mathbf{j}+\mathbf{k})$ N acting through a point with position vector $(4\mathbf{i}-7\mathbf{j}+7\mathbf{k})$, and a third force $\mathbf{F}_3$, is in equilibrium. Find the magnitude of $\mathbf{F}_3$ and a vector equation of its line of action. (*L*)

12 The resultant of a force of magnitude $2F$ N in a direction 090° and a force of magnitude F N in a direction 330° is a force of magnitude 12 N. Calculate the value of F.

It is required to add a third force in a direction 270° so that the resultant of the system is in a direction 000°. Calculate the magnitude of this third force. (*C*)

8.2 Relative velocity

The change of position of a particle observed from a frame of reference will depend on the motion of the frame of reference. Suppose that we wish to determine the velocity of a particle B relative to a moving observer A, given that the velocities of A and B with reference to a fixed origin O are $\mathbf{v}_A$ and $\mathbf{v}_B$ respectively, and that the velocities are constant.

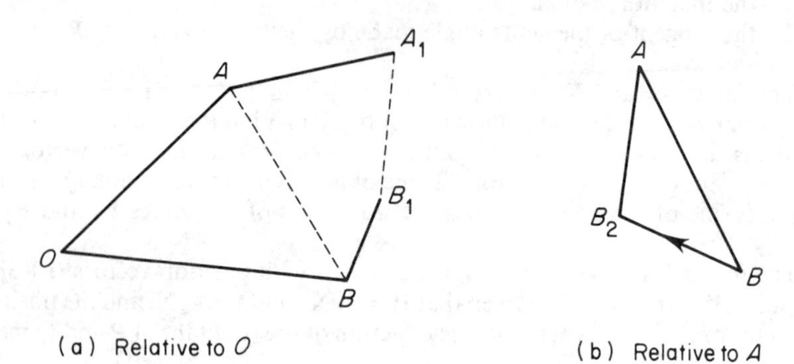

Fig. 8.3

Let the position vectors of A and B with respect to O at time $t = 0$ be $\mathbf{a}$ and $\mathbf{b}$ respectively. At time $t = 0$,

$$AB = \mathbf{b} - \mathbf{a}.$$

At time t, A has moved to A_1, B has moved to B_1, where

$$\overrightarrow{OA_1} = \mathbf{a} + \mathbf{v}_A t \quad \text{and} \quad \overrightarrow{OB_1} = \mathbf{b} + \mathbf{v}_B t.$$

$$\overrightarrow{A_1B_1} = \mathbf{b} + \mathbf{v}_B t - (\mathbf{a} + \mathbf{v}_A t) = \mathbf{b} - \mathbf{a} + (\mathbf{v}_B - \mathbf{v}_A)t$$

and this is represented by the vector $\overrightarrow{AB_2}$ in Fig. 8.4(b).

$$\overrightarrow{AB} = \mathbf{b} - \mathbf{a} \quad \text{and} \quad \overrightarrow{BB_2} = (\mathbf{v}_B - \mathbf{v}_A)t.$$

Thus, relative to A, B moves with velocity $(\mathbf{v}_B - \mathbf{v}_A)$.

The velocity of B relative to a moving observer A, when $\mathbf{v}_B$ and $\mathbf{v}_A$ are their respective velocities relative to the same frame of reference, is

$$(\mathbf{v}_B - \mathbf{v}_A) \qquad 8.7$$

There are many well known practical examples of this principle. If an observer stands at the bow of a liner which is moving with velocity $\mathbf{u}$ and he observes the water which is assumed to be still, the water appears to be moving towards the observer. The velocity of the water relative to the observer is $-\mathbf{u}$, since $\mathbf{v}_B = 0$, $\mathbf{v}_A = \mathbf{u}$.

Similarly, suppose that two trains are moving along two parallel tracks in opposite directions with speeds of u_1 and u_2. The velocities of the two trains are $u_1\mathbf{i}$ and $-u_2\mathbf{i}$ where $\mathbf{i}$ is the unit vector parallel to the track in the direction of motion of the first train moving with speed u_1. To an observer in the train moving with velocity $-u_2\mathbf{i}$, the velocity of approach (or separation) of the other train will be $u_1\mathbf{i}-(-u_2\mathbf{i})$, that is, $(u_1+u_2)\mathbf{i}$.

EXAMPLE 1 *At noon, a motor boat A is at a point 900 m east and 600 m north of the jetty of a port. At the same time another motor boat B is 600 m east and 810 m north of the jetty. Boat A moves with constant velocity* $(12\mathbf{i}+16\mathbf{j})\,\text{ms}^{-1}$, *where* **i** *and* **j** *are unit vectors due east and due north respectively. Boat B moves with a constant velocity of* $(15\mathbf{i}+10\mathbf{j})\,\text{ms}^{-1}$. *Find*

(i) *the vector* $\overrightarrow{AB}$, 30 s *later,*
(ii) *the vector* $\overrightarrow{AB}$, t s *later,*
(iii) *the time when the two boats are nearest to each other and their shortest distance apart.*

Taking the origin at the jetty, at $t=0$,

$$\overrightarrow{OA}=(900\mathbf{i}+600\mathbf{j})\,\text{m}$$

and $$\overrightarrow{OB}=(600\mathbf{i}+810\mathbf{j})\,\text{m}.$$

Hence $$\overrightarrow{AB}=(-300\mathbf{i}+210\mathbf{j})\,\text{m}.$$

The velocity of B relative to A is

$$\mathbf{v}_B-\mathbf{v}_A=(3\mathbf{i}-6\mathbf{j})\,\text{ms}^{-1}.$$

Considering the motion relative to A, at t seconds after noon

$$\overrightarrow{AB}=[\mathbf{-300i+210j+}t\mathbf{(3i-6j)}]\,\mathbf{m}. \quad \text{(ii)}$$

When $t=30$, $$\overrightarrow{AB}=\mathbf{(-210i+30j)\,m}. \quad \text{(i)}$$

At time t, $$|\overrightarrow{AB}|^2=[(3t-300)^2+(210-6t)^2]\,\text{m}^2. \quad \text{(ii)}$$

To find the least value of $|\overrightarrow{AB}|$ we may use calculus. Put

$$y=(3t-300)^2+(210-6t)^2.$$

Then

$$\frac{dy}{dt}=2(3t-300)\times 3+2(210-6t)\times(-6)$$

$$=18t-1800-2520+72t$$

$$=90t-4320$$

$$\frac{dy}{dt}=0 \text{ when } t=48.$$

As $\frac{d^2y}{dt^2}$ is positive, this value of t gives a minimum for y and therefore the boats are nearest to each other when $t=48$. When $t=48$,

$$y=(-156)^2+(-78)^2.$$

So, the shortest distance apart is **174·4 m**. (iii)

Some problems are more easily dealt with by means of a scale drawing.

EXAMPLE 2 *A liner is steaming at a constant velocity of* 30 knots *on a bearing of* 105° *and, at the same time, a tanker is steaming with a constant velocity of* 12 knots *on a bearing of* 220°. *At noon the liner is* 20 nautical miles *due east of the tanker. It is considered dangerous if the tanker approaches within* 1 nautical mile *of the liner. If the two ships pursue their courses with the same constant velocities, estimate whether a situation of danger will arise.* (1 knot = 1 nautical mile h^{-1}.)

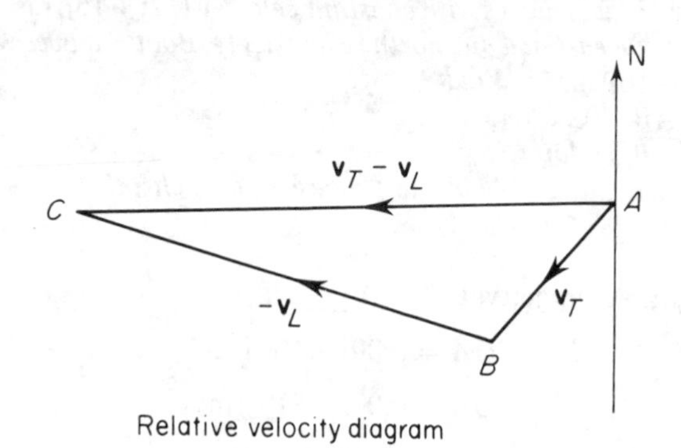

Fig. 8.4

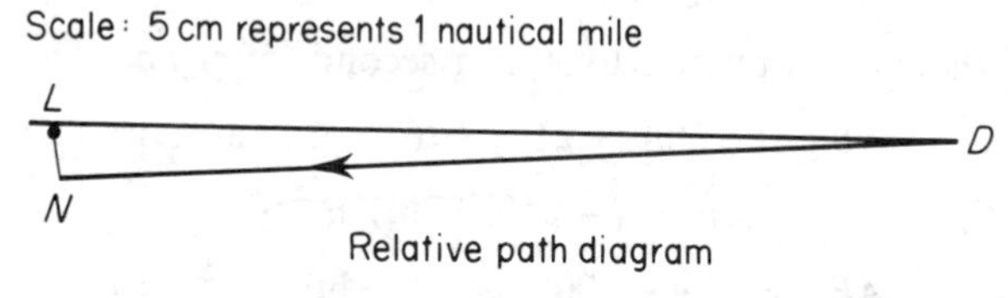

Fig. 8.5

If $\mathbf{v}_T$ and $\mathbf{v}_L$ are the velocities of the tanker and the liner respectively, then the velocity of the tanker relative to the liner is $\mathbf{v}_T - \mathbf{v}_L$ and this is represented on the relative velocity diagram (Fig. 8.4) by adding the vector $(-\mathbf{v}_L)$ to $\mathbf{v}_T$. $\mathbf{v}_T$ is represented by $\overrightarrow{AB}$ and $(-\mathbf{v}_L)$ by $\overrightarrow{BC}$. $\overrightarrow{AC}$ represents the velocity of the tanker relative to the liner. From this diagram it is seen that the velocity of the tanker relative to the liner is 36·8 knots on a bearing of 268°.

The relative path diagram (Fig. 8.5) shows the path $\overrightarrow{DN}$ of the tanker relative to the liner L. Note that $\overrightarrow{DN}$ is parallel to $\overrightarrow{AC}$, the velocity of the tanker relative to the liner. Initially the tanker is at D, 20 nautical miles due east of L. The shortest distance $LN \approx 1$ nautical mile, so the courses are close to a danger situation.

A scale diagram can only give an approximate result. In this case, to resolve the situation greater accuracy could be obtained by using larger scales or by using trigonometry.

EXERCISE 8.3

1 A passenger standing on a ship sees a hovercraft which appears to him to be travelling due north at 40 km h^{-1}. At the same instant, the passenger also sees

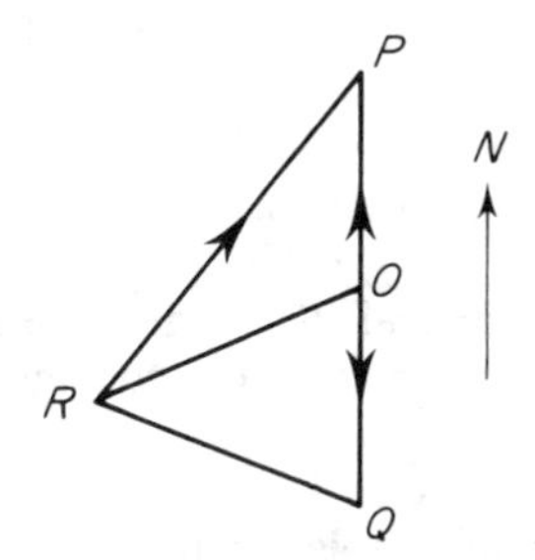

a motor boat which appears to him to be travelling due south at 40 km h^{-1}. The velocity diagram shows $\overrightarrow{OP}$ and $\overrightarrow{OQ}$, representing the apparent velocities of the hovercraft and the motor boat, respectively. The actual velocity of the hovercraft is represented by $\overrightarrow{RP}$. Copy the diagram. Mark clearly and write down on it the vectors which represent the actual velocities of the ship and the motor boat, distinguishing carefully between them.

If the actual direction of motion of the hovercraft and of the motor boat are perpendicular, explain with reasons why the magnitude of the velocity of the ship is 40 km h^{-1}.

If the ship is travelling on a bearing of 060° at this speed, calculate the magnitude and the direction of the actual velocity of (a) the hovercraft and (b) the motor boat. (*L*)

2 Particles A and B move with constant velocities $2\mathbf{i}$ m s^{-1} and $-4\mathbf{j}$ m s^{-1} respectively. At time $t = 0$, A is at the origin and B is at the point with position vector $(5\mathbf{i} + 3\mathbf{j})$ m. Find

(a) the velocity of B relative to A,

(b) the distance between A and B at time $t = 2$ s.

The velocity of a third particle C relative to A is in the direction of the vector $(3\mathbf{i} + 2\mathbf{j})$ and relative to B is in the direction of the vector $(\mathbf{i} + \mathbf{j})$. If the velocity of C is $(a\mathbf{i} + b\mathbf{j})$ m s^{-1}, find the values of a and b. (*L*)

3 At a given instant, a ship P travelling due west at a speed of 30 km h^{-1} is 7 km due north of a second ship Q which is travelling on a bearing of $\theta°$ at a speed of 14 km h^{-1}, where $\tan\theta = \frac{3}{4}$. Show that the speed of Q relative to P is 40 km h^{-1} and find the direction of the relative velocity.

The ships continue to move with uniform velocities. Find correct to three significant figures,

(i) the distance between the ships when they are nearest together,

(ii) the time, in minutes, to attain this shortest distance.

If, initially, the course of Q had been altered to bring the ships as close as possible, the speed of Q and the speed and course of P being unchanged, find the direction of the new course. (*JMB*)

4 A ship P is travelling due east at 30 km h^{-1} and a ship Q is travelling due south at 40 km h^{-1}. Both ships keep constant speed and course. At noon they are each 10 km from the point of intersection of their courses, O, and moving towards O. Find the coordinates, with respect to axes Ox eastwards and Oy northwards, of P and Q at time t hours past noon, and find the distance PQ at this time.

Find the time at which P and Q are closest together.

Find the magnitude and direction of the velocity of Q relative to P, indicating the direction on a diagram.

Show that at the position of closest approach the bearing of Q from P is $\theta° + 90°$, where $\tan\theta = \frac{3}{4}$. *(JMB)*

5 The flight controller at an airport is about to talk down an aircraft A, whose position vector relative to the control tower is $(10\mathbf{i} + 20\mathbf{j} + 5\mathbf{k})$ km and whose constant velocity is $(210\mathbf{i} - 50\mathbf{j})\,\text{km h}^{-1}$.

At this instant a second aircraft B appears on his radar screen with position vector $(-20\mathbf{i} - 10\mathbf{j} + 3\mathbf{k})$ km and constant velocity $(150\mathbf{i} + 250\mathbf{j} + 60\mathbf{k})\,\text{km h}^{-1}$. Find

(i) the velocity of A relative to B,

(ii) the position vector of A relative to B at time t minutes after B first appeared on the radar screen,

(iii) the time that elapses, to the nearest 10 seconds, until the aircraft are nearest one another. *(JMB)*

6 Relative to a harbour H, the position vector of two boats A and B at noon are $\begin{pmatrix} 3 \\ 1 \end{pmatrix}$ and $\begin{pmatrix} 1 \\ -2 \end{pmatrix}$ respectively, the units being km. The velocity of A is $\begin{pmatrix} 10 \\ 24 \end{pmatrix}\text{km h}^{-1}$ and that of B is $\begin{pmatrix} 24 \\ 32 \end{pmatrix}\text{km h}^{-1}$. Determine the time at which the boats are nearest to one another. *(JMB)*

7 A bird flies due north at a speed of $18\,\text{m s}^{-1}$ at a height of 30 m above a horizontal field. A man running on the field at a speed of $6\,\text{m s}^{-1}$ due east passes over a point A on the field 6 s after the bird was vertically above A. Given that the bird passed over A at time $t = 0$, find when the man and the bird were closest together and their distance apart at this time. *(L)*

8 A man cycling at a constant speed u on level ground finds that when his velocity is $u\mathbf{j}\,\text{m s}^{-1}$, the velocity of the wind appears to be $v(3\mathbf{i} - 4\mathbf{j})\,\text{m s}^{-1}$, where $\mathbf{i}$ and $\mathbf{j}$ are unit vectors in the east and north direction respectively. When the velocity of the man is $(u/5)(-3\mathbf{i} + 4\mathbf{j})\,\text{m s}^{-1}$, the velocity of the wind appears to be $w\mathbf{i}\,\text{m s}^{-1}$. Find w and v in terms of u and prove that the true velocity of the wind is

$$(u/20)(3\mathbf{i} + 16\mathbf{j})\,\text{m s}^{-1}.$$

Find the velocity of the wind relative to the man when his velocity is $u\mathbf{i}\,\text{m s}^{-1}$. *(JMB)*

9 To a motorist driving due west along a level road with constant speed u, the wind appears to be blowing in a direction with bearing $\theta°$. When he is driving with the same speed u due east, the apparent direction of the wind has bearing $\phi°$. Show that when he is driving at a speed $2w$ due east, the apparent direction of the wind is on a bearing of $X°$, where

$$2\tan X° = 3\tan\phi° - \tan\theta°.$$

Determine the true direction of the wind. *(L)*

8.3 Acceleration and relative acceleration

If the velocity of a body changes, then the body has an acceleration. This acceleration $\mathbf{a}$ is equal to the derivative with respect to time of the velocity vector.

$$\mathbf{a} = \frac{d\mathbf{v}}{dt}. \qquad 8.8$$

If the frame of reference has an acceleration $\mathbf{a}_1$ with reference to a fixed origin O and the observed particle has an acceleration $\mathbf{a}_2$ relative to O, then by a similar argument to that used for relative velocity, the acceleration of the observed particle relative to the moving frame of reference is $\mathbf{a}_2 - \mathbf{a}_1$.

EXAMPLE 1 *At time t s, the position vector of a moving particle with respect to a fixed origin O is given by*

$$\mathbf{r} = [(20t)\mathbf{i} + (30t - 5t^2)\mathbf{j}]\,\text{m},$$

where $\mathbf{i}$ *is a unit vector due east and* $\mathbf{j}$ *is a unit vector vertically upwards. Find the velocity and acceleration of the particle at time t s. Find also the position vector of the particle when it hits the ground.*

The velocity

$$\mathbf{v} = \frac{d\mathbf{r}}{dt} = [\mathbf{20i} + (\mathbf{30} - \mathbf{10}t)\mathbf{j}]\,\mathbf{ms^{-1}},$$

that is, it has a constant component of its velocity in an easterly direction and a vertical component of $(30 - 10t)$ ms^{-1}.

The acceleration $\qquad \mathbf{a} = \dfrac{d\mathbf{v}}{dt} = -10\mathbf{j}\,\text{ms}^{-2},$

that is, the motion is one of constant acceleration $-10\mathbf{j}$ ms^{-2} or **10 ms^{-2} vertically downwards.**

As $\mathbf{g}$, the acceleration due to gravity $\approx -9{\cdot}81\,\mathbf{j}$ ms^{-2}, this motion can be regarded as an approximation to the motion of a projectile.

At $t = 0$,

$$\mathbf{v} = (20\mathbf{i} + 30\mathbf{j})\,\text{ms}^{-1}.$$

Thus the particle will have been given an initial velocity with components 20 ms^{-1} eastwards and 30 ms^{-1} upwards. The particle is at ground level when the $\mathbf{j}$ component of the position vector is zero, that is, when

$$30t - 5t^2 = 0.$$

This is when $t = 0$ (the start) or when $t = 6$ (when it hits the ground). Thus, the position vector of the particle when it hits the ground is **120i m**, and so the range is 120 m.

8.4 Integration of a vector function of time

We define integration as the inverse process to differentiation, viz it is the determination of $\mathbf{r}$ when $\dfrac{d\mathbf{r}}{dt}$ is a given function of time (or the determination of $\mathbf{v}$ when $\dfrac{d\mathbf{v}}{dt}$ is given). If

$$\frac{d\mathbf{r}}{dt} = \mathbf{v} = v_1\mathbf{i} + v_2\mathbf{j} + v_3\mathbf{k},$$

then
$$\mathbf{r} = \mathbf{i}\int_0^t v_1\,dt + \mathbf{j}\int_0^t v_2\,dt + \mathbf{k}\int_0^t v_3\,dt,$$

and if
$$\frac{d\mathbf{v}}{dt} = \mathbf{a} = a_1\mathbf{i} + a_2\mathbf{j} + a_3\mathbf{k},$$

then
$$\mathbf{v} = \mathbf{i}\int_0^t a_1\,dt + \mathbf{j}\int_0^t a_2 dt + \mathbf{k}\int_0^t a_3\,dt.$$

If the acceleration of a particle at time t s is given as $-10\mathbf{j}$, then

$$\frac{d\mathbf{v}}{dt} = -10\mathbf{j}\,\text{ms}^{-2}.$$

Integrating with respect to the time,

$$\mathbf{v} = [(-10t)\mathbf{j} + \mathbf{c}]\,\text{ms}^{-1},$$

where $\mathbf{c}$ is a constant vector. To find the value of $\mathbf{c}$ it is necessary that the value of $\mathbf{v}$ is known for one value of the time. Thus, if we are given that $\mathbf{v} = 12\mathbf{i}\,\text{ms}^{-1}$ when $t = 0$, $\mathbf{c} = 12\mathbf{i}$, and $\mathbf{v} = [(-\mathbf{10}t)\mathbf{j} + 12\mathbf{i}]\,\text{ms}^{-1}$,

$$\mathbf{r} = [(-5t^2)\mathbf{j} + (12t)\mathbf{i}]\,\text{m} + \mathbf{d},$$

where $\mathbf{d}$ is another constant vector.

If at time $t = 0$, $\mathbf{r} = [2\mathbf{i} + 3\mathbf{j}]\,\text{m}$, then $\mathbf{d} = [2\mathbf{i} + 3\mathbf{j}]\,\text{m}$ and

$$\mathbf{r} = [-5t^2\mathbf{j} + (12t)\mathbf{i} + 2\mathbf{i} + 3\mathbf{j}]\,\text{m}$$

EXAMPLE 1 *At time t s, the velocity of a particle relative to a fixed origin O is given by*

$$\mathbf{v} = [15\mathbf{i} + (8t - 4)\mathbf{j}]\,\text{ms}^{-1}.$$

Initially, the position vector of the particle is $\mathbf{r} = (8\mathbf{i} + 10\mathbf{j})\,\text{m}$. *Find the position vector of the particle and its acceleration when* $t = 10$.

$$\mathbf{v} = \frac{d\mathbf{r}}{dt} = [15\mathbf{i} + (8t - 4)\mathbf{j}]\,\text{ms}^{-1}. \qquad \text{(i)}$$

Integrating with respect to time,

$$\mathbf{r} = [(15t)\mathbf{i} + (4t^2 - 4t)\mathbf{j} + \mathbf{c}]\,\text{m},$$

where $\mathbf{c}$ is a constant vector. Putting $t = 0$, $\mathbf{c} = 8\mathbf{i} + 10\mathbf{j}$. Thus,

$$\mathbf{r}(t) = [(15t)\mathbf{i} + (4t^2 - 4t)\mathbf{j} + 8\mathbf{i} + 10\mathbf{j}]\,\text{m}$$

and at $t = 10$, $\mathbf{r} = \mathbf{(158i + 370j)\,m}$.

$\mathbf{a} = \frac{d\mathbf{v}}{dt}$, so differentiating (i),

$$\frac{d\mathbf{v}}{dt} = 8\mathbf{j}\,\text{ms}^{-2}$$

so the acceleration is constant and equal to $\mathbf{8j\,ms^{-2}}$.

8.5 Composition of velocities

When both the velocity of the frame of reference relative to a fixed origin and the velocity of a particle relative to this moving frame are known, then to find the velocity of the particle relative to the origin, the composition of velocities is required. If $\mathbf{u}$ is the velocity of the frame of reference relative to the origin O and $\mathbf{v}$ the velocity of the particle relative to O, then $(\mathbf{v}-\mathbf{u})$ is the velocity of the particle relative to the moving frame. If $(\mathbf{v}-\mathbf{u})$ and $\mathbf{u}$ are known, then $\mathbf{v}$ may be found from

$$\mathbf{v} = (\mathbf{v}-\mathbf{u}) + \mathbf{u}.$$

This principle is illustrated in the following example.

EXAMPLE 1 *A man runs across the deck of a liner at right angles to its direction of motion. If the liner is moving at* $10\,\text{ms}^{-1}$ *and the man runs at* $4\,\text{ms}^{-1}$ *relative to the liner, find the velocity of the man observed from a stationary point on land.*

In 1 s the liner moves forward 10 m while the man moves 4 m across the deck. Thus the resultant displacement of the man in 1 s is obtained by combining these two displacements as shown in Fig. 8.6. So his velocity relative to a stationary observer on land is of magnitude $\sqrt{116}\,\text{ms}^{-1} \approx$ **10·8 ms^{-1} and in a direction making an angle of tan^{-1}(0·4) ≈ 21·8° with the direction of motion of the liner.**

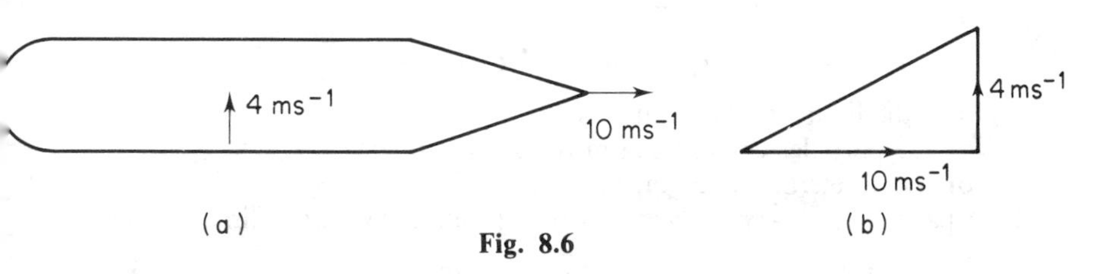

Fig. 8.6

Equation 8.7 and the corresponding result for relative acceleration may also be established by differentiation of position vectors. If $\mathbf{r}_A$ is the position vector of an observer A with respect to a fixed origin O and $\mathbf{r}_B$ the position vector of B with respect to the same origin, then $\overrightarrow{AB} = \mathbf{r}_B - \mathbf{r}_A$. The velocity of B relative to A is

$$\frac{d}{dt}(\overrightarrow{AB}) = \frac{d}{dt}(\mathbf{r}_B - \mathbf{r}_A) = \mathbf{v}_B - \mathbf{v}_A,$$

where $\mathbf{v}_B$ and $\mathbf{v}_A$ are the velocities of A and B with respect to O as in equation 8.7.

Similarly the relative acceleration is

$$\frac{d}{dt}(\mathbf{v}_B - \mathbf{v}_A) = \mathbf{a}_B - \mathbf{a}_A,$$

where $\mathbf{a}_B$ and $\mathbf{a}_A$ are the accelerations of A and B with respect to O.

EXERCISE 8.6

1 Two particles A and B are moving in horizontal straight lines in the plane of the coordinate axes Ox and Oy. At time t s, A has position vector $(t^2\mathbf{i} + 2\mathbf{j})$ m and B has position vector $[t(3\mathbf{i} - 4\mathbf{j})]$ m.
(a) Calculate the distance between A and B when $t = 3$.
(b) Show that the velocity of B is constant and calculate its magnitude and direction.
(c) Calculate the acceleration of A.
(d) Find the magnitude and the direction of the velocity of B relative to A when $t = 2$. (*L*)

2 Two particles P and Q are moving in a horizontal plane. At time t seconds, P and Q have position vectors

$$[(\tfrac{3}{2}t^2 + 3t)\mathbf{i}]\,\text{m} \quad \text{and} \quad [\tfrac{1}{6}t^3\mathbf{i} - (t^2 - 5t)\mathbf{j}]\,\text{m}$$

respectively. Find
(a) the velocity vectors of P and Q when $t = 2$,
(b) the magnitude and direction of the velocity of P relative to Q when $t = 2$,
(c) the acceleration vectors of P and Q when $t = 3$. (*L*)

3 Two particles A and B move so that, at time t s (> 0), they have position vectors $\mathbf{r}_A$ and $\mathbf{r}_B$ respectively, where

$$\mathbf{r}_A = (a \sin wt)\mathbf{i} + (a \cos wt)\mathbf{j},$$
$$\mathbf{r}_B = a(w^2t^2 - 1)\mathbf{i} + (aw^2t^2)\mathbf{j}.$$

Find the least possible value for t for which
(a) the particles are moving in the same direction,
(b) the accelerations of the particles are at right angles. (*L*)

4 A particle of mass m moves so that its position vector $\mathbf{r}$ at time t s is given by

$$\mathbf{r} = \mathbf{a} \cos \omega t + \mathbf{b} \sin \omega t,$$

where $\mathbf{a}$ and $\mathbf{b}$ are constant vectors and ω is a constant. Calculate the velocity $\left(\frac{d\mathbf{r}}{dt}\right)$ and the acceleration $\left(\frac{d^2\mathbf{r}}{dt^2}\right)$ of P. Calculate also, in terms of m, ω, and $\mathbf{r}$, the force exerted on P. Show that P describes a plane curve in periodic time $2\pi/\omega$. (*L*)

5 Two points move in a plane, both starting from the origin, in such a way that after time t s, their position vectors are

$$[3(\cos t - 1)\mathbf{i} + (4\sin t)\mathbf{j}]\,\text{m} \quad \text{and} \quad [(3\sin t)\mathbf{i} + 4(\cos t - 1)\mathbf{j}]\,\text{m}.$$

Find the value of t when the points are first moving in (a) opposite directions, (b) the same direction. (*L*)

6 At time t s, the position vector $\mathbf{r}$ of the point P with respect to the origin O is given by

$$\mathbf{r} = (a\sin pt)\mathbf{i} + a\mathbf{j},$$

where a and p are constants. Show that the vector $\dfrac{\mathrm{d}^2\mathbf{r}}{\mathrm{d}t^2} + p^2\mathbf{r}$ is constant during the motion. (*L*)

7 The position vector $\mathbf{r}$ of a particle at time t seconds is

$$\mathbf{r} = [a(t - \sin t)\mathbf{i} + a(1 - \cos t)\mathbf{j}]\,\text{m},$$

where a is a positive constant. Show that the speed of the particle at time t is $|2a\sin(t/2)|\,\text{ms}^{-1}$. (*L*)

8 At time t s the position vector of a particle of mass m is $[t^2\mathbf{i} + (\sin t)\mathbf{j}]$ m. Find the resultant force acting on the particle when $t = \pi/2$. (*L*)

9 A particle of mass m moves so that its position vector, $\overrightarrow{OP}(= \mathbf{r})$, at time t is given by

$$\mathbf{r} = \mathbf{a}\cos\omega t + \mathbf{b}\sin\omega t,$$

where $\mathbf{a}$ and $\mathbf{b}$ are constant non-parallel vectors. $|\mathbf{a}| > |\mathbf{b}|$ and $\omega(\neq 0)$ is a constant. Show that P lies in a fixed plane which passes through O. Show that the resultant force $\mathbf{F}$ acting on P is directed towards O. If $\mathbf{a}\,.\,\mathbf{b} = 0$, find the greatest and least values of $|\mathbf{F}|$. (*L*)

10 Two particles A and B have position vectors $[(2\sin\omega t)\mathbf{i} + (2\cos\omega t)\mathbf{j}]$ m where $\omega > 0$ and $[2t\mathbf{i} + t^2\mathbf{j}]$ m respectively at time t seconds. Find the Cartesian equations of the paths followed by A and B. Find
(a) the magnitude of the velocity of A relative to B when $t = 0$,
(b) the magnitude of the acceleration of A relative to B when $t = \pi/2\omega$,
(c) the least positive value of t for which the accelerations of A and B are parallel, and (i) in the same sense, (ii) in opposite senses. (*L*)

11 A particle of unit mass moves under the action of a constant force $(\mathbf{i} + 2\mathbf{j})$ N. At time $t = 0$, the particle is stationary at the point with position vector $(2\mathbf{i} + 5\mathbf{j})$ m. Find the velocity at any subsequent time t s and the position vector of the particle at time $t = 2$. (*L*)

12 The force acting at time t seconds $(0 \leqslant t \leqslant 2)$ on a particle of unit mass is $(24t^2\mathbf{i} + 6\mathbf{j})$ N. At time $t = 0$, the particle is at rest at the point with position vector $(-2\mathbf{i} + 3\mathbf{j})$ m. Find the position vector of the particle at time T s $(0 \leqslant T \leqslant 2)$. When $t > 2$, the force acting on the particle is $6\mathbf{j}$ N. Find the position vector of the particle when $t = 3$. (*L*)

MISCELLANEOUS EXERCISE 8

1 ΔABC is equilateral. Forces with magnitudes 5, 7 and 9 N act along $\overrightarrow{AB}$, $\overrightarrow{BC}$ and $\overrightarrow{CA}$ respectively. Find the magnitude and the direction of the resultant of these forces. (*L*)

2 A particle of mass 2 kg moves from rest at the origin under the action of two forces each of magnitude 20 N. One force acts parallel to the vector $4\mathbf{i}-3\mathbf{j}$, and the other parallel to the vector $-3\mathbf{i}+4\mathbf{j}$. Find the acceleration of the particle. *(L)*

3 The position vectors of the vertices B and C of a triangle ABC are respectively $(8\mathbf{i}+3\mathbf{j}+5\mathbf{k})$ m and $(6\mathbf{i}+4\mathbf{j}+9\mathbf{k})$ m. Two forces $(3\mathbf{i}+2\mathbf{j}+\mathbf{k})$ N and $(4\mathbf{i}+5\mathbf{j}+6\mathbf{k})$ N act along AB and AC respectively. A third force $\mathbf{F}$ acts through A. The system of forces is in equilibrium. Find
(a) the magnitude of the force $\mathbf{F}$,
(b) the position vector of A,
(c) the equation of the line of action of $\mathbf{F}$ in vector form. *(L)*

4 (i) Forces $(\mathbf{i}+3\mathbf{j})$ N, $(-2\mathbf{i}-\mathbf{j})$ N, $(\mathbf{i}-2\mathbf{j})$ N act through the points with position vectors $(2\mathbf{i}+5\mathbf{j})$ m, $4\mathbf{j}$ m, $(-\mathbf{i}+\mathbf{j})$ m, respectively. Prove that this system of forces is equivalent to a couple and find its magnitude.
(ii) Three forces are represented in magnitude, direction and line of action by the sides AB, BC, CA of the triangle ABC. Show that this system of forces reduces to a couple and find the magnitude of this couple.
The force acting along the side CA is now reversed in direction. Find completely the resultant of this new system. *(L)*

5 A particle of mass 2 kg starts from rest at the origin and the force $\mathbf{F}_3$, where $\mathbf{F}_3 = (6\mathbf{i}+26\mathbf{j})$ N, acts on it as it moves. Find the position vector of the particle 2 seconds later. *(L)*

6 The position vectors of the vertices A, B, C of a triangle are respectively

$$(4\mathbf{i}+2\mathbf{j}+2\mathbf{k})\,\text{m}, \quad (2\mathbf{i}+\mathbf{j}+4\mathbf{k})\,\text{m}, \quad (-3\mathbf{i}+6\mathbf{j}-6\mathbf{k})\,\text{m}.$$

Find the position vector of G, the centroid of the triangle.
Forces $3\overrightarrow{AB}$ and $2\overrightarrow{AC}$ act along AB and AC. Find the magnitude of their resultant and the position vector of the point P in which the line of action of this resultant meets BC. Show that the area of the triangle GAP is $\frac{1}{2}\sqrt{5}$ and hence find the moment of the resultant about an axis through G perpendicular to the plane ABC. *(L)*

7 A particle of mass m kg is acted upon by a constant force of $21m$ newtons in the direction of the vector $3\mathbf{i}+2\mathbf{j}-6\mathbf{k}$. Initially the particle is at the origin moving with speed 18 ms^{-1} in the direction of the vector $7\mathbf{i}-4\mathbf{j}+4\mathbf{k}$. Find the position vector of the particle 4 seconds later. *(L)*

8 A ship X is steaming due north at 8 knots. An enemy ship Y, 2 nautical miles due east of X, appears to the captain on X to be travelling southwest at 11·3 knots.
(i) Show, by calculation and not by measuring from a scale diagram, that the true velocity of Y is slightly under 8 knots, and hence show that it is travelling approximately due west.
(ii) What is the approximate bearing of X from Y when they are closest together?
(iii) The guns on ship Y can hit anything within 1·5 nautical miles. For how many minutes could X be under attack if both ships maintained their original courses and speeds? *(SUJB)*

9 A ship A whose full speed is 40 kilometres per hour is 20 kilometres due west of a ship B which is travelling uniformly with speed 30 kilometres per hour in a direction due north. The ship A travels at full speed on a course chosen so as to intercept B as soon as possible. Find the direction of this course and calculate to the nearest minute the time A would take to reach B.

When half of this time has elapsed the ship A has engine failure and thereafter proceeds at half speed. Find the course which A should then set in order to approach as close as possible to B, and calculate the distance of closest approach (in kilometres to two decimal places). (*JMB*)

10 A destroyer sights a ship travelling with constant velocity $5\mathbf{j}$ whose position vector at the time of sighting is $2000(3\mathbf{i}+\mathbf{j})$ relative to the destroyer, distances being measured in m and speeds in ms^{-1}. The destroyer immediately begins to move with velocity $k(4\mathbf{i}+3\mathbf{j})$, where k is a constant, in order to intercept the ship. Find k and the time to interception.

Find also the distance between the vessels when half the time to interception has elapsed. (*O&E*)

11 Distances being measured in nautical miles and speeds in knots, a motor boat sets out at 11 a.m. from a position $(-6\mathbf{i}-2\mathbf{j})$ relative to a marker buoy, and travels at a steady speed of magnitude $\sqrt{53}$ on a direct course to intercept a ship. The ship maintains a steady velocity vector $(3\mathbf{i}+4\mathbf{j})$ and at 12 noon is at a position $(3\mathbf{i}-\mathbf{j})$ from the buoy. Find the velocity vector of the motor boat, the time of interception, and the position vector of the point of interception from the buoy. (*L*)

12 Two boats A and B have velocity vectors $\mathbf{u}$ and $\mathbf{v}$ respectively. Show, with the aid of a sketch, how to find the velocity of A relative to B.

The boats A and B are racing when there is a wind of 10 knots blowing from due north. A is sailing at 6 knots on a bearing of 045°. Find the direction of the wind relative to A. (Give bearings correct to the nearest degree.)

At the same time B is sailing at 6 knots on a bearing of 315°. Find the direction of the wind relative to B and the velocity of A relative to B.

The boat A then rounds a buoy and sails on a bearing of 225°. If its new speed relative to B is 10 knots, find its actual speed. (*MEI*)

13 A cruiser sailing due N. at 12 knots sights a destroyer 24 nautical miles due E. sailing at 28 knots on a course α W. of N., where $\cos\alpha = \frac{11}{14}$. Show that the destroyer's course relative to the cruiser is 60° W. of N. and find the relative speed. If the cruiser's guns have a maximum range of 15 nautical miles and both ships maintain course and speed, find for how long the destroyer will be within range.

Immediately the destroyer is sighted an aircraft is despatched from the cruiser and flies in a straight line at a steady speed of 220 knots on an intersection course with the destroyer. Find the course on which the aircraft flies. (A knot is 1 nautical mile per hour.) (*JMB*)

14 A wide river flows from north to south with a speed of $3\,\text{ms}^{-1}$. A boat P crossing the river moves from west to east with a speed of $4\,\text{ms}^{-1}$. Using the west-east direction and the south-north direction as your coordinate axes Ox and Oy respectively, write down in vector form the velocities of the river and the boat. Find the velocity and speed of the boat relative to the river.

A second boat Q has a velocity $(3\mathbf{i}+7\mathbf{j})\,\text{ms}^{-1}$ relative to the river. Calculate the velocity of P relative to Q and find the cosine of the angle between the velocity of Q and the velocity of P.

Initially both boats are on the west bank of the river with Q being 40 m to the south of P. They wish to meet on the river and both boats must sail with the velocities given above. Show that P must sail 2·5 s after Q. (*L*)

15 In this question the unit of distance is the km and the unit of time is the hour. The position vector, referred to axes Ox and Oy, of a port P is $4\mathbf{i}+3\mathbf{j}$. A ship B leaves P and steers with constant velocity vector $3\mathbf{i}+2\mathbf{j}$. At the same instant a ship C sets out from a point Q which has position vector $5\mathbf{i}-12\mathbf{j}$. Write down the vectors to represent
(a) the velocity of ship C if the velocity of ship B relative to ship C is $6\mathbf{i}-2\mathbf{j}$.
(b) vector $\overrightarrow{CB}$ at time t.
Find, by drawing or otherwise, the shortest distance between the ships and the distance BP when the ships are closest together. (*L*)

16 A particle A moving with constant velocity $(3\mathbf{i}-2\mathbf{j})\,\mathrm{m\,s^{-1}}$ passes through a point with position vector $(4\mathbf{i}+4\mathbf{j})\,\mathrm{m}$ at the same instant as a particle B passes through a point with position vector $(-4\sqrt{2}\mathbf{i}+p\mathbf{j})\,\mathrm{m}$. Given that B has a constant velocity $(4\mathbf{i}-\mathbf{j})\,\mathrm{m\,s^{-1}}$ find the velocity of B relative to A and the value of p which ensures that A and B collide.

When $p=4\sqrt{2}$, find the minimum distance between A and B in the subsequent motion. (*L*)

17 Two particles A and B are moving with constant velocity vectors $\mathbf{v}_1=(5\mathbf{i}+3\mathbf{j}-\mathbf{k})\,\mathrm{m\,s^{-1}}$ and $\mathbf{v}_2=(3\mathbf{i}+4\mathbf{j}-3\mathbf{k})\,\mathrm{m\,s^{-1}}$ respectively. Find the velocity vector of A relative to B. At time $t=0$ the particle A is at the point whose position vector is $(-4\mathbf{i}+7\mathbf{j}-6\mathbf{k})\,\mathrm{m}$. If A collides with B when $t=5$, find the position vector of B at $t=0$.

The velocity of A relative to a third moving particle C is in the direction of the vector $2\mathbf{i}+\mathbf{j}-2\mathbf{k}$ and the velocity of B relative to C is in the direction of the vector $2\mathbf{i}+3\mathbf{j}-6\mathbf{k}$. Find the magnitude and direction of the velocity of C. (*L*)

18 One particle A has velocity vector $(4\mathbf{i}+3\mathbf{j}+2\mathbf{k})\,\mathrm{m\,s^{-1}}$ and another particle B has velocity vector $(2\mathbf{i}-\mathbf{j}+4\mathbf{k})\,\mathrm{m\,s^{-1}}$. The velocity of a third particle C relative to A is $(4\mathbf{i}-10\mathbf{j}+4\mathbf{k})\,\mathrm{m\,s^{-1}}$. Find the velocity vector of C and the velocity of C relative to B.
If A is at the origin when B is at the point with position vector $(8\mathbf{i}+16\mathbf{j}-8\mathbf{k})\,\mathrm{m}$, show that A and B will collide. Determine the position vector of the point of collision. (*L*)

19 Particles A and B move with constant velocities $2\mathbf{i}\,\mathrm{m\,s^{-1}}$ and $-4\mathbf{j}\,\mathrm{m\,s^{-1}}$ respectively. At time $t=0$, A is at the origin and B is at the point with position vector $(5\mathbf{i}+3\mathbf{j})\,\mathrm{m}$. Find
(a) the velocity of B relative to A,
(b) the distance between A and B at time $t=2$. (*L*)

20 The radius vector OP of constant length a rotates in a plane with constant angular speed $\omega\,\mathrm{rad\,s^{-1}}$ about the origin O. Find the position vector of P at time t s, given that it is $(a\mathbf{i})\,\mathrm{m}$ when $t=0$. Express in vector form the velocity and the acceleration of P at time t s.

The velocity of a point Q relative to P at time t s is

$$[(a\omega\sin\omega t)\mathbf{i}+(a\omega\cos\omega t)\mathbf{j}]\,\mathrm{m\,s^{-1}},$$

and Q is at the origin when $t=0$. Find the position vector of M, the mid-point of PQ. Find the values of t for which the acceleration of M is parallel to the acceleration of P. (*L*)

9 Planar Motion under a Constant Force.

In this chapter we study the simplest case of planar motion of a particle – that under the action of a constant force. The simplest example of such motion is that of any free body (such as, a stone or a bullet) under the influence of gravity, provided that air resistance is neglected. Most of the work in the chapter is concerned with free motion under gravity, commonly referred to as projectile motion. There are, however, physical agencies other than gravity (for example, a constant electric field) which generate a constant force and in §9.4 the approach used in the earlier sections is generalised to cover the case of motion under the action of a general but constant force.

9.1 Projectile motion

All particle motion is governed by Newton's second law (equation 5.3)

$$m\mathbf{a} = \mathbf{F}, \tag{9.1}$$

where m is the mass of the particle, $\mathbf{F}$ is the force acting on it and $\mathbf{a}$ is its acceleration. For free motion under gravity

$$\mathbf{F} = m\mathbf{g}, \tag{9.2}$$

where $\mathbf{g}$ denotes the acceleration due to gravity. For motion in a vertical plane with the y-axis chosen vertically *upwards* and the x-axis horizontal we have

$$\mathbf{g} = -g\mathbf{j}, \tag{9.3}$$

where g is the magnitude of the acceleration due to gravity. The position vector of a particle with respect to some origin O in the given vertical plane is

$$\mathbf{r} = x\mathbf{i} + y\mathbf{j}, \tag{9.4}$$

where x and y are its Cartesian coordinates. The acceleration is, from equations 8.1 and 8.8, given by

$$\mathbf{a} = \ddot{x}\mathbf{i} + \ddot{y}\mathbf{j}, \tag{9.5}$$

Equation 9.1 then becomes

$$m(\ddot{x}\mathbf{i} + \ddot{y}\mathbf{j}) = -mg\mathbf{j}, \tag{9.6}$$

or, in component form,

$$\ddot{x} = 0, \tag{9.7}$$

$$\ddot{y} = -g. \tag{9.8}$$

Equations 9.7 and 9.8 are independent of each other, with the first referring to horizontal motion and the second to vertical motion. It is generally easier, on a first encounter with projectile theory, to treat vertical and horizontal motion separately rather than look at the combined motion as represented by equation 9.6. The latter approach, based on direct integration of the vector equation 9.6, is given in §9.4. It is also proved in §9.4 that a particle set off in a given vertical plane will always remain in that plane. Here, this result will be assumed without proof.

The constant acceleration formulae of equations 4.4 and 4.5 can be applied immediately to equations 9.7 and 9.8. The equation for horizontal motion is the simpler of the two and it follows from equations 4.4 and 4.5, or just as easily by direct integration, that

$$\dot{x} = u_H, \tag{9.9}$$

$$x = u_H t, \tag{9.10}$$

where u_H is the horizontal component of velocity at time $t = 0$, the origin being chosen at the position of the particle at time $t = 0$. Equation 9.9 shows that the horizontal component of velocity remains constant throughout the motion. Applying equations 4.4 and 4.5 to equation 9.8 gives

$$\dot{y} = u_v - gt, \tag{9.11}$$

$$y = u_v t - \tfrac{1}{2}gt^2, \tag{9.12}$$

where u_v is the initial vertical component of the velocity.

In many cases, such as those involving the motion of a shell, the magnitude and direction of the initial velocity are given rather than the components, and it is, therefore, useful to re-write the above equations in terms of these parameters. For a particle projected from a point O with a speed u in a direction inclined at an angle α to the horizontal (see Fig. 9.1), u_H and u_v can be expressed in terms of u and α by

$$u_H = u\cos\alpha, \quad u_v = u\sin\alpha.$$

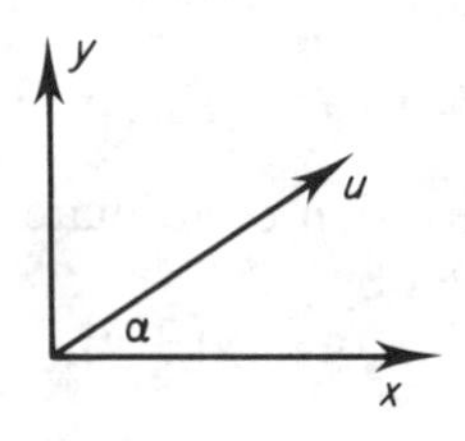

Fig. 9.1

Replacing u_H and u_v by the above expressions in equations 9.9 to 9.12 gives

$$\dot{x} = u \cos \alpha, \tag{9.13}$$

$$x = ut \cos \alpha, \tag{9.14}$$

$$\dot{y} = u \sin \alpha - gt, \tag{9.15}$$

$$y = ut \sin \alpha - \tfrac{1}{2}gt^2. \tag{9.16}$$

Equations 9.9 to 9.12 or, equivalently, equations 9.13 to 9.16 are the basic equations governing projectile motion and are sufficient to solve any projectile problem. In the simplest problems, u and α (or u_H and u_v) are given and any other information required can be obtained from the basic equations. In slightly more complicated problems, u and α are not given explicitly but sufficient information is available to determine them, and hence find x and y for all t.

There are some results concerning projectile motion that occur so often that it is useful to treat them as basic formulae and commit them to memory. These formulae are derived in the next section. It is, however, fairly easy to solve projectile problems directly from first principles without the use of additional formulae, and this approach is illustrated in the following examples.

Again, the S.I. unit system is used in all numerical examples. Also, to avoid including units in displayed equations, in all numerical examples we denote the horizontal and vertical displacements at time t s by x m and y m respectively. The initial speed is denoted by $u\,\mathrm{ms}^{-1}$ (with, equivalently, $u_H\,\mathrm{ms}^{-1}$ and $u_v\,\mathrm{ms}^{-1}$ denoting the horizontal and vertical components of the initial velocity). To simplify calculations, $g\,\mathrm{ms}^{-2}$ will be approximated to by $10\,\mathrm{ms}^{-2}$. Therefore, in all the numerical examples, x, y, u, u_H, u_v and t denote dimensionless quantities which satisfy equations 9.9 to 9.16, with g taken as 10.

EXAMPLE 1 *A particle is projected from a point O on the ground with initial velocity* $(15\mathbf{i} + 20\mathbf{j})\,\mathrm{ms}^{-1}$, *where* **i** *and* **j** *are unit vectors horizontally and vertically upwards respectively. Find the time taken for it to reach the ground again and the distance from the point of projection at which it hits the ground.*

We see that $u_H = 15$, $u_v = 20$ and substituting these into equations 9.9 to 9.12 gives

$$\dot{x} = 15,$$
$$x = 15t,$$
$$\dot{y} = 20 - 10t,$$
$$y = 20t - 5t^2.$$

The path followed by the particle will be as shown in Fig. 9.2. The precise shape of the path will be determined in the next section, but from the observation 'what goes up must come down' we know the path is roughly as in Fig. 9.2. The particle will once again be at ground level when $y = 0$, that is,

$$0 = 20t - 5t^2.$$

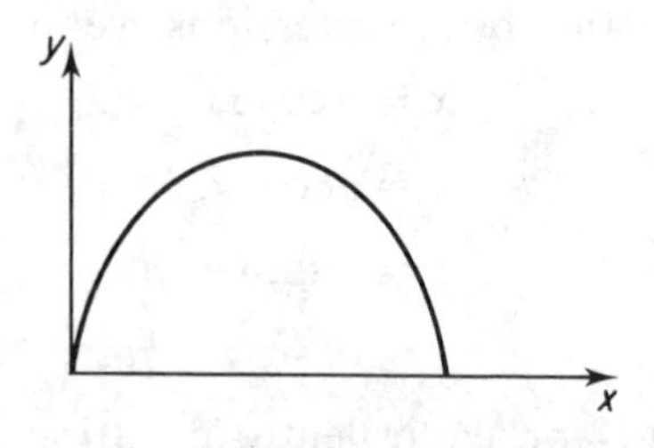

Fig. 9.2

This gives $t = 0$ (the starting point) and $t = 4$. The distance travelled along the ground can be found by substituting $t = 4$ into the expression for x, giving $x = 60$. Therefore, it hits the ground after **4 s** at a distance of **60 m** from O, that is, at the point with position vector $60\mathbf{i}$ m relative to O.

EXAMPLE 2 *Determine the greatest height reached by the particle in example* 1 *and the Cartesian equation of its path.*

The particle stops rising (that is, it reaches its greatest height), when its vertical component of velocity $\dot{y}$ is zero. This gives

$$20 - 10t = 0,$$
$$t = 2.$$

At this time, y is $20 \times 2 - 5 \times 4 = 20$. Therefore, the greatest height reached is **20 m** and at this position, the horizontal displacement is 30 m, so the highest point has position vector $(30\mathbf{i} + 20\mathbf{j})$ m relative to O.

Since both x and y are given in terms of t, the latter can be eliminated to give a relation between y and x. We have that

$$t = \frac{x}{15}$$

and substituting in the expression for y gives

$$y = \frac{4}{3}x - \frac{x^2}{45},$$

which can be re-written as

$$\mathbf{45y = 60x - x^2}.$$

This is the required result, but it can be put in a more useful form by completing the square on the right-hand side to give

$$45y = -(x - 30)^2 + 900$$

or

$$45(y - 20) = -(x - 30)^2.$$

Making the substitutions $Y = y - 20$, $X = x - 30$ (which means the origin of the X, Y coordinates is at $x = 30$, $y = 20$ – the highest point), gives

$$45Y = -X^2.$$

The path is symmetric about the line $X = 0$ (or $x = 30$) since the equation is unchanged when X is replaced by $-X$, and Y can only take negative values. The path is, in fact, a parabola with its vertex (which is at the highest point) upwards.

EXAMPLE 3 *A particle is projected from the ground at an angle of* $30°$ *to the horizontal with speed* $80\,\text{ms}^{-1}$. *Find the distance from the point of projection at which the particle lands on a horizontal plane through the point of projection, and find the angle between the horizontal and the direction of motion of the particle when the latter has been moving for* 1 s.

In this case the angle of projection is given so we use equations 9.13 to 9.16 with $u = 80$ and $\alpha = 30°$ to obtain

$$\dot{x} = 80\cos 30° = 40\sqrt{3},$$
$$x = 40\sqrt{3}t,$$
$$\dot{y} = 80\sin 30° - 10t = 40 - 10t,$$
$$y = 40t - 5t^2.$$

The horizontal plane will be reached when $y = 0$, that is at $t = 8$ s. Substituting this value of t in the expression for x gives the required distance as $\mathbf{320\sqrt{3}\,m}$.

After 1 s, the horizontal and vertical components of velocity are $\dot{x}$, $\dot{y}$ where

$$\dot{x} = 40\sqrt{3}, \quad \dot{y} = 30.$$

The velocity vector is therefore $40\sqrt{3}\mathbf{i} + 30\mathbf{j}$, which makes an angle $\mathbf{\tan^{-1}(\sqrt{3}/4)}$ to the horizontal.

EXAMPLE 4 *One second after a particle P has been projected it has a horizontal component of velocity of* $10\,\text{ms}^{-1}$ *and an upward vertical one of* $30\,\text{ms}^{-1}$. *Find the maximum height reached by P above the point of projection.*

In this question the initial components of velocity are not given directly and have to be found. The horizontal component of velocity is constant and is, therefore, equal to $10\,\text{ms}^{-1}$. Substituting the given conditions in equation 9.11

$$30 = u_v - 10,$$

so that $u_v = 40$. At the highest point $\dot{y} = 0$, and substituting this and the value of u_v into equation 9.11 gives

$$0 = 40 - 10t,$$

so that $t = 4$. Substituting this into equation 9.12 gives $y = 80$, and so the maximum height reached is **80 m**.

EXAMPLE 5 *Find the initial speed and direction of projection of a particle which,* 3 s *after projection, is moving with a speed of* $10\sqrt{2}\,\text{ms}^{-1}$ *in a direction inclined at* $45°$ *above the horizontal.*

The given information shows that after 3 s the horizontal and vertical components of velocity are both $10\,\text{ms}^{-1}$. Equations 9.13 and 9.15 then give

$$u\cos\alpha = 10$$
$$u\sin\alpha - 30 = 10.$$

Therefore, $u \cos \alpha = 10$, $u \sin \alpha = 40$, and squaring and adding gives $u = \sqrt{1700} \approx 41$ and $\alpha = \tan^{-1} 4 \approx 76°$. The initial speed of the particle is **41 m** and the direction of projection is **76°** to the horizontal.

EXERCISE 9.1

Questions 1 to 6 refer to a particle projected at time $t = 0$ s with velocity $(u_H\mathbf{i} + u_v\mathbf{j})$ ms^{-1} from a point with position vector $\mathbf{r}_o$ relative to a given origin O. **i** and **j** are unit vectors directed horizontally and vertically upward respectively. Take $g = 10$ ms^{-2}.

1 $\mathbf{r}_o = \mathbf{0}$, $u_H = 2$, $u_v = 8$; find **r** when $t = 1$ s.
2 $u_H = 4$, $u_v = 11$; find the velocity at time t.
3 $\mathbf{r}_o = 3\mathbf{i} + 5\mathbf{j}$, $u_H = 3$, $u_v = 5$; find **r** when the particle is level with O.
4 $\mathbf{r}_o = \mathbf{i} + 3\mathbf{j}$, $u_H = $ r, $u_v = 6$; find **r** when the particle is at its maximum height above O.
5 At time $t = 3$ s, $\dot{\mathbf{r}} = 21\mathbf{i} + 11\mathbf{j}$; find u_H and u_v.
6 At time $t = 4$ s the velocity is of magnitude 4 ms^{-1} and inclined at an angle 30° below the horizontal; find u_H and u_v.

9.2 Basic projectile formulae

For projection from one point to another on the same horizontal level, simple formulae will now be derived for the horizontal distance between the two points, for the total time taken, and for the total height reached above the point of projection. Use of these formulae would have simplified the calculations in the preceding examples.

The distance between the point of projection and the point reached on the same level is, for obvious reasons, called the range. A particle projected with speed u at an angle α to the horizontal is, at a time t after projection, at a distance y above the point of projection where, from equation 9.16,

$$y = ut \sin \alpha - \tfrac{1}{2}gt^2.$$

It will, therefore, next be at the same level when $y = 0$, that is, when

$$t = \frac{2u \sin \alpha}{g}.$$

This is generally referred to as the time of flight T.

The range R is, from equation 9.14, $uT \cos \alpha$ giving

$$R = \frac{2u^2 \sin \alpha \cos \alpha}{g} = \frac{u^2 \sin 2\alpha}{g}.$$

The upward vertical component $\dot{y}$ of the particle velocity is, from equation 9.17

$$\dot{y} = u \sin \alpha - gt.$$

The maximum height above the point of projection is attained when $\dot{y} = 0$ so that

$$t = \frac{u \sin \alpha}{g}.$$

Substituting this value of t in the expression for y shows H, the maximum height, to be given by

$$H = \frac{u^2 \sin^2 \alpha}{2g}.$$

The time to the highest point is seen to be one half of the total time of flight, and the horizontal displacement when the particle is at its highest point is one half the range. An alternative method of obtaining H would have been to use equation 4.6 ($v^2 = u^2 + 2as$) for vertical motion with u replaced by $u \sin \alpha$, a by $-g$ and v by $\dot{y}$. Setting $\dot{y} = 0$ would then have given the previous result.

Eliminating t between

$$x = ut \cos \alpha$$

and

$$y = ut \sin \alpha - \tfrac{1}{2}gt^2$$

gives

$$y = x \tan \alpha - \frac{1}{2}\frac{gx^2}{u^2} \sec^2 \alpha.$$

This is the Cartesian equation of the path, which, by completing the square on the right-hand side, can be re-written as

$$y = -\frac{1}{2}\frac{g \sec^2 \alpha}{u^2}\left(x - \frac{u^2 \sin \alpha \cos \alpha}{g}\right)^2 + \frac{u^2 \sin^2 \alpha}{2g}$$

or

$$y - H = -\frac{1}{2}\frac{g \sec^2 \alpha}{u^2}(x - \tfrac{1}{2}R)^2.$$

If X and Y are defined by $X = x - \frac{1}{2}R$, $Y = y - H$, so that the origin of (X, Y) is the highest point, then the above equation becomes

$$Y = -\frac{1}{2}\frac{g \sec^2 \alpha}{V^2} X^2.$$

The path is, therefore, a parabola symmetric about the vertical line through the vertex, with its vertex, which is at the highest point, upwards. The general path is as shown in Fig. 9.3.

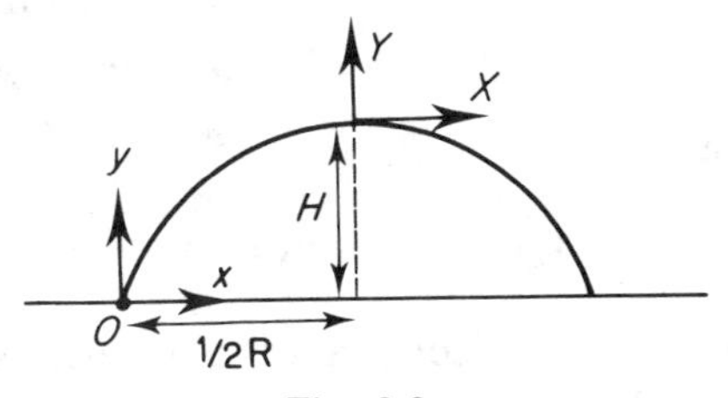

Fig. 9.3

Summary of formulae

The basic formulae of equations 9.9 to 9.15 and those obtained above are, for ease of reference, collected together below. They are written in the two separate forms involving u and α, and u_H and u_v.

$$\dot{x} = u\cos\alpha = u_H, \qquad 9.17$$

$$x = ut\cos\alpha = u_H t, \qquad 9.18$$

$$\dot{y} = u\sin\alpha - gt = u_v - gt, \qquad 9.19$$

$$y = ut\sin\alpha - \tfrac{1}{2}gt^2 = u_v t - \tfrac{1}{2}gt^2, \qquad 9.20$$

$$y = x\tan\alpha - \frac{1}{2}\frac{gx^2\sec^2\alpha}{u^2} = \frac{xu_v}{u_H} - \frac{1}{2}\frac{gx^2}{u_H^2}. \qquad 9.21$$

The inclination θ of the path to the horizontal is given by

$$\tan\theta = \frac{\dot{y}}{\dot{x}} = \frac{u\sin\alpha - gt}{u\cos\alpha} = \frac{u_v - gt}{u_H}. \qquad 9.22$$

$$\text{Maximum height } H = \frac{u^2\sin^2\alpha}{2g} = \frac{u_v^2}{2g}. \qquad 9.23$$

$$\text{Time to highest point} = \tfrac{1}{2}\text{ time of flight} = \frac{u\sin\alpha}{g} = \frac{u_v}{g}. \qquad 9.24$$

$$\text{Range } R = \frac{2u^2\sin\alpha\cos\alpha}{g} = \frac{u^2\sin 2\alpha}{g} = \frac{2u_H u_v}{g}. \qquad 9.25$$

$$\text{Displacement to highest point} = \tfrac{1}{2}R. \qquad 9.26$$

$$\text{Time of flight } T = \frac{2u\sin^2\alpha}{g} = \frac{2u_v}{g}. \qquad 9.27$$

If it is required, for given u^2, to find the values of α which produce a given range R_o then equation 9.25 gives

$$\sin 2\alpha = \frac{gR_o}{u^2}.$$

There are two acute angles α such that $\sin 2\alpha$ has a given value since if α_o is one such value then $\frac{1}{2}\pi - \alpha_o$ is another $[\sin 2(\frac{1}{2}\pi - \alpha_o) = \sin 2\alpha_o]$. Therefore, there are two values of the angle of projection (their sum being $\pi/2$) which produce a given range. The maximum range, for given u^2, is attained when $\sin 2\alpha = 1$, that is, when $\alpha = \pi/4$.

Problem solving

Projectile problems can be regarded, in general, as being of one of two kinds;

(i) those where the velocity is given initially and various quantities, such as the range, have to be found,

(ii) problems where the initial velocity is not given explicitly but sufficient information is given for the initial velocity to be found, and then any other required information can be found as in (i).

Problems of the first type are relatively simple and almost reduce to substituting into formulae. Those of the second kind can be more difficult, and it is important to use the information as directly as possible. Since the equations governing horizontal motion are simpler than those for vertical motion, first extract as much information as possible from the horizontal motion. For example, if the initial components of velocity are given and the time taken to reach a given point is required then substituting the horizontal coordinate into equation 9.18 solves the problem immediately, whereas substituting the vertical coordinate into equation 9.20 gives a quadratic and the correct root has then to be chosen. The quadratic occurs because there are two positions (one going up and one coming down, see Fig. 9.3) when the particle is at a given height.

Sometimes a problem requires, for given speed of projection, the angle of projection to be found such that a projectile passes through a given point (h, k). The method in this case is to substitute the values of h and k into equation 9.21 and replace $\sec^2\alpha$ by $1+\tan^2\alpha$. This gives

$$k = h\tan\alpha - \frac{1}{2}\frac{gh^2}{u^2}(1+\tan^2\alpha),$$

which is a quadratic for $\tan\alpha$, showing that there will be two possible values of α.

In solving projectile problems, it is certainly necessary to be able to recall equations 9.17 to 9.20 (or at least know how to find them quickly from the constant acceleration formulae). If a question does not refer directly to angle of projection, then it is often easier to use u_H and u_v rather than u and α. It is also useful to remember (or at least be able to derive) equations 9.21 and 9.23 and, for projection from point to point on the same horizontal level, the use of equations 9.25 and 9.26 can ease a problem.

In §9.1, example 1, for example, substituting $u_v = 20$ into equation 9.27 gives the time of flight immediately. Similarly, substituting $u_H = 15$ and $u_v = 20$ into equation 9.26 gives the range. Again, setting $u_v = 20$ into equation 9.23 gives the maximum height found in §9.1 Example 2. Some of these basic approaches to solving projectiles are illustrated in the following examples.

EXAMPLE 1 *A particle projected from a given point O is again level with O after 4 s. Find the maximum height reached above O.*

In this case, the time of flight is given and therefore use of equation 9.27 gives

$$\frac{2u_v}{10} = 4,$$

so that $u_v = 20$. Substituting this into equation 9.23 gives the maximum height as $20^2/20$ m = **20 m**. In this particular problem, there was no particular advantage in using $u \sin \alpha$ rather than u_v.

EXAMPLE 2 *The greatest height above O reached by a particle projected from O is* 20 m *and when next level with O, the projectile is at a distance of* 60 m *from O. Find the tangent of the angle between the direction of motion of the projectile and the horizontal* 3 s *after projection.*

In this case, the greatest height and the range are given, and substituting the relevant values into equations 9.23 and 9.25 gives

$$\frac{u_v^2}{2g} = 20,$$

$$\frac{2u_H u_v}{g} = 60.$$

The first equation gives $u_v = 20$ and substituting this value in the second equation gives $u_H = 15$.

These results could equally well have been found by using the formulae in terms of u and $\sin \alpha$, the first equation giving $u \sin \alpha$ and the second $u \cos \alpha$, as before. Another approach would have been to divide the two equations to obtain $\tan \alpha$ and then find $\cos \alpha$ or $\sin \alpha$ and substitute in one of the equations to find u. This is a rather more complicated method and illustrates the point made earlier that, unless information is specifically required about the angle of projection, it is usually better to work directly with the components of velocity.

If θ is the angle between the path and the horizontal after 3 s, then

$$\tan \theta = \frac{\dot{y}}{\dot{x}} = \frac{u_v - 3g}{u_H},$$

using equations 9.17 and 9.19. Substituting for u_H and u_v gives $\tan \theta = -\frac{2}{3}$, showing that the particle is moving downwards.

EXAMPLE 3 *Determine the directions in which a particle has to be projected with speed* 15 ms^{-1} *from a point O on a horizontal plane so that it lands again on the plane at a distance of* 11·25 m *from O.*

In this case the range is given and, as the angle of projection is required, it seems advisable to use the form involving u and α. Equation 9.25 gives

$$\frac{2 \times 15^2 \sin \alpha \cos \alpha}{g} = 11{\cdot}25$$

so that

$$\sin \alpha \cos \alpha = \tfrac{1}{4}.$$

This equation can be solved by using the result $\sin 2\alpha = 2 \sin \alpha \cos \alpha$ (which has been used in deriving one of the alternative forms in equation 9.25). Therefore, $\sin 2\alpha = \frac{1}{2}$ and 2α is either 30° or 150°, so that α is **either 15° or 75°**.

EXAMPLE 4 *A particle projected with speed* $10\sqrt{10}$ ms^{-1} *from a point O on the ground passes through a point P at a distance of* 2·5 m *above O, and the horizontal displacement of P from O is* 50 m. *Find the tangents of the possible angles of projection.*

In this case, the coordinates of one point on the path are known, and the method of solution is that described above of substituting the coordinates into equation 9.21 with $\sec^2\alpha$ replaced by $1+\tan^2\alpha$. This gives

$$25 = 50\tan\alpha - \frac{10\times 2500}{2000}(1+\tan^2\alpha)$$

which simplifies to

$$\tan^2\alpha - 4\tan\alpha + 3 = 0 = (\tan\alpha - 1)(\tan\alpha - 3).$$

The required values of $\tan\alpha$ are, therefore, **1 and 3**.

EXAMPLE 5 *A particle P is projected from the point with position vector* $(7\mathbf{i}+11\mathbf{j})$ m *referred to a fixed origin O, where* **i** *and* **j** *are unit vectors in the horizontal and upward vertical directions respectively, with velocity* $(5\mathbf{i}+6\mathbf{j})$ m s^{-1}. *Determine the position vector,* **r**, *of P when* **r** *next satisfies* $\mathbf{r}.\mathbf{j} = 11$.

As stated this problem looks very unfamiliar, but it is very easily translated into a simple projectile problem. The condition $\mathbf{OP}.\mathbf{j} = 11$ means that the 'y' component of P is again 11, which means that P is again level with O. The horizontal distance OP is therefore the range of the particle projected with horizontal and vertical components of velocity of 5 ms^{-1} and 6 ms^{-1} respectively. Substituting in equation 9.25 gives the range as 6 m so that

$$\mathbf{r} = (13\mathbf{i}+11\mathbf{j})\,\text{m}.$$

EXERCISE 9.2

In numerical questions take $g = 10$ ms^{-2} and give times and distances to three significant figures and angles to the nearest degree.

Questions 1 to 12 refer to a particle projected from a point O with speed u ms^{-1} in a direction inclined at an angle α above the horizontal.

In questions 1 to 4, find the range, greatest height and time of flight.

1 $u = 10$, $\alpha = 30°$.
2 $u = 5{\cdot}5$, $\alpha = 37°$.
3 $u = 10$, $\sin\alpha = 0{\cdot}6$.
4 $u = 12{\cdot}3$, $\tan\alpha = 0{\cdot}25$.
5 $u = 20$, $\alpha = 30°$; find the direction of motion of the particle 0·5 s after projection.
6 $u = 40$, $\alpha = 60°$; find the direction of motion of the particle 7 s after projection.
7 $u = 20$, $\tan\alpha = 0{\cdot}75$; find the direction of motion 3 s after projection.

In questions 8 and 9, find the times at which the stated height h is attained.

8 $u = 24$, $\alpha = 30°$, $h = 7$ m.
9 $u = 60$, $\tan\alpha = 4/3$, $h = 112$ m.
10 If $u = 20$ and the range is 25 m; find the possible angles of projection.
11 If $u = 24$ and the range is 93 m; find the possible angles of projection.

12 $u = \sqrt{5}$ and the particle passes through the point with position vector $(10\mathbf{i} + 15\mathbf{j})$ m, where $\mathbf{i}$ and $\mathbf{j}$ are unit vectors horizontally and vertically upwards. Find the values of $\tan\alpha$.

13 A particle is projected so that the range is three times the greatest height. Find the ratio of the horizontal and vertical components of the velocity of projection and, given that the range is 2400 m, find the time of flight.

14 A particle is projected with speed $13\,\text{ms}^{-1}$ inside a horizontal tunnel of height h m. Find, when (a) $h = 5$ m and (b) $h = 3$ m, the maximum range within the tunnel.

9.3 Further projectile problems

We now consider some rather more complicated problems, in which the point of projection and the point at which the projectile lands are not on the same horizontal level so that the formulae for range and so on cannot be used. It is not really sensible to attempt to give a general method, but the general approach is to use any information given in as simple a way as possible in the basic equations 9.17 to 9.20. The following examples illustrate some of the problems that may be encountered.

EXAMPLE 1 *A particle is projected horizontally with speed* $40\,\text{ms}^{-1}$ *from a point O at a distance* 180 m *above a horizontal plane. Determine the horizontal displacement from O of the point on the plane at which the particle lands.*

From equations 9.18 and 9.20 and the initial condition, the position of the particle is given by

$$x = 40t,$$
$$y = -\tfrac{1}{2}gt^2 = -5t^2.$$

The particle hits the plane when $y = -180$, and substitution in the second equation above gives $t = 6$. Using this value in the first equation shows that the horizontal displacement is **240 m**.

EXAMPLE 2 *A particle P is projected from a point O at a distance of* 15 m *above a horizontal plane. The velocity of P at O has an upward component of* $10\,\text{ms}^{-1}$. *Find the time taken to reach the plane.*

The vertical displacement y m of P from O is given by

$$y = 10t - 5t^2,$$

and it hits the plane when $y = -15$. Therefore

$$-15 = 10t - 5t^2$$

or
$$t^2 - 2t - 3 = 0 = (t-3)(t+1).$$

Taking the positive root gives the time to be **3 seconds**.

EXAMPLE 3 *A particle is projected from a point O on a plane inclined at an angle* $\tan^{-1}(\tfrac{1}{2})$ *to the horizontal so as to land on the plane at a point on the line of*

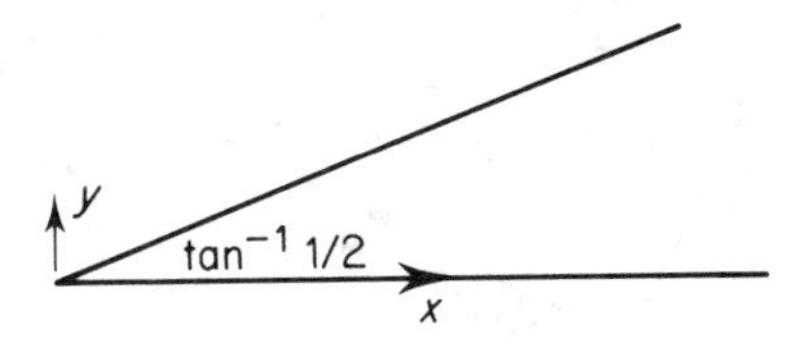

Fig. 9.4

greatest slope through O and above it. The horizontal and vertical components of velocity are 3 ms^{-1} *and* 4 ms^{-1} *respectively. Determine the distance along the plane from O of the point at which the particle lands.*

The displacements x m and y m at time t s are given by

$$x = 3t,$$
$$y = 4t - 5t^2.$$

Obviously the plane has no effect on the equations of motion, but its presence complicates the condition which determines where the particle lands. The coordinates of the points on the line of the greatest slope of the plane through O (Fig. 9.4) satisfy

$$\frac{y}{x} = \frac{1}{2}$$

and substituting for x and y gives

$$\frac{4t - 5t^2}{3t} = \frac{1}{2}.$$

$$t = 2t^2$$

so that $t = 0$ and $\frac{1}{2}$. Therefore, at the point of impact $x = \frac{3}{2}$ and $y = \frac{3}{4}$ so that the distance up the plane is given by $(x^2 + y^2)^{\frac{1}{2}}$ m $= \frac{3}{4}\sqrt{5}$ **m.**

An alternative approach sometimes used in dealing with problems involving range along an inclined plane is to use coordinate axes along and perpendicular to the plane. This method, which simplifies the geometry, means that the equations of motion in the directions along and perpendicular to the plane are equally complicated. The general approach is illustrated in the following example.

EXAMPLE 4 *A particle is projected from a point O on a plane inclined at an angle β to the horizontal so as to land on the plane at a point on the line of greatest slope through O. The initial velocity of projection is of magnitude u in a direction inclined at an angle α ($> \beta$) to the horizontal. Determine the range along the plane.*

Axes OX, OY are chosen along and perpendicular to the plane, as shown in Fig. 9.5, with $\mathbf{p}$ and $\mathbf{q}$ denoting the unit vectors in these directions. The unit vector in the upward vertical direction is, $\mathbf{p}\sin\beta + \mathbf{q}\cos\beta$, so the acceleration due to gravity is given by $-g(\mathbf{p}\sin\beta + \mathbf{q}\sin\beta)$. Equation 9.1 becomes

$$m\mathbf{a} = -mg(\mathbf{p}\sin\beta + \mathbf{q}\cos\beta).$$

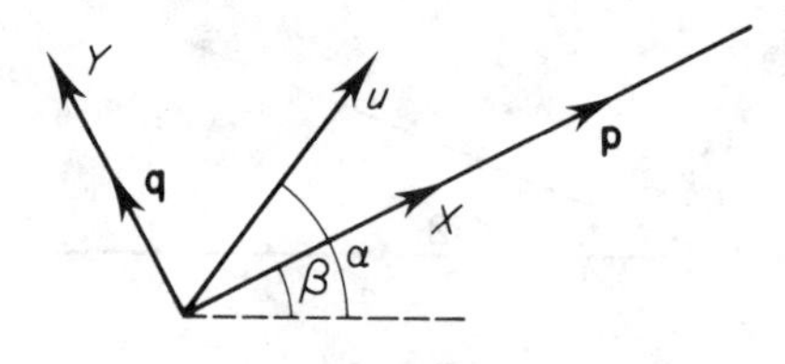

Fig. 9.5

The acceleration **a** is given by

$$\mathbf{a} = \ddot{X}\mathbf{p} + \ddot{Y}\mathbf{q}$$

and, therefore, on taking components,

$$\ddot{X} = -g\sin\beta,$$
$$\ddot{Y} = -g\cos\beta.$$

The accelerations along and perpendicular to the plane are both constant and, therefore, X and Y can be found by using the constant acceleration formulae of equations 4.4 and 4.5. The initial components of velocity along OX and OY are $u\cos(\alpha-\beta)$ and $u\sin(\alpha-\beta)$ respectively, so that

$$X = u\cos(\alpha-\beta)t - \tfrac{1}{2}g\sin\beta t^2,$$
$$Y = u\sin(\alpha-\beta)t - \tfrac{1}{2}g\cos\beta t^2.$$

At the point of impact with the plane $Y = 0$

$$t = \frac{2u\sin(\alpha-\beta)}{g\cos\beta}.$$

The horizontal distance travelled in this time is $ut\cos\alpha$, and if the range up the plane is R then this distance is also equal to $R\cos\beta$. Hence

$$R = \frac{2u^2\cos\alpha\sin(\alpha-\beta)}{g\cos^2\beta}$$

EXERCISE 9.3

In numerical exercises take $g = 10\,\text{ms}^{-2}$ and give times and distances to three significant figures.

1 A particle is projected from a point O at a height of 20 m above a horizontal plane with velocity $42\,\text{ms}^{-1}$ in a direction inclined at an angle 30° above the horizontal. Find the horizontal displacement from O of the point on the plane at which the projectile lands.

2 An aeroplane, flying horizontally with a speed of $150\,\text{ms}^{-1}$ at a height 500 m above horizontal ground, is required to drop a bomb so as to hit a point P on the ground. Determine the point at which the bomb should be released.

Questions 3 to 5 refer to a particle projected with speed $u\,\text{ms}^{-1}$ at an angle α above the horizontal from a point on a plane inclined at an angle β to the horizontal. The ranges up and down the plane are to be found.

3 $u = 20$, $\alpha = 60°$, $\beta = 30°$.

4 $u = 23$, $\alpha = 47°$, $\beta = 23°$.

5 A particle is projected with speed u from a point on a plane inclined at an angle of 30° to the horizontal so as to strike this plane at right angles. Find the range along the plane.

9.4 General planar motion under constant forces

So far we have concentrated entirely on the problem of motion under one particular constant force – gravity. The methods used can, however, be applied equally well to motion under any constant force (for example, a constant electric field acting on a charged particle exerts a constant force on the particle). We now consider the solution of problems where a constant force, other than that of gravity, acts.

The basic equation governing the motion is still equation 9.1 ($m\mathbf{a} = \mathbf{F}$). If the coordinate axes are chosen along and perpendicular to $\mathbf{F}$, then the component of acceleration in the former direction is constant and non-zero whilst that in the latter direction is zero. This is exactly the same as in §9.1 (equations 9.7 and 9.8) and any problem can be solved as in that section. The path will usually be a parabola. For motion under gravity the vertex pointed in the opposite direction to the force and, therefore, for a general force $\mathbf{F}$ the path will be as shown in Fig. 9.6. (If the component of velocity perpendicular to $\mathbf{F}$ is zero, however, the path is not a parabola but degenerates into a straight line.)

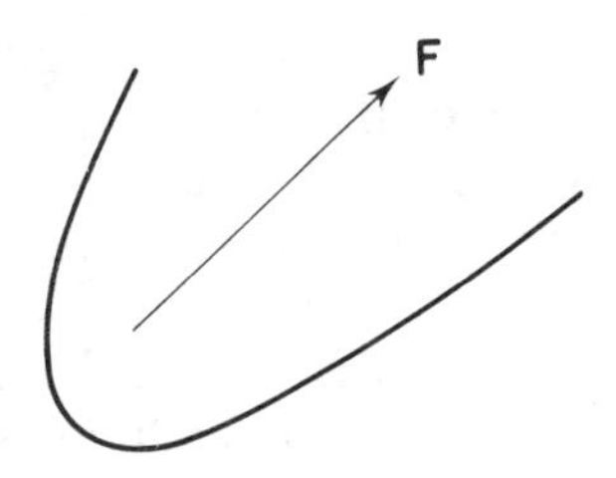

Fig. 9.6

Some problems, in which the given constant force is parallel to one of the basic Cartesian axes used in the problem, can often be solved as for gravitational motion and even the standard formulae, suitably interpreted, can be used. This is illustrated in the following example.

EXAMPLE 1 *A charged particle Q of mass m is free to move in the region $0 \leqslant x \leqslant a$ between two plane charged plates at $x = 0$ and $x = a$. The charge on the plates is such that Q is repelled from the plate $x = a$ by a force of magnitude $4ma/T^2$, where T is a constant. The particle leaves a point O on the plate $x = 0$ with speed a/T at an angle of 60° to the x-axis and returns to the plate at a point P. Find the maximum distance that Q moves away from the plate and also the distance OP.*

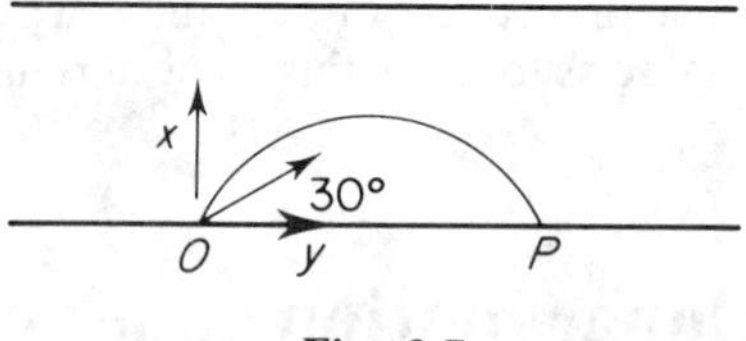

Fig. 9.7

The configuration is as shown in Fig. 9.7, where Oy is taken to be perpendicular to Ox in the plane of initial motion. The force acting is $-(4ma/T^2)\mathbf{i}$ and, therefore, the equation of motion is

$$\frac{\mathrm{d}^2\mathbf{r}}{\mathrm{d}t^2} = \frac{-4a}{T^2}\mathbf{i}.$$

In component form this becomes

$$\ddot{x} = \frac{-4a}{T^2},$$

$$\ddot{y} = 0.$$

These are simply the normal projectile equations with x and y interchanged and g replaced by $4a/T^2$, and the path will be a parabola as illustrated in Fig. 9.7. The maximum distance from $x = 0$ is the 'maximum height'. It is found from equation 9.23 with $u_v = (a/2T)$ [or $u = a/T$ and $\alpha = 30°$], $g = 4a/T^2$, and so it is $\boldsymbol{a/32}$. OP is the 'range' and from equation 9.25 this is $\boldsymbol{a\sqrt{3}/8}$.

If a given force is not parallel to one of the given Cartesian axes in a problem (for example, if it is of the form $a\mathbf{i}+b\mathbf{j}$, where $\mathbf{i}$ and $\mathbf{j}$ are not along the given axes), then it is possible to take new axes and then solve the problem as for motion under gravity. However, changing to new axes involves additional algebra, which can lead to mistakes, and it is generally best not to do so. Instead, the simplest approach is likely to be to integrate the equation of motion in vector form as follows.

Since $\mathbf{F}$ is constant in equation 9.1, the equation can be integrated as in §8.5 to give

$$\frac{m\mathrm{d}\mathbf{r}}{\mathrm{d}t} = \mathbf{F}t + \mathbf{a}, \qquad 9.28$$

where $\mathbf{a}$ is a constant vector. If, at $t = 0$, the particle has velocity $\mathbf{u}$ then substituting into equation 9.28 when $t = 0$ gives $\mathbf{a} = m\mathbf{u}$, so that

$$\frac{m\mathrm{d}\mathbf{r}}{\mathrm{d}t} = \mathbf{F}t + m\mathbf{u}. \qquad 9.29$$

Integration of equation 9.29, assuming that $\mathbf{r} = 0$ when $t = 0$, gives

$$m\mathbf{r} = \tfrac{1}{2}\mathbf{F}t^2 + m\mathbf{u}t. \qquad 9.30$$

Equation 9.30 shows that the particle always stays in the plane through the point of projection and containing the force and the initial speed. This proves that, for free motion under gravity, the projectile always remains in the vertical plane containing the initial velocity vector. Also, for motion under gravity with a particle projected with speed u at an angle α to the horizontal, $\mathbf{u} = u\,(\mathbf{i}\cos\alpha + \mathbf{j}\sin\alpha)$ and $\mathbf{F} = -mg\mathbf{j}$ so that

$$\mathbf{r} = -\tfrac{1}{2}gt^2\mathbf{j} + u\,(\mathbf{i}\cos\alpha + \mathbf{j}\sin\alpha)t$$

Taking the components of this gives equations 9.18 and 9.20.

EXERCISE 9.4

The following questions refer to a particle moving with acceleration $\mathbf{a}\,\text{ms}^{-1}$, which at $t = 0$ s passes through the point $\mathbf{r} = \mathbf{r}_o$ with velocity $\mathbf{u}\,\text{ms}^{-1}$. $\mathbf{i}$ and $\mathbf{j}$ are used to denote perpendicular unit vectors.

1 $\mathbf{u} = 2\mathbf{i} + 4\mathbf{j}$, $\mathbf{a} = 5\mathbf{i} - 7\mathbf{j}$; find the velocity at time t s.

2 $\mathbf{r}_o = 0$, $\mathbf{u} = \mathbf{i} - 3\mathbf{j}$, $\mathbf{a} = 2\mathbf{i} + 12\mathbf{j}$; find $\mathbf{r}$ after time t s.

3 $\mathbf{r}_o = 2\mathbf{i} + 5\mathbf{j}$, $\mathbf{u} = \mathbf{i} + \mathbf{j}$, $\mathbf{a} = 6\mathbf{i} - 4\mathbf{j}$; find $\mathbf{r}$ after time t s.

4 $\mathbf{r}_o = 2\mathbf{i} + 4\mathbf{j}$, $\mathbf{u} = 2\mathbf{i} + 5\mathbf{j}$, $\mathbf{a} = 6\mathbf{i} - 2\mathbf{j}$; find when the particle intersects the line $2\mathbf{i} + 4\mathbf{j} + \lambda(8\mathbf{i} + 3\mathbf{j})$.

5 $\mathbf{r}_o = \mathbf{i} + 2\mathbf{j}$, $\mathbf{u} = 2\mathbf{i} + \mathbf{j}$, $\mathbf{a} = 4(\mathbf{i} + \mathbf{j})$; find the maximum distance of the particle from the line $(\mathbf{i} + 2\mathbf{j}) + \lambda(14\mathbf{i} + 13\mathbf{j})$.

6 A charged particle of mass m moves in the region $0 \leqslant y \leqslant 2a$ and is acted upon by a force of magnitude $4mv^2/a$ acting in the positive y-direction. Given that the particle has a speed v parallel to the plane $y = 0$ when at the point $(0, a)$, find the point at which it hits the plane $y = 2a$.

MISCELLANEOUS EXERCISE 9

The acceleration due to gravity should be taken as $10\,\text{ms}^{-2}$.

1 A cricketer hits a ball so that it strikes the ground at a point P which is at a distance of 70 m from him. At the highest point of its path the ball is at a height of 20 m. Determine the initial speed and the angle of projection.

[You may assume that the ball is at ground level when it is hit by the cricketer].

A fielder starts to run at a speed of $7\,\text{ms}^{-1}$ at the instant the ball is hit. Show that he cannot reach the point P before the ball lands if he starts at more than 29 m from P.

Show that the ball can be hit, with the same initial speed, but at a different angle of projection, so as to land at the same point and find the maximum height in this case. *(L)*

2 A particle is projected with speed V at an angle of elevation α and moves under gravity with negligible air resistance. Show that

(i) the greatest height attained, H, is given by $H = (V^2\sin^2\alpha)/(2g)$;

(ii) the range R on the horizontal plane is given by $R = (V^2\sin 2\alpha)/g$.

Prove that, for a given R and V with $gR < V^2$, there are two possible values of α. If H_1 and H_2 are the greatest heights for these two values of α, show that $H_1 + H_2 = V^2/(2g)$.

In the case in which $H_1 = 2H_2$, show that the smaller angle of projection is given by $\sin^{-1}(1/\sqrt{3})$. If a small alteration is made in this angle (whilst V

remains fixed), causing H_2 to increase by ε, prove that the range increases by approximately $\varepsilon\sqrt{2}$. (*JMB*)

3 A particle is projected from the origin O with speed V at an angle of elevation θ, and it moves freely under gravity. Find $\tan\theta$ if the greatest height reached by the particle above the level of O is equal to the range on the horizontal plane through O.

A second particle projected from O with speed V and elevation α has the same range as the first particle, and $\alpha \neq \theta$. Show that the greatest height of the second particle is one-sixteenth of that of the first particle.

A third particle projected from O has the minimum speed V_1 necessary to achieve the same horizontal range as the first particle. Find V_1 in terms of V (*C*)

4 If a particle is projected with speed u at an angle of elevation α, show that the horizontal range is $u^2 \sin 2\alpha/g$ and the maximum height attained is $u^2 \sin^2\alpha/(2g)$.

A golf ball is struck so that it leaves a point A on the ground with speed 49 m/s at an angle of elevation α. If it lands on the green which is the same level as A, the nearest and furthest points of which are 196 m and 245 m respectively from A, find the set of possible values of α. Find also the maximum height the ball can reach and still land on the green.

There is a tree at a horizontal distance 24·5 m from A and to reach the green the ball must pass over this tree. Find the maximum height of the tree if this ball can reach any point on the green.

(Assume the point A, the green and the base of the tree to be in the same horizontal plane.) (*AEB 1972*)

5 Using any relevant formulae, write down expressions, in terms of g and the initial horizontal and vertical components of velocity u_H and u_v respectively, for the range and maximum height of a projectile.

A cricketer hits a ball so that it first lands on the ground at a point P at a distance of 75 m from him. At the highest point of its path the ball reaches a height of 20 m. Assuming that the ball is projected from ground level determine

(i) the horizontal and vertical components of the velocity of projection,

(ii) the tangent of the angle between the horizontal and the direction of motion of the ball 1 s after it has been hit,

(iii) the furthest distance from P that a fielder, who can run at a speed of 8 ms^{-1}, can stand in order that, starting when the ball is hit, he can arrive at P before the ball lands, (*AEB 1982*)

(iv) the tangent of a different angle of projection which is such that the ball, when projected with the same initial speed, again first lands at P. (*AEB 1982*)

6 A particle is projected at an angle of elevation α from a point A on horizontal ground. When travelling upwards at an angle β to the horizontal, the particle passes through a point B. The line AB makes an angle θ with the horizontal. Show that

$$2\tan\theta = \tan\alpha + \tan\beta. \qquad (AEB\ 1969)$$

The point B is at a horizontal distance of 30 m from point A and is at a height of 20 m above ground. At B, the particle is travelling upwards at an angle arctan (1/3) to the horizontal. Find the angle of projection and the initial speed of the particle. (*L*)

7 A ball is projected with a velocity, whose horizontal and vertical components are u and v respectively, so as just to clear a vertical wall of height b at a distance a from the point of projection. Show that the range R on a horizontal plane is

$$R = a + 2bu^2/ag.$$

If the ball also just clears another wall of height a and distance b from the point of projection, show that

$$R = (a^2 + ab + b^2)/(a + b). \qquad (W)$$

8 A particle is projected with speed u and at an angle of elevation α from a point O. Show that, at time t after projection, the position vector of the particle relative to O is $\mathbf{r}$, where $\mathbf{r} = (u \cos\alpha)t\mathbf{i} + [(u \sin\alpha)t - \frac{1}{2}gt^2]\mathbf{j}$ and $\mathbf{i}$ and $\mathbf{j}$ are unit vectors directed horizontally and vertically upwards respectively.

Given that $u = 40\,\text{ms}^{-1}$ and that the particle strikes a target A on the same horizontal level as O, where $OA = 60\,\text{m}$, find the least possible time, to the nearest tenth of a second, that elapses before the particle hits the target A. (*L*)

9 A particle is to be projected from a point O on a horizontal plane, with a speed of $20\sqrt{10}\,\text{m/s}$ in a direction inclined at an angle α to the horizontal, so as to pass through the point $P(200\,\text{m}, 100\,\text{m})$. The upward vertical direction is Oy and the horizontal direction is Ox.

(i) Show that the only possible values of $\tan\alpha$ are 3 and 1.

(ii) Find, when $\tan\alpha = 3$, the time taken to travel from O to P and the tangent of the angle between the horizontal and the path of the particle at P.

(iii) Find also, when $\tan\alpha = 3$, the total range of the particle on the horizontal plane through O and the horizontal displacement from O of the other point at which the particle is at a height 100 m above O.

(iv) Sketch, for the above two values of $\tan\alpha$, the complete path of the particle until it reaches the plane. State clearly which sketch refers to which value of $\tan\alpha$ and show the positions of O and P on both sketches. (*AEB 1982*)

10 A stone is thrown from a point O on level ground with a speed $13\,\text{ms}^{-1}$ at an angle $\tan^{-1}\left(\frac{12}{5}\right)$ to the horizontal. The stone just misses the top of a pole in its path and then reaches a maximum height of twice the height of the pole. At the instant the stone is thrown, a bird on top of the pole sets off with constant speed of $v\,\text{ms}^{-1}$ away from O in a horizontal line in the vertical plane containing O and the pole. Find

(a) the distance, in metres correct to one decimal place, of the base of the pole from O,

(b) v, correct to one decimal place, given that the stone hits the bird. (*AEB 1983*)

11 A particle is projected with velocity u at an angle α to the horizontal. Show that at any subsequent time the horizontal distance x and the vertical distance y are connected by the relation

$$y = x\tan\alpha - gx^2\sec^2\alpha/2u^2.$$

A ball is thrown with velocity u in a plane perpendicular to a vertical wall whose horizontal distance from the point of projection is a. Show that
(i) the ball will not strike the wall if $u < (ag)^{\frac{1}{2}}$,
(ii) the greatest height above the level of projection at which the ball can strike the wall is

$$(u^4 - a^2g^2)/2gu^2. \qquad (W)$$

12 A particle is projected from a point O on horizontal ground h with a speed of $56\,\text{m s}^{-1}$, and just passes over a vertical mast of height 32 m, the base of which is 256 m from O. If α and β, where $\alpha < \beta$, are the two possible angles of projection, show that $\tan\alpha = \frac{3}{4}$, and find $\tan\beta$.

Two particles are projected from O, each with speed $56\,\text{m s}^{-1}$ and with angles of projection α and β respectively. Find the times t_1 s and t_2 s taken by them to reach the top of the mast, showing that the ratio $t_2 : t_1$ is $\sqrt{65} : 5$.

Find the distance between the two points at which the particles hit the ground. *(C)*

13 Axes Ox and Oy are horizontal and vertically upwards respectively. Particles are to be projected from O with speed $\sqrt{(3ga)}$ in the plane of these axes at various angles of elevation θ above Ox. A vertical screen is to be erected along part of the line $x = a$ to intercept all particles whose range along Ox would otherwise be more than 24/25 of the maximum possible range along Ox. Show that the screen must intercept all particles for which

$$\tfrac{3}{4} \leqslant \tan\theta \leqslant \tfrac{4}{3}$$

and find the y coordinates of the pair of points between which the screen should be fixed. *(JMB)*

14 A particle is projected from a point O with speed V at an angle of elevation α so as to strike a smooth vertical wall, and after rebounding the particle passes through O. Show that the time taken by the particle to reach the wall is

$$\left(\frac{2e}{e+1}\right)\frac{V\sin\alpha}{g},$$

where e is the coefficient of restitution between the wall and the particle.

Find the angle of inclination to the horizontal at which the particle rebounds from the wall. *(O)*

15 A particle is projected with velocity V at an angle of elevation α from a point A of a smooth horizontal floor and rebounds from a point B of a smooth vertical wall to strike the floor for the first time at A. It then rebounds from the floor and next strikes it at a point C. The distances of A from the wall, of B above the floor and of C from A are denoted by d, h and c respectively, and the coefficient of restitution at both the wall and the floor is e.

(i) Show that the time taken from the instant of projection for the particle to return to A is $(2V\sin\alpha)/g$.
(ii) Prove that $c = de(1+e)$.
(iii) Find h in terms of d, α and e.
(iv) If the times of flight from A to B, B to A and A to C form an arithmetic progression, find e. *(JMB)*

16 A ball is projected from ground level with velocity V at an angle α to the horizontal ground. Derive expressions for the range R and the greatest height H attained in its subsequent flight.

The ball is projected from ground level in a large room with a smooth horizontal ceiling at height h. The speed of projection is V [$> \sqrt{(2gh)}$], and the ball subsequently just touches the ceiling. Show that its range is

$$2\sqrt{\left\{\frac{2h(V^2-2gh)}{g}\right\}}.$$

The ball is projected with velocity $3\sqrt{(gh)}$ at an angle of 45° with the horizontal. When it hits the ceiling its horizontal component of velocity is unaltered, but its vertical component of velocity is halved and reversed. Show that the angle made with the horizontal by its trajectory on its subsequent impact with the ground is

$$\tan^{-1}\left(\sqrt{\left(\frac{7}{12}\right)}\right). \qquad (C)$$

17 A and B are points distant a apart on horizontal ground and a ball is thrown from A towards B with velocity u at an angle 2α (< 30) to the horizontal, and simultaneously a second ball is thrown from B towards A with velocity v at an angle α to the horizontal. If the balls collide, show that $v = 2u\cos\alpha$ and hence that they collide after a time

$$t = a/u(2\cos 2\alpha + 1).$$

Show also that
(i) the two trajectories would have the same maximum height above AB,
(ii) the range of the ball from B would always exceed the range of the ball from A. *(W)*

18 A particle is projected with initial velocity $u\mathbf{i} + v\mathbf{j}$, $\mathbf{i}$ and $\mathbf{j}$ being unit vectors in the horizontal and upward vertical directions respectively, and moves with constant gravitational acceleration of magnitude g. Establish the formulae

$$\dot{\mathbf{r}} = u\mathbf{i} + (v - gt)\mathbf{j}$$

and

$$\mathbf{r} = ut\mathbf{i} + (vt - \tfrac{1}{2}gt^2)\mathbf{j},$$

where the position vector of the particle at time t is relative to the point of projection.

The particle is projected from the origin with initial velocity $28\mathbf{i} + 100\mathbf{j}$ towards an inclined plane whose line of greatest slope has the equation $\mathbf{r} = 480\mathbf{i} + \lambda(2\mathbf{i} + \mathbf{j})$ where λ is a parameter. Show that the particle strikes the plane after 20 seconds and determine the distance along the line of greatest slope from the point where $\lambda = 0$ to the point of impact.

Show also that α, the acute angle between the direction of motion at impact and the inclined plane, is given by $5\sqrt{5}\cos\alpha = 2$.
(Velocity components are measured in ms^{-1} and displacements in m. Take g as $9{\cdot}8\,\text{ms}^{-2}$.) *(JMB)*

19 An aircraft is flying with speed V in a direction inclined at an angle α above the horizontal. When the aircraft is at height h, a bomb is dropped. Show that the horizontal distance R, measured from the point vertically below the point at which the bomb is dropped to the point where the bomb hits the ground, is given by

$$gR = \tfrac{1}{2}V^2\sin 2\alpha + V(2gh + V^2\sin^2\alpha)^{1/2}\cos\alpha.$$

(L)

20 A particle is projected from a point O on a cliff at a speed of 29 m/s and moves freely under gravity. The point O is at a vertical height of 60 metres above the sea and the particle strikes the sea at a point A seconds after the instant of its projection. Calculate

(i) the sine of the angle of elevation at which the particle is projected,
(ii) the horizontal distance between O and A.

At the instant when the particle is at the point B, where OB is horizontal, a second particle is projected from O and strikes the sea at A simultaneously with the first particle. Calculate

(iii) the time of flight of the second particle,
(iv) the magnitude and the direction of the velocity of projection of the second particle. (*AEB* 1979)

21 An aircraft, flying in a straight line and at a constant height h above horizontal ground with constant speed V, releases a container at time $t = 0$. On impact with the ground, the horizontal component of the velocity of the container is unaltered whereas the vertical component is reversed in direction and its magnitude is halved. Neglecting air resistance, show that
(a) the first impact with the ground occurs after a time $\sqrt{(2h/g)}$,
(b) the horizontal distance travelled by the container up to the second impact with the ground is $2V\sqrt{(2h/g)}$,
(c) the maximum height reached between the first and second impact with the ground is $h/4$. (*L*)

22 (i) A stone is thrown horizontally out of a window at a height h above level ground. Show that when half the time for the fall has elapsed the height of the stone is $\frac{3}{4}h$ above the ground.
(ii) Another stone is lobbed from level ground; it just clears a vertical wall which is 9 m high and 30 m distant. It hits the ground 10 m the other side of the wall. Show that the greatest height of the stone was 12 m. Hence, or otherwise, find the magnitude and direction of the initial velocity of this second stone. (*AEB 1979*)

23 A plane is inclined at an angle α to the horizontal. A particle is projected from the foot of this plane in the vertical plane containing the line of greatest slope. It strikes the inclined plane at right angles at a distance d from the point of projection. If air resistance can be neglected, show that its greatest distance from the inclined plane during its flight is $\frac{1}{4}d \cot \alpha$. (*O*)

24 A particle is projected with given speed u from a point of a plane of inclination α so that its range r up the plane is as large as possible. Show that

$$r = u^2 (1 - \sin \alpha)/g \cos^2 \alpha.$$

Show that the greatest distance of the particle from the plane, measured vertically (*not* perpendicular to the plane, during the motion is $\frac{1}{4}r$. (*O&C*)

25 A particle P is projected horizontally, with speed V, from a point O on a plane which is inclined at an angle β to the horizontal. The particle hits the plane at a point A which is on the line of greatest slope through O. Show that the time of flight is

$$\frac{2V}{g} \tan \beta.$$

Find the tangent of the acute angle between the horizontal and the direction of motion of P when P reaches A.

A second particle Q is projected from O, with speed V, in a direction perpendicular to the plane. Find the time taken for Q to return to the plane and show that Q hits the plane at A. *(JMB)*

26 An aeroplane is travelling horizontally with speed $63\,\text{ms}^{-1}$ at a height 1200 m above horizontal ground. When it is vertically above a point O on the ground, a parachutist of mass 80 kg steps out of the aeroplane and falls freely under gravity with negligible air resistance. After falling x m, he opens his parachute, and is then subject to a resistive force with vertical component 1120 N and horizontal component 240 N. When the parachutist reaches the ground, the vertical component of his velocity is zero. Show that $x = 360$, and that the resultant speed of the parachutist just before he opens his parachute is $105\,\text{ms}^{-1}$ Find also the total time taken for the fall.

Find the distance from O of the point where the parachutist lands, and his speed just before landing. [Take $g = 9.8\,\text{ms}^{-2}$.] *(L)*

27 A particle starts from the origin O and moves in a horizontal plane with a constant acceleration $(-2\mathbf{i}+\mathbf{j})\,\text{cm/s}^2$, $\mathbf{i}$ and $\mathbf{j}$ being unit vectors in the directions of the coordinates axes Ox and Oy respectively, the unit of distance being 1 cm on each axis. The initial velocity of the particle is $(9\mathbf{i} - 4\mathbf{j})\,\text{cm/s}$. By considering the resolved parts of the motion of the particle in the directions Ox and Oy, or otherwise, show that the position vector of the particle after 5 s is $20\mathbf{i}-(15/2)\mathbf{j}$.

Find the position vector of the particle after t s and hence show that the particle moves along the curve whose equation is

$$x^2+4xy+4y^2-8x-18y=0. \qquad \textit{(AEB 1971)}$$

28 The forces $\mathbf{F}_1$ and $\mathbf{F}_2$, where $\mathbf{F}_1 = (5\mathbf{i}-12\mathbf{j})\,\text{N}$ and $\mathbf{F}_2 = (3\mathbf{i}+4\mathbf{j})\,\text{N}$, both act through the point P. Determine the cosine of the angle between $\mathbf{F}_1$ and $\mathbf{F}_2$. Find also the resultant, $\mathbf{F}$, of the two forces $\mathbf{F}_1$ and $\mathbf{F}_2$ and show that its magnitude is $8\sqrt{2}\,\text{N}$.

The position vector of P is $(-8\mathbf{i}+8\mathbf{j})\,\text{m}$ and a particle of mass 8 kg is initially at rest at P. The particle is acted upon throughout its subsequent motion by the force $\mathbf{F}$ and no other force. Show that the particle reaches the origin 4 s after leaving P and find its velocity at that instant. *(L)*

29 $\mathbf{i}$ and $\mathbf{j}$ are unit vectors parallel to the x- and y-axes respectively.

(a) At time $t = 0$ a particle of mass 2 kg is at the origin and has velocity given by $(2-\mathbf{j})\,\text{ms}^{-1}$. From $t = 0$ until $t = 4$ s a force given by $(\mathbf{i}+3\mathbf{j})\,\text{N}$ acts on the particle. Determine the particle's position and velocity when $t = 4$ s.

(b) At a certain instant the particle A is at the origin and the particle B is at the point $(10, 0)$. The velocities of A and B are constant, $(2\mathbf{i}+2\mathbf{j})$ and $(\mathbf{i}-\mathbf{j})$ units per second respectively. Show the velocity of B relative to A, and hence, or otherwise, show that the least distance apart of A and B is $3\sqrt{10}$ units. *(C)*

10 Motion in a Circle

The study of planar motion is extended in this chapter to possibly the next simplest class of such motion – that of a particle constrained to move in a circle. The earlier parts of the chapter are concerned with motion at constant speed in a horizontal circle but, in the last section, the work is extended to include motion at varying speed in a vertical circle.

10.1 Basic results

In this section we derive the basic expressions for the velocity and acceleration of a particle P which is constrained to move in a circle of radius r and centre O (Fig. 10.1(a)). The position of the particle at any instant is completely specified by the angle θ between OP and a fixed direction. The latter direction is chosen to be in the direction of Ox, and Oy is taken to be perpendicular to Ox, so that $\theta = \pi/2$ corresponds to a point on the positive y-axis. The x and y coordinates of P are $r \cos \theta$ and $r \sin \theta$ respectively, and

$$\mathbf{OP} = \mathbf{r} = r\,(\mathbf{i} \cos \theta + \mathbf{j} \sin \theta). \qquad 10.1$$

The velocity $\mathbf{v}$ of P is given by (cf, §8.1)

$$\mathbf{v} = \frac{\mathrm{d}\mathbf{r}}{\mathrm{d}t} = r\dot{\theta}\,(-\mathbf{i} \sin \theta + \mathbf{j} \cos \theta). \qquad 10.2$$

The vector $-\mathbf{i} \sin \theta + \mathbf{j} \cos \theta$ is a unit vector perpendicular to OP and is in the direction shown in Fig. 10.1(b), that is, it is in the sense of θ

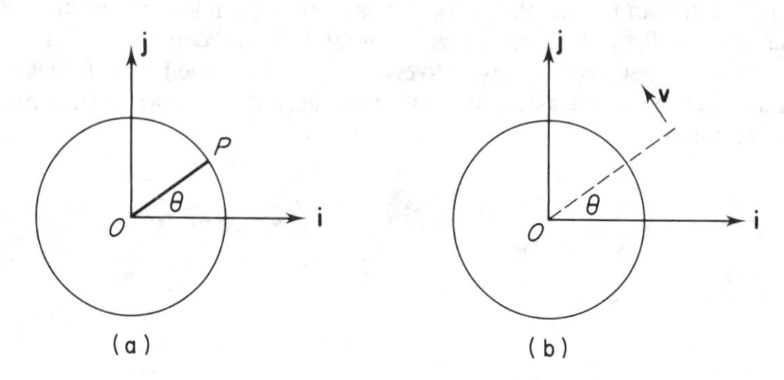

Fig. 10.1

increasing. The velocity is of magnitude $r|\dot{\theta}|$ and $|\dot{\theta}|$ is called the *angular speed* of P. It is sometimes incorrectly referred to as the angular velocity but, as $\dot{\theta}$ is a scalar, this would not be consistent with the vector concept of velocity. (In this book we shall not use the concept of angular velocity. This term is used for the vector whose magnitude is $|\theta|$ and whose direction is out of the paper. More precisely, in the standard vector notation defined in *Pure Mathematics for Advanced Level*, §12.3, the angular velocity $\boldsymbol{\omega}$ is defined by $\boldsymbol{\omega} = \dot{\theta}\mathbf{k}$.)

We consider first the case when P has a constant speed, that is, $\dot{\theta}$ is constant. Henceforth, ω will be assumed to denote the constant angular speed and the sense of rotation will be chosen so that $\dot{\theta}$ is positive. The angular speed is the rate of change of angle with respect to time, and in the m.k.s. system the unit of angular speed is rad s^{-1}.

The acceleration $\mathbf{a}$ is found by differentiating equation 10.2 again with respect to t giving

$$\mathbf{a} = \frac{\mathrm{d}^2\mathbf{r}}{\mathrm{d}t^2} = -r\omega^2(\mathbf{i}\cos\theta + \mathbf{j}\sin\theta). \qquad 10.3$$

The vector $(\mathbf{i}\cos\theta + \mathbf{j}\sin\theta)$ is parallel to OP and, therefore, equation 10.3 shows that the acceleration is directly towards O and of magnitude $\omega^2 r$. Since $v = \omega r$, the magnitude of the acceleration can also be written as v^2/r. Therefore, for motion at constant speed, the velocity and acceleration are of magnitudes ωr and $\omega^2 r$, and are as shown in Fig. 10.2(a), where the positive sense of rotation is counter-clockwise.

Since the angular speed is constant we have that

$$\theta = \omega t + \alpha$$

where α is the value of θ when $t = 0$. The particle, therefore, returns to any particular position at intervals $2\pi/\omega$ and the period T of the motion is given by

$$T = \frac{2\pi}{\omega}. \qquad 10.4$$

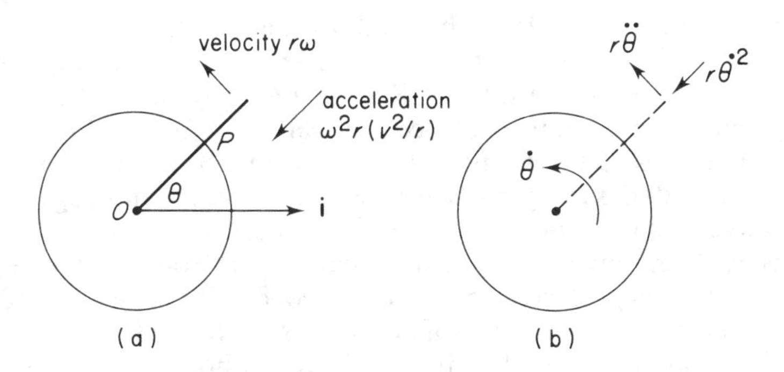

Fig. 10.2

At this point it is convenient to generalise equation 10.3 to cover the case when $\dot{\theta}$ is not constant. Straightforward differentiation of equation 10.2 now gives

$$\frac{\mathrm{d}^2\mathbf{r}}{\mathrm{d}t^2} = -r\dot{\theta}^2(\mathbf{i}\cos\theta + \mathbf{j}\sin\theta) + r\ddot{\theta}(-\mathbf{i}\sin\theta + \mathbf{j}\cos\theta). \qquad 10.5$$

The first term is effectively that obtained in equation 10.3, and the second represents a vector of magnitude $r|\ddot{\theta}|$ perpendicular to OP in the sense of θ increasing. Therefore, for non-uniform motion the acceleration has components as illustrated in Fig. 10.2(b).

10.2 Motion at constant speed in a horizontal circle

The above results will now be applied to problems of motion at constant speed in a horizontal circle. The basic equation governing all particle motion is equation 5.3

$$\mathbf{F} = m\mathbf{a}$$

and the general method of solution is, as for projectiles, to consider the components of the above equation in particular directions. The most obvious directions for circular motion are along and perpendicular to the radius in the plane of motion and perpendicular to the plane. There is no acceleration perpendicular to the radius vector in the plane of motion and, therefore, there can be no force in that direction. Since the motion is horizontal there is no acceleration perpendicular to the plane and so the total vertical component of the forces acting will be zero – that is, the vertical components of the forces acting are in equilibrium. Therefore, the radial component of equation 5.3 is

$$m\omega^2 r = \frac{mv^2}{r} = F_r, \qquad 10.6$$

where F_r is the magnitude of the total force acting towards the centre.

It is possible to interpret equation 10.6 as a statical equilibrium equation where the inward force F_r is balanced by the 'outward' force $m\omega^2 r$, which is referred to as the 'centrifugal' force. We shall not use this approach as it is rather artificial to try and reduce dynamics to statics when, in fact, statics is a particular case of dynamics! Centrifugal force is effectively a fictitious concept but it can be used, if sufficient care is taken, to solve problems involving circular motion. Apart from the care required and the illogicality referred to above, the concept of centrifugal force is difficult to generalise to the case of non-uniform motion. It is usually wiser to use Newton's laws for all motion problems without confusing them by introducing 'fictitious' forces.

To solve any problem involving motion at constant speed in a horizontal circle all that is necessary is to
(a) consider vertical equilibrium,
(b) calculate the magnitude of the force F_r acting towards the centre and then use equation 10.6.
These principles are illustrated in the following examples.

EXAMPLE 1 *A particle of mass* 2 kg *is attached to one end of a light string of length* 0·3 m *and is made to move at constant speed on a smooth horizontal table. The string is such that it will break when the tension in it exceeds* 5·4 N. *Find the maximum angular speed of the particle.*

The situation is as shown in Fig. 10.3, where the only radial force acting on the particle is the tension which is taken to be T newtons. Equating the tension to the mass times the radial acceleration gives

$$T = 0{\cdot}6\omega^2,$$

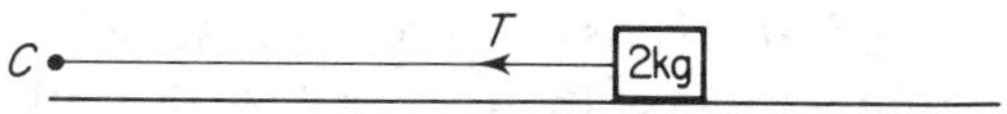

Fig. 10.3

where ω rad s^{-1} is the angular speed. The maximum value of the tension is 5·4 N which means that ω_{max}, the maximum value of ω, satisfies

$$0{\cdot}6\omega_{max}^2 = 5{\cdot}4,$$

that is,

$$\omega_{max} = 3.$$

So the maximum angular speed is **3 rad s^{-1}**.

EXAMPLE 2 *A car moves without skidding at a constant speed of* 20 ms^{-1} *in a horizontal circle of radius* 100 m. *Determine the minimum value of the coefficient of friction between the road and the car tyres.*

The radial force F N inward (see Fig. 10.4) satisfies

$$F = \frac{mv^2}{r} = 4m$$

where m kg is the mass of the car. Vertical equilibrium gives

$$R = mg,$$

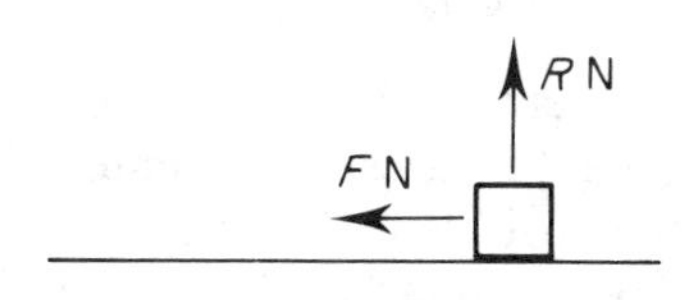

Fig. 10.4

where the reaction of the plane is denoted by R N. Since the car is not skidding, the ratio F/R must be less than μ, the coefficient of friction. Approximating to g by 10 ms^{-2} gives $\mu \geqslant \mathbf{0{\cdot}4}$.

EXAMPLE 3 *A particle of mass m, constrained to move at constant speed in a horizontal circle of radius 2a, is acted upon by a force directed towards the centre of the circle and of magnitude 6mv/T, where v is its speed and T is a constant. Determine the speed of the particle.*

(This is not an entirely artificial example as a force of the above form would be produced on an electron by a magnetic field perpendicular to the plane.)
The equation of motion gives

$$\frac{mv^2}{2a} = \frac{6mv}{T},$$

so that $v = \mathbf{12a/T}$.

The conical pendulum

A slightly more complicated class of problems arises when considering the motion, in a horizontal circle of radius a, of a particle attached to the end of a light string, the other end of which is attached to a fixed point vertically above the centre of the circle (Fig. 10.5). The string describes the surface of a cone, hence the term conical pendulum.

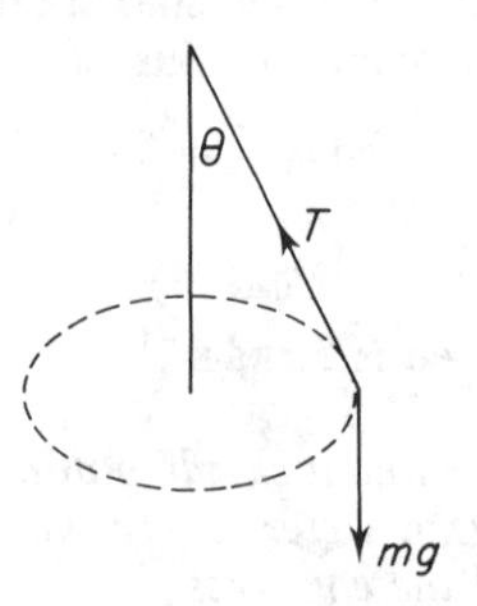

Fig. 10.5

Taking the particle to be of mass m we have, using the notation of Fig. 10.5, that vertical equilibrium gives

$$T\cos\theta = mg, \qquad 10.7$$

whilst the radial component of Newton's law gives

$$T\sin\theta = m\omega^2 a. \qquad 10.8$$

The above equations govern the motion in any problem of the conical pendulum type but have to be supplemented by other conditions, depending on the type of problem. In one of the simpler variants the length l of the string will be given rather than the radius of the circle, so the relation $a = l\sin\theta$ has to be used. The string may also be elastic, so that the vertical equilibrium equation determines the tension and, therefore, the length of the string. A further variant occurs when a particle is constrained to move on the surface of a cone. In this case, the normal reaction and (for a rough cone) possibly the tangential reaction have to be taken into account in setting up the basic equations. It is often necessary to use the condition $|\cos\theta| < 1$ or $|\sin\theta| < 1$ to establish whether a particular motion is possible.

In all problems, the basic procedure followed should be to mark all the forces clearly in a diagram, obtain the equation of vertical equilibrium for each particle and take the radial component of Newton's law for each particle. This procedure is illustrated in the following examples.

EXAMPLE 4 *A particle P of mass* 2 kg *is attached to one end of a light inextensible string of length* 0·4 m, *the other end of which is held fixed at a point O. P describes, with constant speed, a horizontal circle whose centre is directly below O. During the motion the string is inclined at an angle of* 30° *to the vertical. Find the time for one complete revolution of the particle.*

We can use the diagram of Fig. 10.5, with $\theta = 30°$ so that

$$\frac{T\sqrt{3}}{2} = 2g,$$

$$\tfrac{1}{2}T = 2\omega^2 a.$$

The radius a is equal to 0·4 sin 30° m = 0·2 m and, therefore, eliminating T gives

$$\frac{0{\cdot}2\omega^2}{g} = \frac{1}{\sqrt{3}},$$

where $g = 10\ \text{ms}^{-2}$ and the angular speed is $\omega\ \text{rad s}^{-1}$. This gives $\omega \approx 5{\cdot}37$, and therefore the period is found from equation 10.4 ($T = 2\pi/\omega$) to be approximately **1·2s**.

EXAMPLE 5 *A particle of mass m is attached to one end of a light elastic string, of modulus* 8 *mg and natural length* 4*l*. *The other end of the string is attached to a fixed point O and the particle is made to describe, at constant speed, a horizontal circle whose centre is directly below O. The string is of length* 5*l during this motion. Find the angular speed of the motion and the angle between the string and the downward vertical.*

Using the notation of Fig. 10.5 we have

$$T\cos\theta = mg$$
$$T\sin\theta = m\omega^2 a.$$

Hooke's law shows that $T = 2mg$. Also, the radius a is equal to $5l\sin\theta$ so that

$$T = 5lm\omega^2 = 2mg$$

giving $\boldsymbol{\omega = (2g/5l)^{\frac{1}{2}}}$. Substituting for T in the equation for vertical equilibrium gives $\cos\theta = \frac{1}{2}$, and so $\boldsymbol{\theta = 60°}$.

EXAMPLE 6 *Determine the minimum value of the modulus of elasticity so that the motion described in the previous example could occur.*

Hooke's law gives

$$T = \frac{\lambda}{4},$$

where λ denotes the elastic modulus and, therefore, substituting in the first equation gives

$$\cos\theta = \frac{4mg}{\lambda}.$$

Therefore, as $\cos\theta < 1$, $\lambda > \mathbf{4mg}$.

EXAMPLE 7 *A particle describes a horizontal circle with angular speed 2ω on the inner surface of a smooth cone of semi-angle β, placed with its axis vertical and vertex O downwards. Find the height above O at which the motion takes place.*

The configuration is as shown in Fig. 10.6, with R denoting the normal reaction of the cone. Vertical equilibrium gives

$$R\sin\beta = mg,$$

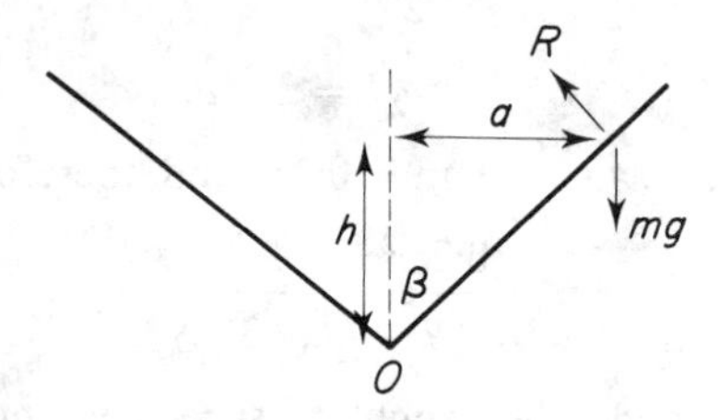

Fig. 10.6

where m is the mass of the particle. The inward force radially is $R\cos\beta$ and, therefore, Newton's law gives

$$R\cos\beta = 4m\omega^2 a.$$

The required height h is such that $a = h\tan\beta$. Eliminating R and substituting for a gives

$$h = \frac{\mathbf{g}}{\mathbf{4\omega^2}}\mathbf{\cot^2\beta}.$$

EXERCISE 10.2

In numerical exercises, take $g = 10\,\text{ms}^{-2}$ and give answers correct to three significant figures.

Questions 1 to 4 refer to a particle of mass m attached to a light string, the other end of which is fixed at a point O. The particle is describing a circle, at constant speed V, on a smooth horizontal plane through O.

1 $m = 2$ kg, the string is inextensible of length 3 m and $v = 9\,\text{ms}^{-1}$. Find the tension in the string.

2 $m = 3$ kg, and the string is inextensible of length 2 m and can sustain a maximum force of $294\pi^2$ N without breaking. Find the maximum number of revolutions per second possible without breaking the string.

3 $m = 1{\cdot}5$ kg and the string is elastic of unstretched length 0·5 m and modulus 350 N. Find the extension if the particle makes 1 complete revolution per second.

4 The string is elastic of unstretched length 0·3 m and when the particle is suspended from it is extended a distance 0·02 m. Find the period of revolution when the particle describes a circle of radius 0·33 m at the end of the string.

5 Find the distance above the earth at which a satellite must be placed in order that it remains permanently above the same point on the earth's surface. Assume that the earth is a sphere of radius $6{\cdot}4 \times 10^6$ m, and that the acceleration due to gravity varies as the inverse square of the distance from the centre of the earth and is equal to $10\,\text{ms}^{-2}$ at the earth's surface.

6 A car travelling on level ground describes a circle of radius 25 m at a speed of $9\,\text{ms}^{-1}$. Find the least value of the coefficient of friction so that the car does not slip.

7 A particle is threaded on a rough rod which can rotate about a vertical axis through a point O of itself. The particle is attached to O by a light elastic string of modulus $10\,mg$ and natural length l. Given that the coefficient of friction is 0·4, find the maximum angular speed at which the rod can rotate with the particle stationary relative to it and the string of length $1{\cdot}2l$.

8 A smooth hollow cone is fixed with its axis vertical and vertex downwards. A particle moving on the inner surface of the cone describes a horizontal circle with speed v at a height h above the vertex. Find v in terms of g and h.

Questions 9 to 13 refer to a particle of mass m attached to the end of a light string, the other end of which is attached to a point O. The particle describes, with constant speed v, a horizontal circle in a horizontal plane below O, the centre of the circle being a distance h directly below O.

9 The string is inextensible and of length 1·5 m, and is inclined at an angle $\tan^{-1} 3/4$ to the downward vertical. Find v.

10 $m = 3$ kg, and the string is inextensible and of length 2 m. Find the tension in the string when the particle describes 3 revolutions per second.

11 Given that the period of one revolution is 2 s find h.

12 $m = 4$ kg and the string is elastic of natural length 0·5 m. Find the modulus, given that when the particle describes 4 revolutions per second the string is of length 0·6 m.

13 $m = 0{\cdot}5$ kg, the string is elastic of natural length 0·75 m and modulus 100 N, and is inclined at an angle of 30° to the downward vertical. Find the period.

14 A particle is attached by two light inextensible strings of equal length to two points A and B, which are at a distance a apart with A being directly above B. The particle describes a horizontal circle with uniform angular speed ω. Show that $\omega > (2g/a)^{1/2}$ and find the ratio of the tensions in the strings when $\omega = 3(g/a)^{\frac{1}{2}}$.

15 Two particles A and B of masses 0·4 kg and 0·1 kg respectively are connected by a light inextensible string of length 0·6 m which passes through a small smooth fixed ring R. Particle A remains in equilibrium at a distance 0·4 m below R whilst particle B describes a horizontal circle whose centre is in the line AR. Find the angular speed of B.

16 A smooth hemispherical bowl is held with its rim, which is of radius a, horizontal, and a particle moving on the inner surface of the bowl describes a horizontal circle of radius b with uniform speed. Find the period of rotation.

10.3 Banking of tracks

A car or train moving round a circular bend of a horizontal road or track will have an acceleration towards the centre of the bend and so there must be a force acting on the car towards the centre of the bend. The principal source of such a force is friction. Therefore, on relatively smooth roads this force is small and (since the force is proportional to the square of the speed) the speed at which the car can take a bend without slipping is also relatively low. If, however, the road is banked, the reaction of the road has a component in the inward direction and the possible maximum speed increases. Therefore, banking a bend increases the maximum speed at which it can be taken, and the basic equation 10.6 is used in practical road design problems. We now examine the approaches used in such problems.

First, we consider a vehicle on a bend which is banked at an angle θ to the horizontal, and assume that the vehicle is at a point on the road such that it is moving in a horizontal circle of radius a. We also assume that there is no friction acting along the line of greatest slope of the bank, and so the situation is as shown in Fig. 10.7. (We could not assume that the road is smooth since this would mean the vehicle could not move.) From vertical equilibrium, the normal reaction R satisfies

$$R\cos\theta = mg,$$

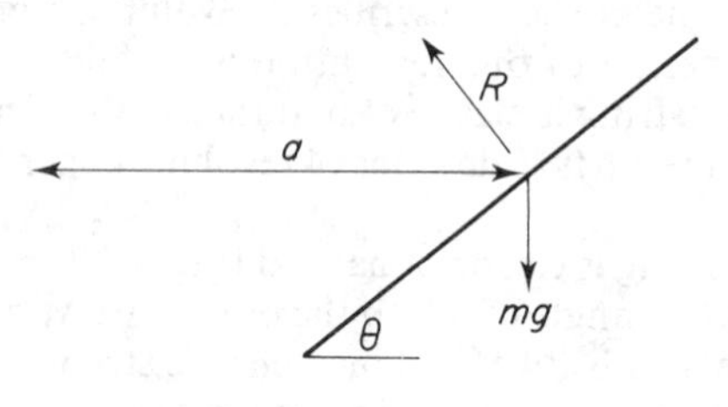

Fig. 10.7

where m is the mass of the vehicle, and Newton's law radially gives

$$R\sin\theta = mv^2/a.$$

Eliminating R gives

$$v^2 = ag \tan \theta.$$

The value of v given by this equation is known as the 'self-steering' speed because in theory, if the steering wheel had been set properly, a vehicle travelling at this speed would steer round the bend without adjustment. Present practice is that banking (called superelevation by traffic engineers) should be such that a 'self-steering' vehicle is travelling at the average speed of the traffic using the road.

When a vehicle is going faster than the 'self-steering' speed, there has to be a further force acting inwards. This is provided by friction, whose magnitude is restricted by the value of the coefficient of friction. In practice, the latter has to be determined from empirical data, and using this data, bankings on bends have to be constructed such that the bend can be taken safely up to a certain maximum speed. The basis of this calculation is as follows.

The same configuration as in Fig. 10.7 is assumed with the addition of a friction force F acting down the plane. The acceleration mv^2/a has components $(mv^2/a) \sin \theta$ parallel to R and $(mv^2/a) \cos \theta$ parallel to F. Taking the components of Newton's law in these directions gives

$$R - mg \cos \theta = \frac{mv^2}{a} \sin \theta$$

$$F + mg \sin \theta = \frac{mv^2}{a} \cos \theta.$$

Applying the condition $F \leqslant \mu R$

$$\frac{v^2}{ag} \leqslant \frac{(\mu \cos \theta + \sin \theta)}{(\cos \theta - \mu \sin \theta)}, \qquad 10.9$$

which gives the maximum speed that can be attained without the vehicle slipping up the bank. In practice $\tan \theta$ never exceeds 0·07, because otherwise there is a danger that very slow moving vehicles might slide down relatively smooth banking (in §2.6 example 2 it was shown that a body would slide down a slope if $\tan \theta > \mu$.) Measurements of μ suggest that it is of the order of 0·3, so that $\mu \tan \theta$ is considerably less than unity. Therefore, equation 10.9 can be approximated to

$$\frac{v^2}{ag} \leqslant \tan \theta + \mu, \qquad 10.10$$

which is the relation used in practice.

Equation 10.10 can be applied to determine the minimum curve radius for any speed. In designing roads, the maximum possible values of μ are not used for safety reasons. Further, investigations have shown that the main factor controlling vehicle speed on a curve is the feeling of discomfort felt by motorists negotiating the bend. This sensation is related to the quantity $v^2/(ag)$, and experiments show that motorists tend to become very uncomfortable when $v^2/(ag)$ is about 0·3. Taking this into account, the maximum value of $v^2/(ag)$ used in design is about 0·2, which corresponds to μ of about 0·13 for the maximum value of $\tan\theta$. Even this value is less than most values occurring in practice. (Most road surfaces are such that μ is at least 0·25.)

For steeply banked tracks (which would not occur in practice) it is possible to carry out calculations similar to the above to determine the least speed that a vehicle can travel at without slipping down the banking. In this case, F would act up the banking, and replacing F by $-F$ gives the minimum speed v to be defined by

$$\frac{v^2}{ag} = \left(\frac{\sin\theta - \mu\cos\theta}{\cos\theta + \mu\sin\theta}\right).$$

EXERCISE 10.3

In numerical exercises take $g = 10\,\text{ms}^{-2}$ and give answers to two significant figures.

1 A car describes a horizontal circle of radius 100 m at $18\,\text{ms}^{-1}$ on a track which is banked at an angle α to the horizontal. Determine $\tan\alpha$ so that there is no lateral force acting on the car. Find, for this value of α, the least coefficient of friction such that the car can go round the track without slipping at a speed of $28\,\text{ms}^{-1}$.

2 A railway curve is an arc of a circle of radius r and the track is banked at an angle α so that there is no lateral force on the rails when the train is moving at speed v. Find the lateral force when a train of mass m traverses the curve at speed $1{\cdot}5\,v$.

10.4 Motion in a vertical circle

The problems on circular motion that we have examined so far have been relatively simple in that the only force acting in the plane of the motion has been radial and so the angular speed has remained constant. We now extend our considerations to another class of problems – those involving motion in a vertical circle – where the angular speed is no longer constant.

Fig. 10.8 shows a vertical circular wire, of centre O and radius a, on which a bead P of mass m is threaded. The angle between OP and the downward vertical is denoted by θ. The wire is assumed to be smooth so that the only force it exerts on P is radial and of magnitude R, and this is shown in Fig. 10.8 as acting radially inwards. The only other force acting on P is the force of gravity vertically downwards, which has a component

$mg \cos\theta$ acting outwards and a component $mg \sin\theta$ tangentially in the direction of decreasing θ. The components of the acceleration of P radially

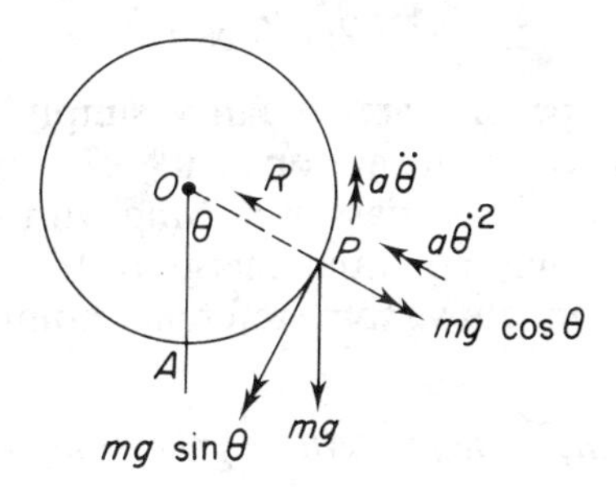

Fig. 10.8

inwards and tangentially are, as derived in §10.1, $a\dot{\theta}^2$ and $a\ddot{\theta}$ respectively. The radial and tangential components of Newton's law give

$$R - mg\cos\theta = ma\dot{\theta}^2,$$

$$ma\ddot{\theta} = -mg\sin\theta. \qquad 10.11$$

The second equation is the differential equation relating θ with t, and the first step is to try and integrate this equation. It is not immediately obvious how to do this as the derivative is with respect to t whilst the right-hand side depends on θ. The difficulty can be overcome by using equation 4.1

$$\frac{\mathrm{d}^2x}{\mathrm{d}t^2} = v\frac{\mathrm{d}v}{\mathrm{d}x},$$

with x replaced by θ and therefore v by $\dot{\theta}$. This gives

$$\ddot{\theta} = \dot{\theta}\frac{\mathrm{d}\dot{\theta}}{\mathrm{d}\theta} = \frac{\mathrm{d}}{\mathrm{d}\theta}(\tfrac{1}{2}\dot{\theta}^2).$$

Equation 10.11 then becomes

$$\frac{a\mathrm{d}}{\mathrm{d}\theta}(\tfrac{1}{2}\dot{\theta}^2) = -g\sin\theta$$

which can be integrated to give

$$a\dot{\theta}^2 = c + 2g\cos\theta,$$

where c is a constant. (This method of integrating equation 10.11 is a particular case of a general approach described in §16.1.)

Since the speed v is equal to $a\dot{\theta}$, the above equation can be re-written as

$$\frac{v^2}{a} = c + 2mg\cos\theta. \qquad 10.12$$

If it is assumed that P has a speed u at its lowest point A (when $\theta = 0$), c can be found by substituting into equation 10.12. This gives

$$v^2 = u^2 + 2ga\,(\cos\theta - 1). \qquad 10.13$$

This equation can be interpreted fairly simply by generalising the concept of energy conservation developed in §6.2. For general motion, the kinetic energy of a particle of mass m moving with speed v is defined as $\frac{1}{2}mv^2$ and the gravitational potential energy of the particle is defined as mgh, where h is the height above some reference point. Equation 10.13 can be re-arranged as

$$\tfrac{1}{2}mv^2 - mga\cos\theta = \tfrac{1}{2}mu^2 - mga,$$

that is, the sum of the potential and kinetic energies at a general position is equal to that at the initial position or, equivalently, the energy is conserved. The above analysis is valid for any motion on a smooth circle and, therefore, establishes that for any such motion the total energy is conserved. (The conditions under which energy is conserved in more general circumstances are examined in §11.3.) Replacing v by $\dot{\theta}$ in equation 10.13 gives

$$a^2\dot{\theta}^2 = u^2 + 2ga\,(\cos\theta - 1) \qquad 10.14$$

and substituting for $\dot{\theta}^2$ in the equation involving R gives

$$R = \frac{mu^2}{a} + mg\,(3\cos\theta - 2). \qquad 10.15$$

It is possible, by separating the variables in equation 10.14, to obtain a relation between θ and t, but this relation involves a rather complicated integral and so we shall not pursue it further.

Most problems involving motion in a vertical circle essentially reduce to determining, from given initial conditions, the forms of equations 10.14 and 10.15 relevant to a particular situation and then making some inferences from these equations. The variation in the types of problems that can occur are produced, almost entirely, by varying the way in which the particle is constrained. If, for example, the particle is attached to the end of a light inextensible string (or moves on the inside of a smooth cylinder) then R in Fig. 10.8 represents the tension in the string or the reaction of the cylinder). Motion is only possible for R positive and the conditions for the motion to be possible, or the value of θ for which the motion ceases to be possible (that is the particle leaves the circle), are often sought. For problems involving motion on the outer surface of a cylinder or sphere, the reaction is radially outward so that, in the notation of Fig. 10.8, R has to be negative. Again, this condition may have to be applied to determine whether or not the motion is feasible.One may also be required, in particular cases, to determine whether or not complete revolutions are described. Equation 10.14 shows that $\dot{\theta}^2$ decreases as θ increases and, therefore, complete revolutions require the smallest value of $\dot{\theta}^2$ (that is, when $\theta = \pi$) to be positive. It is, of course, necessary to make certain first that the motion has not already ceased to be possible before $\theta = \pi$.

Problem solving

The first step is, as usual, to show in a diagram the forces acting and, in particular, to make sure that (when relevant) they are in the physically sensible sense (for example, the tension in a string is always radially inwards). Once the forces have been determined, the next step is to write down the radial and transverse equations of motion. For a single particle, the tangential equation can be integrated as above to obtain a relation between $\dot{\theta}$ and θ, though it is generally simpler to write down the equation of energy directly. The expression obtained for $\dot{\theta}$ can then be used to express the radial force in terms of θ. In this way, equations corresponding to equations 10.14 and 10.15 will be obtained, and often to these only need to be manipulated in order to answer the demands of a particular question.

EXAMPLE 1 *A particle free to move on the inner surface of a smooth hollow sphere of internal radius 2a is given a horizontal velocity of magnitude* $(2ga)^{1/2}$, *from the lowest point of the cylinder. Determine the maximum height the particle rises above the point of projection.*

The forces acting are as shown in Fig. 10.9. The equations of motion are

$$R - mg\cos\theta = 2ma\dot{\theta}^2,$$
$$-mg\sin\theta = 2ma\ddot{\theta}.$$

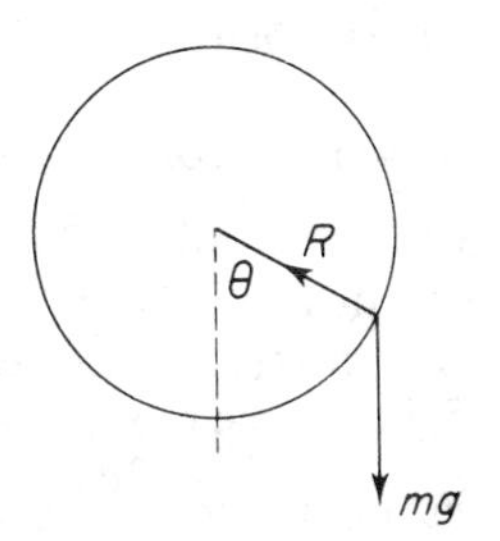

Fig. 10.9

The equation of energy (which is just an integral of the tangential equation of motion) gives

$$\tfrac{1}{2}mv^2 - 2\,mga\cos\theta = \tfrac{1}{2}m(2a)^2\dot{\theta}^2 - 2mga\cos\theta = \text{constant}.$$

The initial conditions give the constant to be $-mga$ so that

$$2a^2\dot{\theta}^2 = ga(2\cos\theta - 1)$$

$\dot{\theta}$ vanishes when $\cos\theta = \frac{1}{2}$ so that the greatest height reached above the lowest point is a.

In this case, it is obvious that R is positive when θ is acute and so the particle does not leave the cylinder.

EXAMPLE 2 *A particle P is free to move on the outer surface of a smooth circular cylinder of radius a, and is released from rest at a depth of a/10 below the highest point. Find the height above the centre of the cylinder of the point at which P leaves the cylinder, and the vertical distance that P falls, after leaving the cylinder, whilst moving a horizontal distance a.*

The situation is depicted in Fig. 10.10, with R denoting the reaction of the cylinder acting outwards. The energy equation gives

$$\tfrac{1}{2}a^2\dot{\theta}^2 + ga\cos\theta = \text{constant.}$$

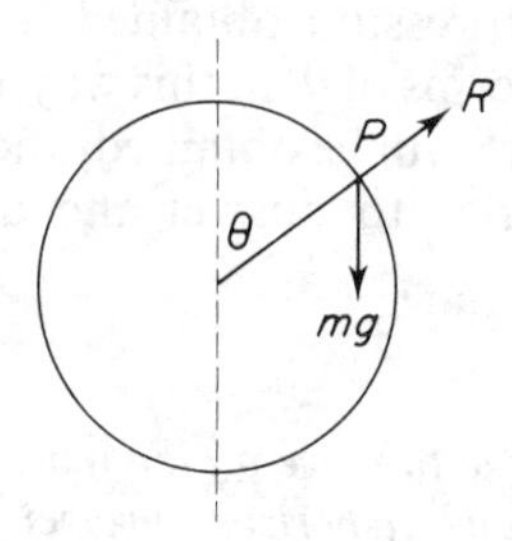

Fig. 10.10

Initially, $\dot{\theta} = 0$ when $\cos\theta = 9/10$ so that the constant is equal to $9ga/10$. Therefore,

$$5a^2\dot{\theta}^2 + 10ga\cos\theta = 9ga.$$

The radial equation of motion gives

$$ma\dot{\theta}^2 = mg\cos\theta - R$$

so that

$$5R = 15mg\cos\theta - 9ga$$

This vanishes when $\cos\theta = 3/5$, that is, when P is at a distance $\mathbf{2a/5}$ above the centre. At this instant the speed of P is $(3ga/5)^{\frac{1}{2}}$.

Once P has left the cylinder it moves as a projectile and its initial velocity is $(3ga/5)^{\frac{1}{2}}$ at an angle $\cos^{-1}(3/5)$ below the horizontal. The horizontal component of velocity, which remains constant, is $\frac{3}{5} \times \left(\frac{3ga}{5}\right)^{\frac{1}{2}}$ and the time taken to move a distance a horizontally is $(5^3 a/3^3 g)^{\frac{1}{2}}$. If y is measured upward from the point of projection then the standard projectile formulae give

$$y = -\left(\frac{3ga}{5}\right)^{\frac{1}{2}}\frac{4}{5}t - \frac{1}{2}gt^2$$

and substituting the value of t into this expression gives the distance dropped to be $\mathbf{(197a/54)}$.

EXERCISE 10.4

In numerical exercises take $g = 10\,\mathrm{ms}^{-2}$ and give answers correct to two significant figures.

Questions 1 and 2 refer to a particle of mass m, attached to one end of a light inextensible string of length l, describing a complete vertical circle of radius l about the other end of the string, which is held fixed.

1 Given that the speed at the highest point is $8gl$, find the speed at the lowest point.

2 $l = 0{\cdot}8\,\mathrm{m}$ and the greatest and least tensions are in the ratio 3:1. Find the greatest speed of the particle.

3 A particle of mass 0·4 kg, attached to a light inextensible string, swings as a pendulum through an arc of 28°. Find the tension in the string when the particle is at its lowest point.

4 A particle of mass $2m$ is attached to one end of a light inextensible string of length a whose other end is fixed at a point O. When in equilibrium at its lowest position, the particle is projected horizontally with speed $(nga)^{\frac{1}{2}}$. Find, while the string remains taut, the tension in the string when it has turned through an angle θ. Determine the range of n so that the string is never slack.

5 Find, in the previous question, the value of n such that the string becomes slack at a distance $\frac{1}{2}a$ above O.

6 A bead of mass m is threaded on a smooth circular loop of wire of radius a which is fixed in a vertical plane. The bead is released from rest at the end of a horizontal diameter. Find the reaction of the wire when the bead has turned through an angle θ.

7 A particle is projected horizontally with speed $\frac{1}{2}(ga)^{\frac{1}{2}}$ from the highest point of a smooth sphere of radius a. Find the height above the centre at which the particle leaves the surface of the sphere.

8 A particle is released from rest at a depth $\frac{1}{2}a$ below the highest point of a smooth sphere of radius a. Find the height above the centre at which the particle leaves the surface of the sphere.

MISCELLANEOUS EXERCISE 10

Unless otherwise stated, in numerical exercises g should be taken as $10\,\mathrm{ms}^{-2}$.

1 A straight horizontal wire is free to rotate about a vertical axis through one end O. A bead of mass m is threaded on the wire and is attached by a light thread, passing through a small smooth ring at O, to a particle of mass $8m$ which is free to move in a vertical plane. The wire is made to rotate with constant angular speed ω about the vertical axis through O so that the hanging particle remains stationary and the bead remains at rest relative to the wire at a distance a from O.

(i) Given that the wire is smooth calculate

(a) the angular speed ω,

(b) the speed of the bead,

(c) the time for one complete rotation of the wire.

(ii) Given that the wire is rough with coefficient of friction $\frac{1}{4}$, determine the permissible range of ω,

(iii) Find the permissible range of ω if the wire is rough as in (ii) and the bead is of mass $8m$ and the hanging particle of mass m. *(AEB 1982)*

2 A smooth rod rotates in a horizontal plane about its mid-point, which is at a vertical distance h below a fixed point A, at a constant angular speed $\sqrt{(2g/h)}$. A light elastic string of modulus $5mg$ and natural length h has one end attached at A and its other end attached to a ring of mass m which is free to slide along the rod. The ring is stationary relative to the rod with the string inclined at an angle $\alpha\ (>0)$ to the vertical. Show that $\cos\alpha = 3/5$ and find
(a) the tension in the string,
(b) the force exerted by the ring on the rod.

The speed of rotation is slowly increased and the string breaks when the tension is $10mg$. Find the angular speed of the rod at the instant when the string is just about to break. *(L)*

3 One end of a light inelastic string of length a is attached to a particle P of mass m, and the other end of the string is attached to a fixed point O. P moves with constant speed in a horizontal circle whose centre C is a fixed point vertically below O. If the angular velocity at which CP rotates is ω, show that $OC = g/\omega^2$.

A second string, also of length a, is fixed to P and C, and the new system rotates about OC with uniform angular velocity Ω and with both strings taut. Find expressions in terms of m, a, ω and Ω for the tensions in OP and PC, and deduce that $\Omega > \omega\sqrt{2}$. *(C)*

4 A solid right circular cone of semi-vertical angle α is fixed with its base on a horizontal table and its vertex V upwards. One end of a light string is attached to a particle of mass m. The other end of the string is attached to a pivot at V. The particle moves on the smooth outer surface of the cone in a horizontal circle with constant angular velocity ω, the length of the string being a. Find in terms of m, g, a, α and ω the tension T in the string and the reaction R between the cone and the particle.
Show that the particle will remain on the cone only if $\omega^2 \leqslant g/(a\cos\alpha)$.
(i) If $\omega^2 = g/(2a\cos\alpha)$ and the string is elastic with modulus mg, find its natural length.
(ii) Show that there is a value of ω for which T and R are equal only if $\alpha > 45°$. If $\sin\alpha = \frac{4}{5}$, find the equal values of T and R. *(C)*

5 A conical pendulum consists of a light string of length l together with a bob of mass m attached to its free end. The bob describes a horizontal circle with constant angular speed ω and the string makes an angle θ with the vertical. Find an expression for ω^2 in terms of g, l and θ. *(JMB)*

6 State the magnitude and the direction of the acceleration of a particle moving at constant speed u in a horizontal circle of radius a and centre O. Prove this statement.

A smooth hemispherical bowl of internal radius R is held with its rim horizontal and uppermost. A particle describes a horizontal circle of radius $3R/5$ on the inner surface of the bowl. Find the period of revolution of the particle.

One end of a light elastic string, of natural length $3R/4$ and modulus of elasticity $2mg$, is attached to the revolving particle whose mass is m. The other end of the string is fixed at C, where C is the centre of the circular rim of the bowl. Given that the path and the speed of the particle are unchanged, find the tension in the string and the normal reaction between the particle and the surface of the bowl. *(AEB 1977)*

7 Particles of mass $3m$ and $2m$ are attached to the ends A and B respectively of a light inextensible string of length a which passes through a small smooth vertical ring at a fixed point P. The ring is free to rotate about the vertical axis through P.

Given that the particle at B describes a horizontal circle about A, which remains at rest, determine the cosine of the angle of inclination of BP to the vertical and the angular speed of rotation of B.

When both particles describe horizontal circles, with the same angular speed, about the vertical axis through P so that at any instant A, B, P lie in a vertical plane, determine

(i) the ratio $\dfrac{AP}{BP}$,

(ii) the cosine of the angle between BP and the vertical when AP is inclined at an angle $\cos^{-1}\left(\dfrac{1}{4}\right)$ to the vertical. (*AEB 1982*)

8 A particle P moves in a circle, centre O and radius r, so that OP has constant angular speed ω. Prove that the acceleration of the particle is $r\omega^2$ and is directed towards O.

One end of a light inextensible string of length $5a$ is tied at a fixed point A which is at a distance $3a$ above a smooth horizontal table. A particle of mass m, which is tied at the other end of the string, rotates with constant speed in a circle on the table. If the reaction between the particle and the table is R, find the tension in the string when (i) $R = 0$, (ii) $R = 3mg/4$.

Show that the respective times of one revolution for these two values of R are in the ratio $1:2$. (*AEB 1976*)

9 A particle of mass m is attached to one end of a light inelastic string of length l, whose other end A is fixed. The particle describes a horizontal circle with constant angular velocity ω. If h is the height of A above the plane of the circle, prove that the tension in the string is mgl/h, and that $h\omega^2 = g$.

Two particles, of masses m_1 and m_2, are attached to the ends of a light inelastic string, which passes through a small fixed horizontal smooth ring. The particles describe horizontal circles with the same angular velocity. Prove that the ring divides the string in the ratio $m_2 : m_1$. (*O*)

10 A light elastic string AB of natural length a and modulus of elasticity mg is joined to a light inextensible string BC of length a. The ends A and C of the string are fastened to two fixed points with A vertically above C and $AC = a$. A particle of mass m is fixed at B and rotates with speed v in a horizontal circle. Show that if AB makes an angle $\pi/6$ with the downward vertical then

$$v^2 = \tfrac{1}{2}\sqrt{3ag},$$

and find the tension in BC. (*O*)

11 The figure shows a particle P of mass $2m$ which is attached to a fixed point O by a light inextensible string of length l. A ring R of mass $3m$, which is attached to P by another light inextensible string of length l, is free to slide on a smooth vertical wire passing through O. The plane OPR rotates about the wire with constant angular velocity ω, where $\omega^2 > 4g/l$. Show that $OR = 8g/\omega^2$.

What happens if $\omega^2 \leqslant 4g/l$? (*C*)

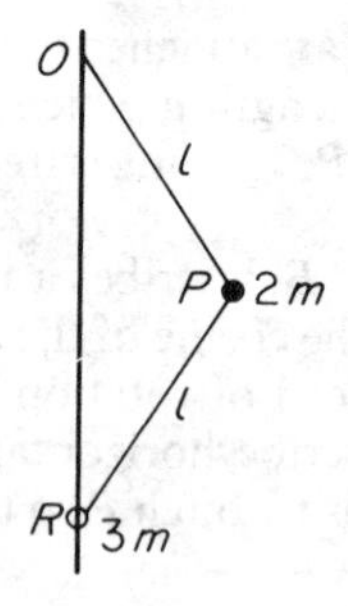

12 A particle is moving at constant speed u in a horizontal circle of radius a and centre O. Prove that the acceleration of the particle is of magnitude u^2/a and is directed towards O.

A smooth ring R of mass m is threaded on to a smooth fixed vertical pole on which it is free to move. One end of a light inextensible string of length $2l$ is tied to the pole at the uppermost point A and the other end is tied to R. A particle P of mass m is attached to the string at its midpoint. With each half of the string taut, P moves in a horizontal circle at constant speed v, and the distance $RA = 6l/5$. It may be assumed that, in this motion, the string does not wrap itself around the pole and that, at any instant, the triangle APR is vertical.

(i) Calculate the tensions, in terms of mg, in the two halves of the string.
(ii) Show that $v^2 = 16gl/5$.
(iii) Find the time, in terms of l and g, for P to complete one revolution.

(*AEB 1979*)

13 A circular cone of semi-vertical angle α is fixed with its axis vertical and its vertex, A, lowest, as shown in the diagram. A particle P of mass m moves on the inner surface of the cone, which is smooth. The particle is joined to A by a light inextensible string AP of length l. The particle moves in a horizontal circle with constant speed v and with the string taut. Find the reaction exerted on P by the cone.

Find the tension in the string and show that the motion is possible only if

$$v^2 > gl \cos \alpha. \qquad (JMB)$$

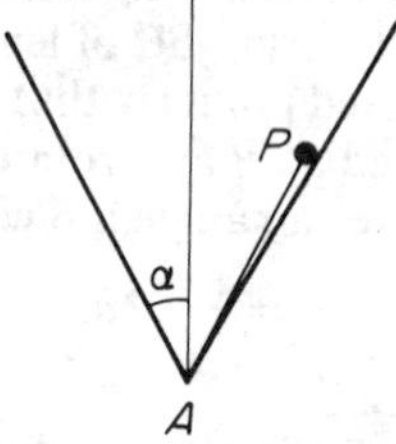

14 A hollow cone with semi-vertical angle α is fixed with its axis vertical and its apex downwards. A particle describes horizontal circles on the inner surface of the cone with constant speed v at a vertical height h above the apex. Find, in terms of g and h, the value of v such that the motion is possible without there being any frictional force acting on the particle.

Given that the coefficient of friction between the particle and the cone is $\frac{2}{3}$ and that $\tan\alpha = \frac{3}{4}$ show that the ratio of the maximum constant speed with which a circle can be described to the minimum constant speed with which a circle can be described is $(51)^{\frac{1}{2}}$.

15 A particle which moves on the inner surface of a fixed, smooth hollow sphere of radius a is attached to the topmost point of the sphere by a light inextensible string of length b, where $a\sqrt{2} < b < 2a$. If the particle describes a circle with angular velocity ω in a horizontal plane, with the string taut, prove that

$$2ga/b^2 \leqslant \omega^2 \leqslant 2ga/(b^2 - 2a^2). \qquad (O)$$

16 The maximum speed at which a car can travel around a horizontal circular bend of radius 100 m without skidding is 63 km/h. Show that the coefficient of friction between the wheels of the car and the road is 5/16.

Calculate the least angle at which the road should be banked in order that the car can negotiate the bend without skidding at 84 km/h, assuming that the coefficient of friction remains unchanged. (*AEB 1972*)

17 (a) A car travels, without skidding, at $63\,\text{km}\,\text{h}^{-1}$ round a circular bend of radius 80 m on a horizontal surface. Show that the coefficient of friction between the wheels and the road is at least 25/64.

(b) A circular bend of radius r is banked at an angle θ such that the maximum speed at which a car can travel around it without skidding is v. If the coefficient of friction between the wheels and the road is μ $(= \tan\lambda)$, show that this maximum speed is given by

$$v^2 = rg\tan(\theta + \lambda). \qquad (W)$$

18 The end O of a light inelastic string OP of length a is fixed and a particle of mass m is attached at the other end P. The particle is held with the string taut and horizontal and is given a velocity u vertically downwards. When the string becomes vertical it begins to wrap itself around a small smooth peg, A, at a depth $a/2$ below O. Find the tension in the string when AP subsequently makes an angle θ with the downward vertical and show that if $u = 0$ the string becomes slack when $\cos\theta = -\frac{2}{3}$.

Find the minimum value of u in order that the particle makes complete revolutions about A. (*AEB 1974*)

19 A rigid wire ABC is fixed in a vertical plane. The portion AB, of length b, is straight and horizontal and BC is a smooth circular arc of centre O (vertically above B), radius a and length $\frac{1}{2}\pi a$. A bead P of mass m is threaded on the wire and projected from B with speed u towards C. Denoting by θ the angle BOP

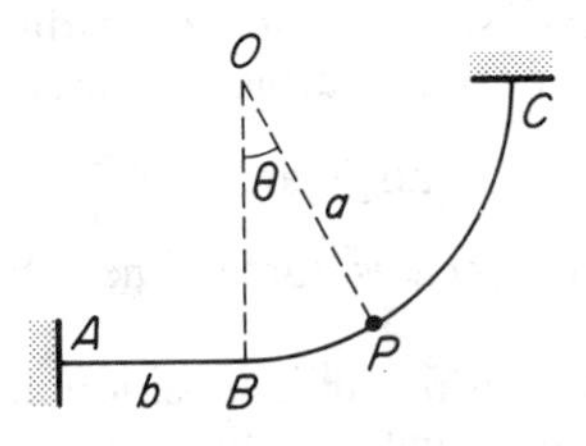

when P is between B and C, write down an equation of motion relating $d^2\theta/dt^2$ to θ.

Show that, while P is moving from B to C,

$$\left(\frac{d\theta}{dt}\right)^2 = \frac{u^2}{a^2} - \frac{2g}{a}(1-\cos\theta).$$

Find the reaction of the wire on the bead in terms of m, g, a, u and θ, indicating clearly its direction.

The bead collides with a fixed stop at C at which the coefficient of restitution is e. Find its speed when it returns to B.

On the straight portion BA the motion is resisted by a constant force F. Show that the bead will reach A if

$$Fb \leqslant \tfrac{1}{2}me^2u^2 + m(1-e^2)ga. \qquad (JMB)$$

20 A particle of mass m is just displaced from the top of a fixed sphere of radius a. If the particle experiences a resisting force of magnitude $\frac{1}{2}mg\sin\theta$ when its angular displacement is θ, find its speed at the instant when it leaves the sphere. (*O*)

21 A smooth circular wire of radius a is fixed in a vertical plane, and a bead A is fixed to the wire at the same horizontal level as the centre. A bead B, which is free to move on the wire, is given a velocity u at the bottom point of the wire, so that it travels towards A. The coefficient of restitution between the beads is e. Prove that if

$$2ga < u^2 < 2ga(1+e^{-2}),$$

B will oscillate back and forth along the wire, passing through the lowest point each time; but that, if

$$2ga(1+e^{-2}) < u^2 < 2ga(1+e^{-4}),$$

B eventually comes to rest touching A. (*O*)

22 A particle P, of mass m, is attached by an inextensible string of length a to a fixed point O and performs complete vertical circles of radius a. At the top of the circle, the speed of P is $\sqrt{(2ag)}$. Show that the string must be able to withstand a tension of magnitude $7mg$.

The particle P is now replaced by another particle Q of mass $2m$. Particle Q is held at a height a above O and given a horizontal velocity of magnitude $\sqrt{(2ag)}$. Given that the string breaks under a tension of magnitude $8mg$, show that the string breaks when it is horizontal.

Given that O is at a height $2a$ above horizontal ground and that the coefficient of restitution between Q and the ground is $1/\sqrt{2}$, show that the greatest height of Q above the plane after impact is $2a$. (*L*)

23 A particle P is projected horizontally with speed u, where $u^2 < ag$, from the highest point A of a fixed smooth sphere of radius a and whose centre is at O. The particle slides on the outer surface of the sphere. Show that it leaves the sphere when PO makes an angle $\arccos\left(\frac{2}{3}+\frac{u^2}{3ag}\right)$ with the vertical.

At this instant the speed is three times the initial speed. Show that $u = \sqrt{(ag/13)}$. (*L*)

24 A particle, of mass m, is at rest in its position of stable equilibrium inside a smooth hollow spherical bowl of radius a and centre O. The particle is now projected with a horizontal velocity u, where $u^2 = \left(2+\frac{3}{\sqrt{2}}\right)ga$. Show that when the particle leaves the bowl its speed v is given by $v^2 = \dfrac{ga}{\sqrt{2}}$.

Axes Ox and Oy are now taken in the plane of the motion such that Ox is horizontal and points in the direction of the initial velocity u, and Oy points vertically upwards. Find the coordinates of the point at which the particle leaves the bowl and show that the particle now moves along the parabola

$$y = \frac{a}{\sqrt{2}} + x - \sqrt{2}\frac{x^2}{a}.$$

Verify that the particle next meets the bowl at the point $\left(-\dfrac{a}{\sqrt{2}}, -\dfrac{a}{\sqrt{2}}\right)$.

(*AEB 1982*)

25 A particle of mass m is attached to the end A of a light inextensible string AB of length a. The end B is attached to a fixed point. Initially the end A is held vertically above B with the string taut and the particle is projected horizontally with a speed V, where $V > \sqrt{(ag)}$. Show that the speed of A when it is directly below B is

$$\sqrt{(V^2 + 4ag)}.$$

At this instant the particle collides and coalesces with a second particle, also of mass m, which is stationary. Find the tension when the string attached to the composite particle is at an angle θ to the vertical. Show also that if $V^2 > 4ag$ then the composite particle makes complete revolutions. (*L*)

26 A smooth narrow tube is bent into the form of a circle, centre O and radius a, fixed in a vertical plane. Two particles A and B, of masses m and λm ($\lambda > 1$) respectively, are connected by a light inextensible string of length πa. The particles are initially at rest in the tube at opposite ends of the horizontal diameter. The string is taut and occupies the upper half of the tube.

The particles are released and after time t the diameter AOB makes an angle θ with the horizontal. Given that the string remains taut during the motion,

(i) show that $a\,\dot{\theta}^2 = \dfrac{2(\lambda - 1)g \sin\theta}{(\lambda + 1)}$,

(ii) find in terms of θ

(a) the force exerted by the tube on A,

(b) the tension in the string. (*AEB 1980*)

11 General Particle Motion.

In this chapter, the formal definitions of impulse, momentum and energy are extended to cover general particle motions. The concepts of impulse and momentum are generalised in §11.1, and applied in §11.2 to problems involving oblique impact with smooth planes. In §11.3, the definitions of kinetic energy, power and work appropriate for non-rectilinear motion are introduced, and in §11.4, some of the conditions under which energy is conserved are established. In §11.5, the solution of simple problems involving vector quantities which have three non-zero independent components is illustrated.

11.1 Impulse and linear momentum

The motion of a particle of mass m moving under the action of a force $\mathbf{F}$ is governed by Newton's second law (equation 6.3) which may be written as

$$m\frac{\mathrm{d}^2\mathbf{r}}{\mathrm{d}t^2} = \mathbf{F}, \qquad 11.1$$

where $\mathbf{r}$ is the position vector of the particle. Integrating this equation from time $t = t_1$ to time $t = t_2$ gives

$$m\left(\frac{\mathrm{d}\mathbf{r}}{\mathrm{d}t}\right)_{t=t_2} - m\left(\frac{\mathrm{d}\mathbf{r}}{\mathrm{d}t}\right)_{t=t_1} = \int_{t_1}^{t_2} \mathbf{F}\,\mathrm{d}t. \qquad 11.2$$

In Chapter 5, the integral with respect to time of the component of force along a line was defined as the component of the impulse of the force, and this is now generalised for non-rectilinear motion by defining the impulse $\mathbf{I}$ of a force $\mathbf{F}$ acting from $t = t_1$ to $t = t_2$ by

$$\mathbf{I} = \int_{t_1}^{t_2} \mathbf{F}\,\mathrm{d}t. \qquad 11.3$$

Again by analogy with rectilinear motion, the linear momentum of the particle is defined to be $m\,\dot{\mathbf{r}}$ so that equation 11.2 becomes

change in linear momentum = impulse of force. 11.4

This is the same result as that obtained previously for rectilinear motion. For problems involving impact and collisions, the force will not be known but the impulse can be found from the change of momentum, as for rectilinear motion.

Linear momentum as defined above is more obviously a vector than as defined in Chapter 5. This is because in Chapters 4 and 5 it was assumed that a specific reference direction was always used, so that $\dot{x}$, for example, was the component of velocity in the direction of x increasing. The actual velocity is $\dot{x}\mathbf{i}$ but, to avoid complications, the unit vector $\mathbf{i}$ is omitted when dealing with problems of motion in a straight line. Similarly, the linear momentum is $m\dot{x}\mathbf{i}$, and for planar motion we have

$$\text{linear momentum} = m(\dot{x}\mathbf{i} + \dot{y}\mathbf{j}).$$

When $\mathbf{F}$ is a relatively simple function of time, equation 11.2 can be integrated directly twice to determine $\mathbf{r}$ as a function of time. The particular case of $\mathbf{F}$ constant has already been onsidered in §9.4.

EXAMPLE 1 *Find the impulse of the force* $(\mathbf{i}e^{-t} + \mathbf{j}\cos t)$ N *acting from* $t = 0$ s *to* $t = (\pi/2)$ s.

The impulse is the time integral of the force, that is

$$\int_0^{\pi/2} (\mathbf{i}e^{-t} + \mathbf{j}\cos t)\,dt \quad \text{Ns} = [\mathbf{i}(1 - e^{-\pi/2}) + \mathbf{j}]\ \text{Ns}.$$

EXAMPLE 2 *A particle P of mass* 2 kg *is moving under the action of a force, which at time t* s *is* $8(3t\mathbf{i} + 5t^3\mathbf{j})$ N. *At time* $t = 0$ *the particle is moving through the origin with velocity* $(\mathbf{i} + 2\mathbf{j})$ ms^{-1}. *Find its position at any subsequent time.*

The equation of motion is

$$\frac{2\mathrm{d}^2\mathbf{r}}{\mathrm{d}t^2} = 8(3t\mathbf{i} + 5t^3\mathbf{j}),$$

where the position vector of P is denoted by $\mathbf{r}$ m. Integrating this and using the initial velocity gives

$$\frac{\mathrm{d}\mathbf{r}}{\mathrm{d}t} = (\mathrm{i} + 2\mathrm{j}) + 6t^2\mathbf{i} + 5t^4\mathbf{j}.$$

A further integration gives

$$\mathbf{r} = (\mathbf{i} + 2\mathbf{j})t + 2t^3\mathbf{i} + t^5\mathbf{j}.$$

EXERCISE 11.1

Questions 1 to 6 refer to a particle of mass m kg acted upon by a force $\mathbf{P}$ N for time t s. The initial velocity of the particle is $\mathbf{u}$ ms^{-1} and its velocity after t s in $\mathbf{v}$ ms^{-1}.

1 $\mathbf{P} = \mathbf{i} + 2\mathbf{j}$, $m = 0{\cdot}2$, $t = 1$, $\mathbf{u} = 3\mathbf{i} - \mathbf{j}$; find $\mathbf{v}$.

2 $m = 0{\cdot}2$, $\mathbf{u} = 5\mathbf{i} + 3\mathbf{j}$, $\mathbf{v} = 2\mathbf{i} - \mathbf{j}$; find the impulse of the force.

3 $\mathbf{P} = t\mathbf{i} - 3t^2\mathbf{j}$, $m = 0{\cdot}2$, $\mathbf{u} = 5\mathbf{i} + \mathbf{j}$, $t = 2$; find $\mathbf{v}$.

4 $\mathbf{P} = 6t^2\mathbf{i} + 4t^3\mathbf{j}$, $\mathbf{u} = \mathbf{i}$, $\mathbf{v} = 9\mathbf{i} + 4\mathbf{j}$, $t = 1$; find m.

5 The impulse of the force is $11\mathbf{i} + 4\mathbf{j}$ Ns, $\mathbf{u} = 2\mathbf{i} - 14\mathbf{j}$, $m = 0{\cdot}5$; find $\mathbf{v}$.

6 $\mathbf{P} = 6t\mathbf{i} + 12t^2\mathbf{j}$, $m = 2$, $\mathbf{u} = 2\mathbf{i} + 3\mathbf{j}$; find the position vector, relative to its initial position, of the particle after 2 s.

11.2 Oblique impact with smooth planes

We are now in a position to extend the work of §7.2 to cover small spheres colliding obliquely with a smooth plane. In such a collision, the impulse of the plane has, by the definition of a smooth plane, to be normal to its surface. It follows from equation 11.4 that there is no change in momentum (and, hence, velocity) parallel to the plane. Newton's experimental law will still be satisfied normal to the surface and, therefore, for oblique collisions the conditions to apply are

(i) $\dfrac{\text{speed normal to the plane after impact}}{\text{speed normal to the plane before impact}} = e,$

(ii) velocity parallel to the wall is unchanged.

EXAMPLE 1 *A smooth sphere, moving with speed* $10\,\text{ms}^{-1}$, *collides with a smooth wall. The coefficient of restitution for collisions between the sphere and the wall is* $\frac{1}{2}$, *and at the instant of collision the sphere is moving at an angle of* 30° *to the wall. Find the speed and the direction of motion of the sphere after collision.*

Fig. 11.1 shows the sphere moving after collision with speed $v\,\text{ms}^{-1}$ at an angle θ to the wall. The velocity component parallel to the wall is unchanged so that

$$v\cos\theta = 10\cos 30^\circ = 5\sqrt{3}.$$

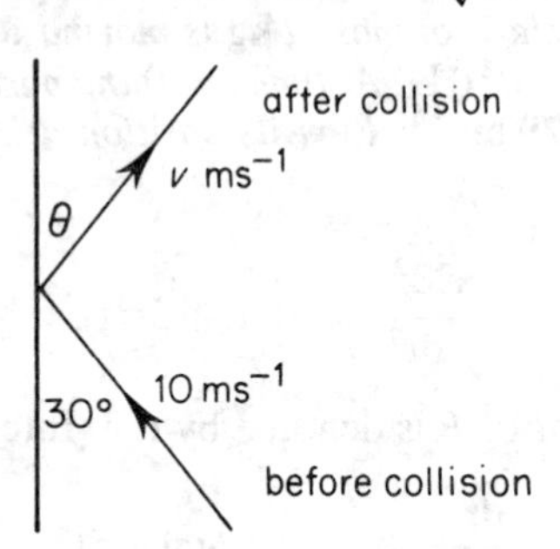

Fig. 11.1

Applying Newton's law to the motion normal to the wall gives

$$v\sin\theta = \tfrac{1}{2}\times 10\sin 30^\circ = \tfrac{5}{2}.$$

Dividing the equations gives $\tan\theta = 1/(2\sqrt{3}) \Rightarrow \theta = \mathbf{16{\cdot}1^\circ}$, whilst squaring and adding the equations gives $v = \mathbf{5\sqrt{13}/2}$.

EXAMPLE 2 *A small smooth sphere moving in the x, y plane collides with a smooth wall passing through the line x = 0. At the instant of collision the sphere has velocity* $(4\mathbf{j} - 3\mathbf{i})\,\text{ms}^{-1}$. *Given that the coefficient of restitution is* $\frac{1}{3}$ *find the velocity of the sphere immediately after collision.*

Fig. 11.2 shows the sphere moving after collision with velocity components of magnitude $v\,\text{ms}^{-1}$ and $u\,\text{ms}^{-1}$ parallel and perpendicular to the wall respectively.

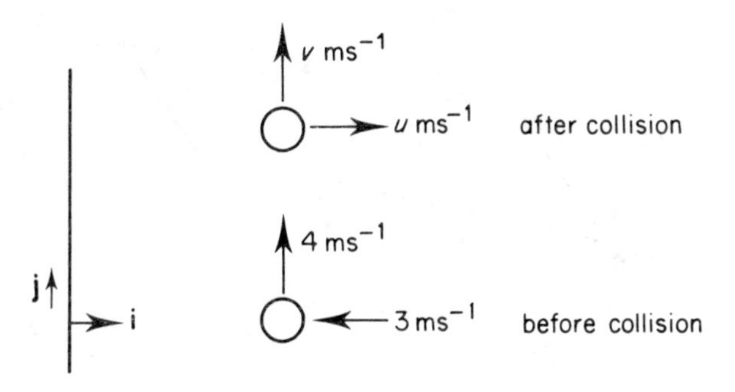

Fig. 11.2

Newton's law gives $u = 1$ and v remains as 4, so that the velocity after collision is $(\mathbf{i} + 4\mathbf{j})$ $\mathbf{ms^{-1}}$.

EXERCISE 11.2

1 A ball travelling at $16\,\text{ms}^{-1}$ strikes a smooth wall at an angle of $40°$. The coefficient of restitution is 0·3. Show that immediately after collision the ball is moving at an angle of approximately $14°$ to the wall with a speed of approximately $12{\cdot}48\,\text{ms}^{-1}$.

2 A ball which strikes a smooth wall is moving at an angle of $60°$ to the wall before collision and at an angle of $30°$ to the wall immediately after collision. Find the coefficient of restitution.

3 A ball strikes a smooth horizontal plane when moving at an angle of $70°$ to the plane, the coefficient of restitution being 0·5. Find, to the nearest degree, the angle its direction of motion makes with the plane immediately after the first and the second bounces.

4 A small smooth sphere of mass 2 kg moving in the x, y plane collides with a smooth wall containing the line $x = 0$. At the instant of collision the sphere has velocity $(-8\mathbf{i} + 7\mathbf{j})\,\text{ms}^{-1}$. Given that the coefficient of restitution is 0·25 find the velocity of the sphere immediately after collision and also the impulse given to the sphere by the wall.

5 A small smooth sphere moving in the x,y plane collides with a smooth wall perpendicular to the vector $3\mathbf{i} + 4\mathbf{j}$. At the instant of collision the sphere has velocity $(5\mathbf{i} - 15\mathbf{j})\,\text{ms}^{-1}$ and the coefficient of restitution is $\frac{1}{3}$. Find the velocity of the sphere immediately after collision.

11.3 Work, power, energy

When the point of application of a constant force $\mathbf{F}$ is moved (see Fig. 11.3) from A to B, where $\overrightarrow{AB} = \mathbf{d}$, then the work done is *defined* to be $\mathbf{F}\,.\,\mathbf{d}$. This means that, when the angle between $\mathbf{F}$ and $\mathbf{d}$ is θ, the work done is equal to $Fd\cos\theta$. Note that $F\cos\theta$ is the magnitude of the component of $\mathbf{F}$ along $\mathbf{d}$. Therefore, the work done is the product of the displacement and the component of the force in the direction of the displacement. This

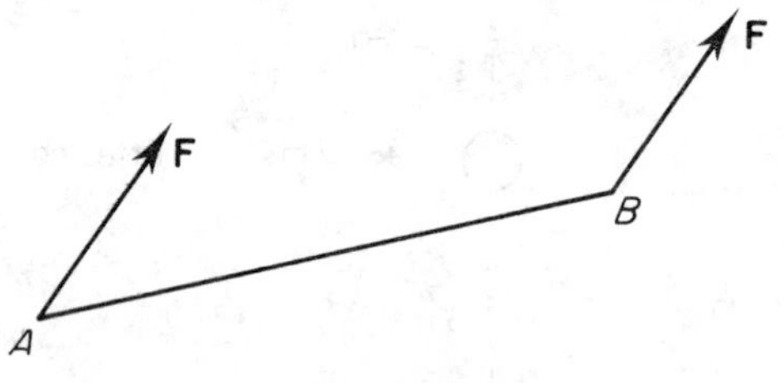

Fig. 11.3

is consistent with the definition given in Chapter 5. In fact, this definition of work using the scalar product is simply a generalisation of that for rectilinear motion.

When the force is no longer constant and an infinitesimal displacement $\delta\mathbf{r}$ is made then, from the above definition, the work δW done during this displacement is given by $\delta W = \mathbf{F}.\delta\mathbf{r}$. The total work done in moving the point of application by a finite distance could be worked out by adding up the work done in each infinitesimal displacement. However, it requires a more complicated definition of an integral than is available at this level and so an indirect approach will be used. If the displacement $\delta\mathbf{r}$ takes place in time δt then, from the above definition, the rate of working $\mathrm{d}W/\mathrm{d}t$, which is the power, is given by

$$\text{power} = \frac{\mathrm{d}W}{\mathrm{d}t} = \lim_{\delta t \to 0} \frac{\delta W}{\delta t} = \lim_{\delta t \to 0} \mathbf{F}.\frac{\delta \mathbf{r}}{\mathrm{d}t} = \mathbf{F}.\dot{\mathbf{r}} = \mathbf{F}.\mathbf{v} \qquad 11.5$$

This definition of power is again a generalisation of that in Chapter 5. When $\mathbf{F}$ and $\mathbf{v}$ are known as functions of time, then equation 11.5 can be integrated over any time interval to find the work done in that interval.

The kinetic energy of a particle was defined for rectilinear motion as one half the product of the mass of the particle and the square of its speed. This is also the definition of the kinetic energy for general motion. The speed is now $|\dot{\mathbf{r}}|$ so that

$$\text{kinetic energy} = \tfrac{1}{2}m\dot{\mathbf{r}}^2 = \tfrac{1}{2}m\left(\frac{\mathrm{d}\mathbf{r}}{\mathrm{d}t}\right)^2 \qquad 11.6$$

Differentiating equation 11.6 with respect to time gives, on using the product rule for differentiating vectors (§8.1)

$$\frac{\mathrm{d}}{\mathrm{d}t}(\text{kinetic energy}) = m\dot{\mathbf{r}}.\ddot{\mathbf{r}} = \mathbf{F}.\dot{\mathbf{r}}. \qquad 11.7$$

It also follows from this equation and the definition of power that

$$\frac{\mathrm{d}}{\mathrm{d}t}(\text{kinetic energy}) = \text{power} = \frac{\mathrm{d}W}{\mathrm{d}t}. \qquad 11.8$$

Integrating this equation over any time interval gives

$$\begin{array}{lcl}\text{change in kinetic energy} & = & \text{work done, over that interval,} \\ \text{over any interval} & & \text{by all the forces acting.}\end{array} \qquad 11.9$$

This result is the same as that deduced in Chapter 5 for rectilinear motion. When several particles are connected together, equation 11.9 holds for each particle separately provided all the forces, including those of interaction between particles, are taken into account. Therefore, on adding the equations for the separate particle, equation 11.9 holds provided that the right-hand side is the total work done by all the applied and interactive forces. This result is effectively the basis of the principle of conservation of energy which will be discussed in the following section.

EXAMPLE 1 *Find the work done by the force* $(3\mathbf{i}+7\mathbf{j})$ N *when its point of application is moved from the point with position vector* $(5\mathbf{i}+4\mathbf{j})$ m *to that with position vector* $(11\mathbf{i}+6\mathbf{j})$ m.

The displacement of the point of application is

$$(11\mathbf{i}+6\mathbf{j})\,\text{m} - (5\mathbf{i}+4\mathbf{j})\,\text{m} = (6\mathbf{i}+2\mathbf{j})\,\text{m}.$$

Therefore the work done is $(3\mathbf{i}+7\mathbf{j}).(6\mathbf{i}+2\mathbf{j})$ J = **32 J**.

EXAMPLE 2 *The point of application of a time varying force* $(t\mathbf{i}+3t^2\mathbf{j})$ N *where t is the time in seconds, moves so that, at time t* s, *its position vector relative to a fixed origin O is* $(3t^2\mathbf{i}+t\mathbf{j})$ m. *Find the work done in* 1 s *in moving from the origin to the point with position vector* $(3\mathbf{i}+\mathbf{j})$ m.

This problem has to be solved by first finding the rate of working and then integrating it. The velocity of the point of application is $(6t\mathbf{i}+\mathbf{j})\,\text{ms}^{-1}$ and therefore the power is $(t\mathbf{i}+3t^2\mathbf{j}).(6t\mathbf{i}+\mathbf{j})\text{W} = 9t^2\,\text{W}$. Integrating this from $t = 0$ to $t = 1$ gives the work done to be **3 J**.

EXAMPLE 3 *A particle of mass* 2 kg *is acted on by a variable force so that its position vector, at time t* s, *relative to a fixed origin O is* $(t^3\mathbf{i}+t^2\mathbf{j})$ m. *Find the work done by the force in the time interval* $0 \leqslant t \leqslant 2$.

The force is not given, but as the position vector of the particle is given, it could be found from Newton's law. In this case it is, however, much easier to find the kinetic energy and then use equation 11.9. As t varies from 0 to 2, the velocity changes from zero to $(12\mathbf{i}+4\mathbf{j})\,\text{ms}^{-1}$ and, therefore, the change in the kinetic energy is $\frac{1}{2}.2(12\mathbf{i}+4\mathbf{j})^2$ J = **160 J**, which is the work done.

EXERCISE 11.3

1 Find the kinetic energy at time T s of a particle of mass m kg whose position vector at time t s, relative to a fixed origin, is $\mathbf{r}$ m when
(a) $m = 2$, $\mathbf{r} = 3t\mathbf{i}+7t\mathbf{j}$.
(b) $m = 2{\cdot}5$, $\mathbf{r} = 3t^2\mathbf{i}+t^4\mathbf{j}$, $T = 2$.
(c) $m = 4$, $\mathbf{r} = e^t\mathbf{i}+2e^{-t}\mathbf{j}$, $T = \ln 2$.

2 Find the work done by a force $\mathbf{F}$ N moving from $\mathbf{r} = \mathbf{r}_1$ m to $\mathbf{r} = \mathbf{r}_2$ m when
(a) $\mathbf{F} = 2\mathbf{i}+\mathbf{j}$, $\mathbf{r}_1 = 3\mathbf{i}+4\mathbf{j}$, $\mathbf{r}_2 = 7\mathbf{i}+9\mathbf{j}$.
(b) $\mathbf{F} = 4\mathbf{i}-3\mathbf{j}$, $\mathbf{r}_1 = \mathbf{i}-\mathbf{j}$, $\mathbf{r}_2 = 3\mathbf{i}+\mathbf{j}$.
(c) $\mathbf{F} = 5\mathbf{i}-3\mathbf{j}$, $\mathbf{r}_1 = \mathbf{i}+7\mathbf{j}$, $\mathbf{r}_2 = 2\mathbf{i}+3\mathbf{j}$.

3 Find the work done when the force $\mathbf{F}$ N, whose point of application at time t s has position vector, relative to a fixed origin, of $\mathbf{r}$ m, moves from $\mathbf{r} = \mathbf{r}_1$ to $\mathbf{r} = \mathbf{r}_2$ when
(a) $\mathbf{F} = \mathbf{i}+\mathbf{j}$, $\mathbf{r} = t\mathbf{i}+2t\mathbf{j}$, $\mathbf{r}_1 = \mathbf{i}+2\mathbf{j}$, $\mathbf{r}_2 = 3\mathbf{i}+6\mathbf{j}$.
(b) $\mathbf{F} = 2\mathbf{i}+t\mathbf{j}$, $\mathbf{r} = 3t^2\mathbf{i}+4t\mathbf{j}$, $\mathbf{r}_1 = \mathbf{0}$, $\mathbf{r}_2 = 27\mathbf{i}+12\mathbf{j}$.
(c) $\mathbf{F} = 2t\mathbf{i}+\mathbf{j}$, $\mathbf{r} = \mathbf{i}t + 12t^3\mathbf{j}$, $\mathbf{r}_1 = \mathbf{i}+12\mathbf{j}$, $\mathbf{r}_2 = 2\mathbf{i}+96\mathbf{j}$.

4 A particle of mass 2 kg is acted on by a variable force so that its position vector, at time t s, relative to a fixed origin, is $(3t^2\mathbf{i}+t^6\mathbf{j})$ m. Find the work done by the force as the particle moves from the origin to $12\mathbf{i}+64\mathbf{j}$.

11.4 Conservation of energy

It was shown in Chapter 5 that for rectilinear motion, where the force depends solely on displacement, a first integral of the equation of motion was given by the equation of energy conservation. The derivation of this equation was possible only because, under the above circumstances, the work done by the forces acting was a function of position only. By defining potential energy as the work done by the forces of the system in moving to some standard position, the work done by the forces in moving from one position to another is minus the change in potential energy. Therefore, it follows from equation 11.9 that the total energy change is zero. We now examine some of the conditions under which a potential energy can exist for general motion, and determine some of the conditions under which energy conservation occurs. The general conditions are rather complicated and so, we will only consider some of the simpler examples of forces for which the principle of conservation of energy holds.

Force of gravity

The force of gravity on a particle of mass m is $-mg\mathbf{j}$, where $\mathbf{j}$ is the unit vector in the upward vertical direction. The work done by the force of gravity in a displacement from $x_1\mathbf{i}+y_1\mathbf{j}$ to $x_2\mathbf{i}+y_2\mathbf{j}$ is

$$-mg\mathbf{j}\,.\,[(x_2-x_1)\mathbf{i}+(y_2-y_1)\mathbf{j}] = -mg(y_2-y_1).$$

Defining the potential energy, as in Chapter 5, as mg times the height above some standard reference position, we see that the work done is (potential energy at y_2 − potential energy at y_1). So from equation 11.9,

$$\begin{aligned}\text{kinetic energy at } y_2 - {} &= -\text{potential energy at } y_2 + \\ \text{kinetic energy at } y_1 \quad &\quad\ \ \text{potential energy at } y_1.\end{aligned}$$

Therefore, energy is conserved when gravity is the only force acting.

Reactions on smooth surfaces and at smooth hinges

The reaction of a smooth surface on a particle is, by definition, perpendicular to that surface and, therefore, as the particle moves along the surface the work done is zero. (The force is always perpendicular to the displacement so the scalar product is zero.) Hence, reactions of smooth surfaces make no contribution to the right-hand side of equation 11.9.

Similarly, for a particle fixed to a light rod which can move about a smooth hinge, the reaction of the hinge on the rod does no work (it does not move) and makes no contribution to the right-hand side of equation 11.9. Therefore, reactions of smooth surfaces and at smooth hinges can be regarded as corresponding to zero potential energy.

Forces in light inextensible strings

For a system involving particles connected by a light inextensible string, the work done by the tensions at both ends of the string have to be taken into account in equation 11.9. It will now be demonstrated that this work is also zero, so that forces in such strings can be treated as corresponding to zero potential energy.

Fig. 11.4 shows a light inextensible string, not necessarily straight, with the string parallel at its ends A and B to unit vectors **a** and **b** respectively, in the senses shown. The string, if it is to be taut and not straight, must be in contact with some smooth surface such as a peg, pulley or inclined plane. Contact with smooth surfaces only ensures that the tension has the same value T throughout its length.

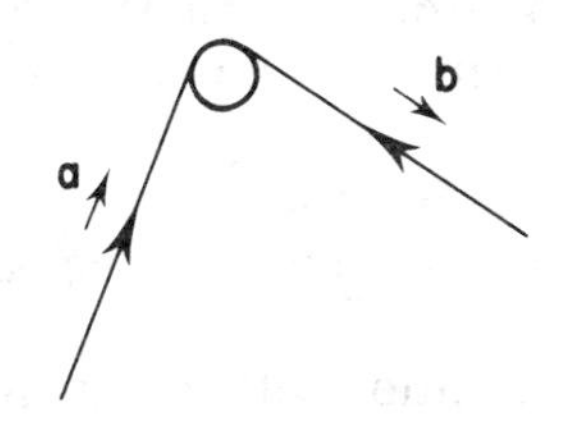

Fig. 11.4

When B is displaced, the component of its displacement $\mathrm{d}x$ parallel to **b** must, since the string is inextensible, be equal to the component of the displacement of A parallel to **a**. Since the tensions at A and B are parallel to **a** and **b** respectively, only the components of displacement along **a** and **b** at A and B respectively need be taken into account in determining the work done. This is, therefore, equal to

$$T\mathbf{a}\,.\,\mathbf{a}\mathrm{d}x - T\mathbf{b}\,.\,\mathbf{b}\mathrm{d}x = 0,$$

as stated above.

Forces in light extensible strings (or springs)

For an extensible string, the displacements of A and B parallel to **a** and **b** respectively will be of the form $\mathrm{d}x\mathbf{a}$ and $\mathrm{d}y\mathbf{b}$, where $\mathrm{d}x$ and $\mathrm{d}y$ are not obviously related to one another, though $\mathrm{d}y-\mathrm{d}x$ will be the change $\mathrm{d}\varepsilon$ in the extension of the string. The work done is $T\mathbf{a}.\mathbf{a}\mathrm{d}x - T\mathbf{b}.\mathbf{b}\mathrm{d}y = -T\mathrm{d}\varepsilon$. For a string of elastic modulus λ and natural length l, the tension is $\lambda\varepsilon/l$ and so the work done is $-\lambda\varepsilon\,\mathrm{d}\varepsilon/l$. Therefore, the total work done as the extension changes from ε_1 to ε_2 is

$$-\int_{\varepsilon_1}^{\varepsilon_2} \frac{\lambda\varepsilon}{l}\,\mathrm{d}\varepsilon = \frac{-\lambda}{2l}(\varepsilon_2^2-\varepsilon_1^2).$$

Defining, as in Chapter 5, the potential energy as $\lambda\varepsilon^2/2l$ shows that the work done is

potential energy at extension ε_1 − potential energy at extension ε_2.

Therefore, equation 11.9 becomes

kinetic energy at extension ε_2 − kinetic energy at extension ε_1 = potential energy at extension ε_1 − potential energy at extension ε_2.

Hence, energy is again conserved.

Summary For systems involving particles moving on smooth surfaces, possibly under the action of gravity, and of forces exerted by light extensible or inextensible strings or possibly in contact with smooth surfaces, the total kinetic and potential energy is conserved with
(a) the gravitational potential energy of a particle of mass m being defined as mg times its height above some reference position,
(b) the potential energy of a light elastic string, of modulus λ and natural length l, being defined as λ (extension)$^2/2l$,
(c) the other forces mentioned not making any contribution to the potential energy.

The principle of conservation of energy is particularly useful in problems involving the above systems when some relation is required between speed and position.

EXAMPLE 1 *A light rod OA of length $2a$ is free to rotate in a vertical plane about a smooth horizontal axis through O. A particle of mass m is attached to the end A and the rod released from rest with OA horizontal. Find the speed of the particle when OA is vertical.*

Fig. 11.5 shows the rod in the position where OA makes an angle θ to the downward vertical. The conditions for energy to be conserved are satisfied and, taking the zero of potential energy at the position where OA is horizontal, we have that

$$\tfrac{1}{2}mv^2 - mga\cos\theta = \text{constant},$$

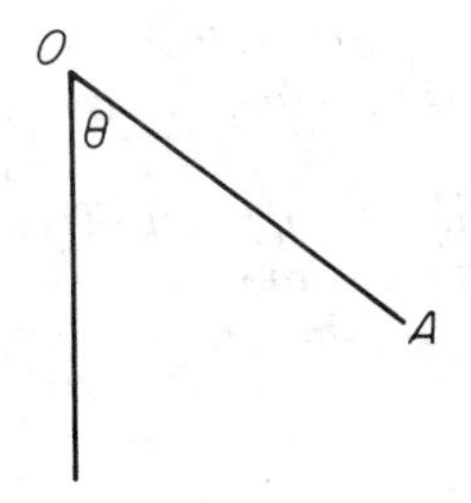

Fig. 11.5

where v is the particle speed. When $\theta = \frac{1}{2}\pi$ the speed is zero, so the constant is zero, and, therefore, when $\theta = 0$ the speed is $(2ga)^{\frac{1}{2}}$.

EXAMPLE 2 *A light rod OA of length 4a is free to rotate on a smooth horizontal table about a fixed vertical axis through O and a particle of mass m is attached at the mid-point of the rod. A light elastic string, of modulus mg and natural length a, is attached to A and the other end is fastened to a fixed point B at a distance 4a from O. Initially, the rod is held with the angle BOA = 60°. The rod is then released. Find its angular speed when the string just becomes slack.*

Fig. 11.6 shows the configuration when angle $BOA = 2\theta$. The triangle BOA is isosceles, so $AB = 8a \sin\theta$. Energy is conserved and, therefore,

$$\tfrac{1}{2}mv^2 + \frac{mg}{2a}(8a\sin\theta - a)^2 = \text{constant},$$

where v is the speed of the particle P.

Initially $v = 0$ and $\theta = 30°$, so the constant is $9mga/2$. The string becomes slack when $\sin\theta = \frac{1}{8}$, and the speed is $(9ga)^{\frac{1}{2}}$. P is describing a circle of radius $2a$ and, therefore, the angular speed of the rod is $\dfrac{3}{2}\left(\dfrac{a}{g}\right)^{\frac{1}{2}}$.

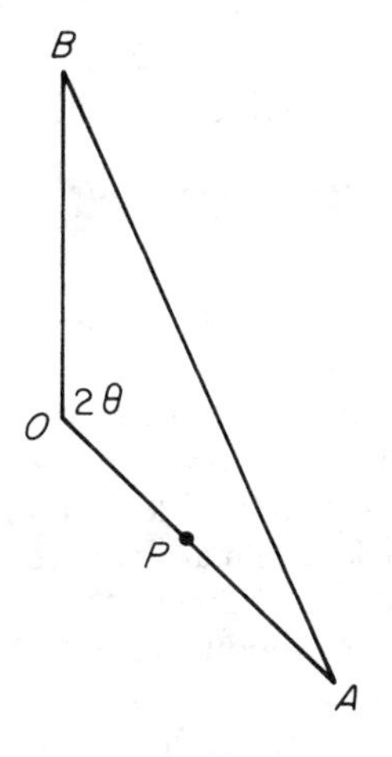

Fig. 11.6

EXAMPLE 3 *Fig. 11.7 shows particles P and Q, of masses m and 3m respectively, connected by a light inextensible string which passes over a smooth circular pulley of radius a, which is fixed with its axis horizontal. Initially, the system is at rest with P held at the point on the pulley directly opposite the centre O, and Q hanging freely. P is then released and after time t the radius vector* $\overrightarrow{OP}$ *has turned through an angle* θ. *Assuming that P remains on the cylinder, find its speed in terms of a, g and* θ.

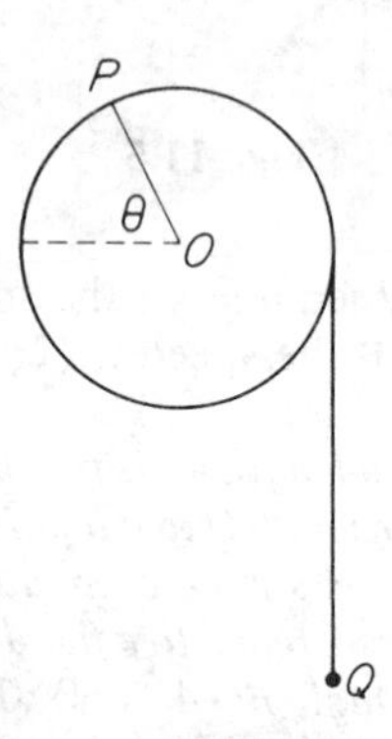

Fig. 11.7

The only force that does work is gravity, so energy is conserved. The potential energy of P, taking the zero point to be the initial position, is $mga \sin \theta$. When the radius vector has turned through an angle θ, a length $a\theta$ of string has left the cylinder, and so Q has dropped a distance $a\theta$ and its potential energy is $-3mga\theta$. Both particles have the same speed v and, therefore,

$$2mv^2 + mga \sin \theta - 3mga\theta = \text{constant}.$$

The constant is found, from the initial conditions, to be zero, so that

$$2v^2 = ga(3\theta - \sin \theta).$$

$$v = \tfrac{1}{2}[ga(3\theta - \sin \theta)]^{\frac{1}{2}}$$

EXERCISE 11.4

In numerical exercises, take $g = 10\,\text{ms}^{-2}$ and give answers correct to three significant figures.

1 A particle of mass 0·2 kg, threaded on a smooth curved piece of wire as shown in Fig. 11.8(a), is released from rest at the point A level with the point O. Find its speed when at the point B on the wire, 0·45 m directly below O.

2 Find the speed at B in question 1 when the particle is also attached to a light spring of modulus 20 N and natural length 0·2 m, given that OA is 0·65 m.

3 Fig. 11.8(b) shows a particle A of mass 4 kg free to slide down a smooth tube, which is inclined at an angle of 30° to the horizontal. The particle is also attached to a light spring of modulus 200 N and natural length 0·5 m. The other end of the spring is attached to a fixed point O and OA and the tube lie in the same vertical plane. At the point B, where $OB = 0{\cdot}6$ m, A has speed $2\,\text{ms}^{-1}$. Find its speed at C where $OC = 0{\cdot}7$ m and $BC = 1{\cdot}2$ m.

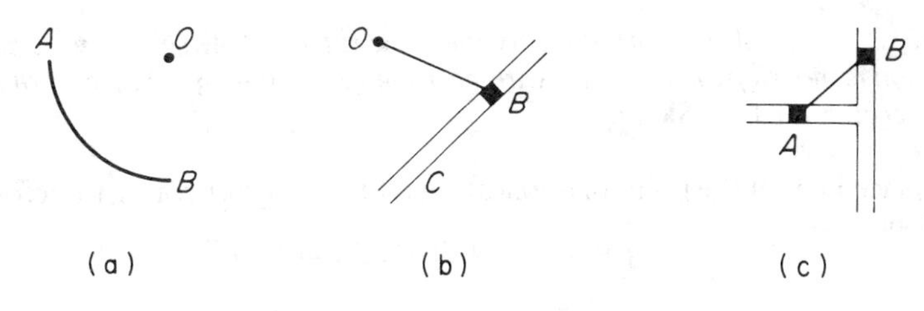

Fig. 11.8

4 A ring R of mass 6 kg can slide on a vertical rod and is also attached to a light spring of elastic modulus 80 N and natural length 0·2 m whose other end is attached to a fixed point O. Initially, the ring is held at rest with OR horizontal and of length 0·3 m. When the ring is released, find its speed after dropping a distance 0·4 m.

5 Fig. 11.8(c) shows two particles A and B of equal mass, free to slide in smooth horizontal and vertical guides as shown. They are joined by a light rod of length 0·4 m and initially held with AB inclined at an angle 30° to the horizontal. After they are released, find the speed of A when AB first becomes horizontal.

6 A heavy uniform inextensible string of length l is held with a length $3l/4$ hanging freely over an edge of a smooth horizontal table. The string all lies in one vertical plane. After it is released, find its speed when just clear of the table.

11.5 Further examples

In this section, we conclude with some further examples of particle motion in which the forces are given in a form involving non-zero components in three independent directions. The principles involved are exactly as in the preceding examples, the only difference being that general three-dimensional vectors are now involved.

EXAMPLE 1 *Find the impulse of the force* $(2t\mathbf{i}+3t^2\mathbf{j}+\mathbf{k})$ N, *where t is the time in seconds, acting from* 0 s *to* 1 s.

By equation 11.3, the impulse is

$$\int_0^1 (2t\mathbf{i}+3t^2\mathbf{j}+\mathbf{k})\mathrm{d}t \text{ Ns} = \mathbf{(i+j+k)\ Ns}.$$

EXAMPLE 2 *A smooth sphere moving with velocity* $(-5\mathbf{i}+4\mathbf{j}+3\mathbf{k})\,\text{ms}^{-1}$ *collides with a smooth plane occupying the region* $x=0$. *The coefficient of restitution is* $\frac{1}{5}$. *Find the velocity after collision.*

Before collision the component of velocity of the sphere normal to the wall is $5\,\text{ms}^{-1}$ towards it. Therefore, by Newton's law, the normal component is $1\,\text{ms}^{-1}$ away from the wall after collision. The other velocity components are unchanged by the collision, and so the velocity after collision is $\mathbf{(i+4j+3k)\ ms^{-1}}$.

EXAMPLE 3 *Find the work done by the force* $(2\mathbf{i}+3\mathbf{j}+4\mathbf{k})$ N *when its point of application is moved from the point with position vector* $(\mathbf{i}+4\mathbf{j}+3\mathbf{k})$ m *to that with position vector* $(5\mathbf{i}+\mathbf{j}+8\mathbf{k})$ m.

The displacement of the point of application is $(4\mathbf{i}-3\mathbf{j}+5\mathbf{k})$ m and, therefore, the work done is

$$(4\mathbf{i}-3\mathbf{j}+5\mathbf{k}).(2\mathbf{i}+3\mathbf{j}+4\mathbf{k})\,\text{J} = \mathbf{19\,J}.$$

EXAMPLE 4 *A particle of mass* 4 kg *is moving under the action of the force* $16(2\mathbf{i}+\mathbf{j}+\mathbf{k})$ N. *The particle passes through the origin with velocity* $(\mathbf{i}+2\mathbf{j}+5\mathbf{k})\,\text{ms}^{-1}$. *Find its kinetic energy and position* 1 s *later.*

The basic equation of motion is

$$\frac{4\mathrm{d}^2\mathbf{r}}{\mathrm{d}t^2} = 16(2\mathbf{i}+\mathbf{j}+\mathbf{k}),$$

where $\mathbf{r}$ m denotes the position vector at time t s. Integrating this equation gives

$$\frac{\mathrm{d}\mathbf{r}}{\mathrm{d}t} = 4(2\mathbf{i}+\mathbf{j}+\mathbf{k})t+\mathbf{A},$$

where $\mathbf{A}$ is a constant vector which is found, from the initial condition, to be $(\mathbf{i}+2\mathbf{j}+5\mathbf{k})$. Therefore,

$$\frac{\mathrm{d}\mathbf{r}}{\mathrm{d}t} = 4(2\mathbf{i}+\mathbf{j}+\mathbf{k})t+(\mathbf{i}+2\mathbf{j}+5\mathbf{k}).$$

When $t = 1$, the velocity is $(9\mathbf{i}+6\mathbf{j}+9\mathbf{k})\,\text{ms}^{-1}$ and, therefore, the kinetic energy is

$$2\times(9\mathbf{i}+6\mathbf{j}+9\mathbf{k})^2\,\text{J} = \mathbf{396\,J}.$$

Integrating the equation for $\dot{\mathbf{r}}$ and using the initial condition gives

$$\mathbf{r} = 2(2\mathbf{i}+\mathbf{j}+\mathbf{k})t^2+(\mathbf{i}+2\mathbf{j}+5\mathbf{k})t.$$

Therefore, when $t = 1$ the position vector is $\mathbf{(5i+4j+7k)\,m}$.

EXERCISE 11.5

1 Find the impulse of the force $(3t^2\mathbf{i}+4t^3\mathbf{j})$ N, where t is the time in s, acting from $t = 1$ to $t = 2$.

2 Find the work done by a force $\mathbf{F}$ N moving from $\mathbf{r} = \mathbf{r}_1$ m to $\mathbf{r} = \mathbf{r}_2$ m where
(a) $\mathbf{F} = \mathbf{i}+2\mathbf{j}+\mathbf{k}$, $\mathbf{r}_1 = 3\mathbf{i}+4\mathbf{j}+7\mathbf{k}$, $\mathbf{r}_2 = 11\mathbf{i}+5\mathbf{j}+8\mathbf{k}$;
(b) $\mathbf{F} = \mathbf{i}-3\mathbf{j}-\mathbf{k}$, $\mathbf{r}_1 = 4\mathbf{i}+\mathbf{j}+\mathbf{k}$, $\mathbf{r}_2 = -\mathbf{i}+2\mathbf{j}-\mathbf{k}$;
(c) $\mathbf{F} = \mathbf{i}+\mathbf{j}-\mathbf{k}$, $\mathbf{r}_1 = \mathbf{i}+\mathbf{j}+2\mathbf{k}$, $\mathbf{r}_2 = \mathbf{i}+3\mathbf{j}-4\mathbf{k}$.

3 Find the work done by a force $\mathbf{F}$ N, whose point of application at time t s has position vector $\mathbf{r}$, in moving from $\mathbf{r} = \mathbf{r}_1$ to $\mathbf{r} = \mathbf{r}_2$ where
(a) $\mathbf{F} = \mathbf{i}+\mathbf{j}+2\mathbf{k}$, $\mathbf{r} = t\mathbf{i}+2t\mathbf{j}+t\mathbf{k}$, $\mathbf{r}_1 = \mathbf{0}$, $\mathbf{r}_2 = \mathbf{i}+2\mathbf{j}+\mathbf{k}$;
(b) $\mathbf{F} = \mathbf{i}+\mathbf{j}-4t\mathbf{k}$, $\mathbf{r} = t^2\mathbf{i}+3t\mathbf{j}+t^2\mathbf{k}$, $\mathbf{r}_1 = \mathbf{i}+3\mathbf{j}+\mathbf{k}$, $\mathbf{r}_2 = 4\mathbf{i}+12\mathbf{j}+16\mathbf{k}$;
(c) $\mathbf{F} = t\mathbf{i}+t^2\mathbf{j}+t\mathbf{k}$, $\mathbf{r} = t^3\mathbf{i}+t^2\mathbf{j}-t^3\mathbf{k}$, $\mathbf{r}_1 = \mathbf{0}$, $\mathbf{r}_2 = \mathbf{i}+\mathbf{j}-\mathbf{k}$.

4 Find the kinetic energy at time T s of a particle of mass m kg whose position vector at time t s relative to a fixed origin O is $\mathbf{r}$ m where
(a) $m = 4$, $\mathbf{r} = t(\mathbf{i}+\mathbf{j}+\mathbf{k})$;
(b) $m = 2$, $\mathbf{r} = t^2\mathbf{i}+4t\mathbf{j}+3t^3\mathbf{k}$, $T = 2$;
(c) $m = 6$, $\mathbf{r} = \mathbf{i}+4t^2\mathbf{j}+t^3\mathbf{k}$, $T = 3$.

5 A small ball moving with velocity $(6\mathbf{i}-8\mathbf{j}-4\mathbf{k})\,\mathrm{ms}^{-1}$ collides with a smooth plane normal to the vector $\mathbf{i}+2\mathbf{j}+2\mathbf{k}$. The coefficient of restitution is 0·5. Find the velocity immediately after collision.

MISCELLANEOUS EXERCISE 11

Unless otherwise stated, the acceleration due to gravity should be taken as $10\,\mathrm{ms}^{-2}$.

1 At time t the momentum of a particle is parallel to the vector $(\cos 2t)\mathbf{i} + (\sin 2t)\mathbf{j}$, and is of magnitude $2\,\mathrm{kg\,ms}^{-1}$. When $t = 0$, the particle passes through the point with position vector $2\mathbf{i}-2\mathbf{j}$ with acceleration of magnitude $8\,\mathrm{ms}^{-2}$. Show that the mass of the particle is 0·5 kg, and find its position vector when $t = \pi/2$. *(L)*

2 A particle of mass m has the velocity vector $3\mathbf{i}+2\mathbf{j}$ when at the point P whose position vector is $\mathbf{i}+\mathbf{j}$; two seconds later it has the velocity vector $7\mathbf{i}-10\mathbf{j}$ and is at Q. The particle is constrained to move along the curve $\mathbf{r} = (a_1t^2+a_2t+1)\mathbf{i}+(b_1t^2+b_2t+1)\mathbf{j}$ where t seconds is the time since the particle was at P.

Find the values of the constants a_1, a_2, b_1 and b_2 and show that the force acting on the particle is constant.

Calculate, giving your answer as a vector, the moment of the force about the origin when the particle is at P. *(AEB 1970)*

3 A particle of mass 4 units moves on a smooth horizontal table which is in the $\mathbf{i}, \mathbf{j}$ plane. The particle which is at rest at the point $\mathbf{r} = \mathbf{0}$ at time $t = 0$, is acted upon by a force $\mathbf{F}$, where $\mathbf{F} = 24(\mathbf{i}+t\mathbf{j})$ at time t. Find
(i) the Cartesian equation of the path of the particle,
(ii) the velocity of the particle at time $t = 2$ and hence or otherwise, the vector equation of the tangent to the path at this instant.

When $t = 2$ an impulse $-8(\mathbf{i}+\mathbf{j})$ is applied to the particle. As a result the particle splits up into two fragments of equal mass, which move in the directions $\mathbf{i}$ and $-\mathbf{i}+\mathbf{j}$. Calculate the speeds of the fragments after the impulse has been applied. *(AEB 1979)*

4 A particle of mass m is free to move on a horizontal table which is in the $\mathbf{i}, \mathbf{j}$ plane. At time $t = 0$ the particle has position vector $\mathbf{r} = 0$ and velocity vector $\mathbf{v} = u\mathbf{i}$. If a force $2mu(e^t\mathbf{i}+e^{-t}\mathbf{j})$ is acting on the particle, show that the velocity of the particle is $u(3\mathbf{i}+\mathbf{j})$ when $t = \log_e 2$ and find its position vector at time t. Find also the cosine of the angle between the velocities of the particle at times $t = 0$ and $t = \log_e 2$.

Find the magnitude of the moment of the force about the origin at time $t = \log_e 2$.

If, at time $t = \log_e 2$, the force stops acting on the particle, find an expression for the position vector of the particle at time t when $t > \log_e 2$. *(AEB 1976)*

5 The position vector of a particle of unit mass at time t is given by $\mathbf{r} = \mathbf{i}e^{-t}\sin t + \mathbf{j}e^{-t}\cos t$. If the particle is moving under the action of a force $\mathbf{F}$ and of a frictional resisting force given by $-2\dot{\mathbf{r}}$, show that $\mathbf{F} = \lambda\mathbf{r}$ where λ is a scalar constant.

Evaluate $\dot{\mathbf{r}}.\ddot{\mathbf{r}}$, and show that the component of acceleration in the direction of motion at any instant is the negative of the speed at that instant. *(JMB)*

6 A small sphere P of mass m is moving with speed u on a smooth horizontal plane along a straight line AO. At O the sphere collides with a smooth vertical wall which makes an acute angle θ with AO. Show that, if the coefficient of restitution at O is e, the sphere leaves O at an angle ϕ to the wall, where

$$\tan\phi = e\tan\theta.$$

Find the value of θ for which the component, perpendicular to AO, of the velocity after collision is greatest.

A sphere Q of mass $2m$, with the same radius as P, is at rest on the plane and OQ is perpendicular to AO. Show that, if $\cot\theta = \sqrt{e}$, P strikes Q directly and at speed $u\sqrt{e}$. Given that the spheres are perfectly elastic, find the impulse exerted by P on Q. *(JMB)*

7 A smooth horizontal rectangular plane $ABCD$ is surrounded by smooth vertical walls. The lengths of AB and BC are a and b, respectively, and P is a point on AB such that $AP = ka$, where $0 < k < 1$. It is required to project a particle horizontally from P so that it strikes the walls BC, CD and DA in turn and then returns to P. The coefficient of restitution between the particle and the walls is e. Show that the required motion takes place if the angle θ between the direction of the initial motion and PB is such that

$$\cot\theta = \frac{a}{b}\left[1 + \frac{k(1-e)}{e}\right]. \qquad (JMB)$$

8 A small ball of mass $\frac{1}{2}$ kg moving on a smooth plane with velocity $(4\mathbf{i} - 5\mathbf{j})\,\text{ms}^{-1}$ strikes a fixed object. After the impact, which lasts for $\frac{1}{10}$ second, the ball proceeds with velocity $(-2\mathbf{i} + 3\mathbf{j})\,\text{ms}^{-1}$. Assuming the impulsive force during the impact acts uniformly, find this force and the loss of energy due to the impact. *(W)*

9 In this question the units of mass, time and length are the kilogram, second and metre respectively. A particle of unit mass is acted upon at time t by a force defined by $\mathbf{F} = 2\mathbf{i} - 4t\mathbf{j} + 2t\mathbf{k}$. When $t = 0$ the velocity of the particle is $-4\mathbf{i} + 3\mathbf{k}$. Determine its velocity at time t and hence find an expression for the power of $\mathbf{F}$ at time t. *(JMB)*

10 A particle of mass 2 units moves under the action of a force $\mathbf{F}$ so that its position vector at time t is given by $\mathbf{r} = 3\mathbf{i} + t^2\mathbf{j} + \frac{1}{3}t^3\mathbf{k}$. Find $\mathbf{F}$ and the power exerted by $\mathbf{F}$ at time t. *(JMB)*

11 At time t a particle is in motion with velocity $\mathbf{v}$ and is being acted upon by a variable force $\mathbf{F}$. Write down expressions for

(i) the power at time t,

(ii) the work done by $\mathbf{F}$ during the time interval $0 \leqslant t \leqslant T$.

The particle, of mass m, moves in a plane where $\mathbf{i}$ and $\mathbf{j}$ are perpendicular unit vectors so that its position vector at time t is given by

$$\mathbf{r} = 2a\cos 2t\mathbf{i} + a\sin 2t\mathbf{j},$$

where a is a positive constant. Derive expressions for the velocity $\mathbf{v}$ and the force $\mathbf{F}$ at time t.

Obtain an expression in terms of t for the power at time t and show that the work done by $\mathbf{F}$ during the interval $0 \leqslant t \leqslant T$ is $3ma^2(1 - \cos 4T)$.

If T varies, find the maximum value of the work done by $\mathbf{F}$ and determine also the smallest value of T for which this maximum value is reached.

(*JMB*)

12 The position vector $\mathbf{r}$, with respect to a fixed origin O, of a particle P_1 at time t is given by

$$\mathbf{r} = \frac{a}{T}\left[(2t + T)\mathbf{i} + \left(\frac{t^2}{T} + \frac{t^3}{T^2}\right)\mathbf{j}\right],$$

where a and T are constants. Show that, when $t = 0$, the particle is passing through the point $(a, 0)$ with velocity $2a\mathbf{i}/T$. Show further that the acceleration of P_1 at that instant has magnitude $2a/T^2$ and state the direction of this acceleration.

A second particle P_2 has constant acceleration $a(\mathbf{i} + 10\mathbf{j})/T^2$. At time $t = 0$, particle P_2 passes with velocity $a(-\mathbf{i} + 3\mathbf{j})/T$ through the point with position vector $a(11\mathbf{i} - 36\mathbf{j})/2$. Show that, when $t = 3T$, P_1 and P_2 are at the same point and are moving with the same velocity. (*L*)

13 (i) A particle P of mass m moves so that its position vector at time t is given by

$$\mathbf{r} = (t - t^2)\mathbf{i} + (t \sin \pi t)\mathbf{j}.$$

Find
(a) the momentum of P when $t = 1$,
(b) the force acting on P when $t = 2$,
(c) the kinetic energy of P when $t = 1$.

(ii) A particle falling freely starts from rest at a point A which is 35 m above ground level. It eventually strikes the ground at B which is distant 40 m from a fixed point O on the ground. Obtain the angular speed of the particle about O when it has descended a distance 5 m. (*L*)

14 A smooth hemispherical bowl is fixed with its circular rim horizontal and uppermost. Two particles A and B, of masses $2m$ and m respectively, are released from rest at opposite ends of a diameter of the rim. The particles collide at N, the lowest point of the bowl, when each is moving with speed u. The particle A is brought to rest by the impact. Show that the coefficient of restitution between the particles is $\frac{1}{2}$ and find the speed of B immediately after the collision. Also find, in terms of m and u,
(a) the magnitude of the impulse of the blow received by A,
(b) the kinetic energy lost in this collision.

Show that, immediately after the second collision of the particles, the speed of A is $\frac{1}{2}u$. By considering conservation of energy, find, in terms of u and g, the height above N to which A rises after this second collision.

(*L*)

15 A light rod AB of length l is hinged at A to a fixed point in such a way that it is free to rotate about A in a vertical plane which contains a small smooth fixed ring C level with A and at a distance l from A. A particle of mass $2m$ is attached to the rod at B. A light inextensible string passes through the ring at C, and has one end attached to B and the other to a particle P of mass m

which hangs freely. The system is released from rest with B at C, and during the subsequent motion the string is sufficiently long for P not to reach C. Show that the upward velocity of P is $2l\cos\theta\dfrac{d\theta}{dt}$ where 2θ is the angle CAB at time t.

Write down the energy equation for the system in terms of θ and $\dfrac{d\theta}{dt}$; and show that the system comes to rest when $\theta = \dfrac{\pi}{3}$. *(JMB)*

16 A nylon climbing rope of length 40 metres is stretched by a force F. The relation between the force F and the corresponding extension x is given in the following table:

x (metres)	0	1	2	3	4	5	6	7	8	9	10
F (newtons)	0	1 600	2 400	3 000	3 600	4 100	4 600	5 200	5 700	6 400	7 900

Use Simpson's rule toestimate the energy stored in the rope when it is extended by 10 metres.

A climber of mass 64 kg is climbing on a vertical rock-face h metres above the point to which he is securely attached by the rope of length 40 m described above. The climber slips and falls freely, being brought to rest for the first time by the rope when it has been extended by 10 metres. Calculate the value of h, giving your answer correct to two significant figures.

Determine also the speed of the climber at the moment the rope becomes taut. Give your answer correct to two significant figures. (Take $g = 9{\cdot}8\ \text{ms}^{-2}$.) *(JMB)*

17 Prove that the work done in increasing the length of a light elastic string, of natural length l and modulus of elasticity λ, from $l+x_1$ to $l+x_2$ is $\lambda(x_2^2-x_1^2)/(2l)$.

A small ring of mass m is threaded on a smooth wire bent into the form of a circle of radius a and fixed in a vertical plane. The ring is connected to the highest point of the wire by a light elastic string of natural length $\frac{1}{4}a\sqrt{3}$ and modulus of elasticity $\frac{1}{3}mg$. The ring is held at rest at a height $\frac{1}{2}a$ above the lowest point of the wire and then released. Show that the ring will come to instantaneous rest before reaching a height a above the lowest point of the wire. *(AEB 1971)*

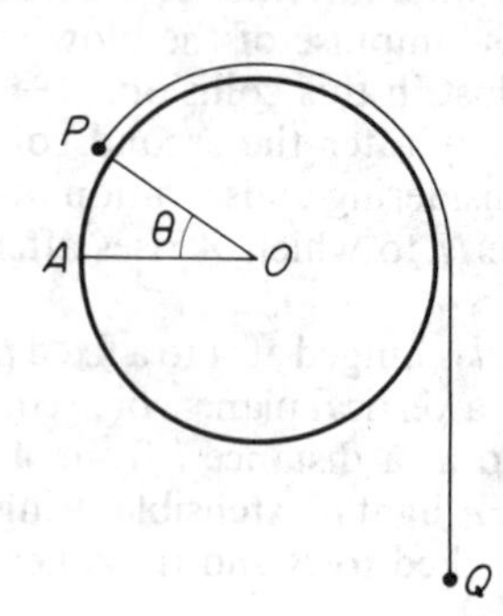

18 A particle P of mass m is at rest at a point A on the smooth outer surface of a fixed sphere of centre O and radius a, OA being horizontal. The particle is attached to one end of a light inextensible string which is taut and passes over the sphere, its other end carrying a particle Q of mass $2m$ which hangs freely. The string lies in the vertical plane containing OA. The system is released from rest, and after time t the angle POA is θ, as shown in the diagram. Show that, while P remains in contact with the sphere,

$$3a\left(\frac{d\theta}{dt}\right)^2 = 2g(2\theta - \sin\theta).$$

Find the tension in the string and the reaction of the sphere on P in terms of m, g and θ. (*JMB*)

19 State Hooke's law for a stretched elastic string.

By integration prove that the work done in stretching an elastic string of natural length c and modulus of elasticity λ to a length $c+x$ is $\lambda x^2/(2c)$.

A particle of mass m is attached at one end of an elastic string of natural length a and modulus of elasticity $4mg$. The other end of the string is attached at a fixed point O. The particle is initially held at O and then allowed to fall vertically. Using the conservation of energy principle, or otherwise, show that the particle falls a distance $2a$ before coming to instantaneous rest. Find the greatest speed of the particle during this fall, stating the distance from O at which it occurs. (*AEB 1974*)

20 A light spring obeys Hooke's Law. A force of 20 N extends the spring by 0·01 m. Show that the work done in extending the spring by b m from the unstretched state is $10^3 b^2$ J.

This spring is placed in a long smooth straight cylindrical tube with one end fixed to the tube. The tube is fixed in a vertical position with the free end of the spring uppermost. The dimensions of the tube and of the spring are such that the spring can only move vertically and the spring always remains inside the tube. A particle of mass 4 kg is firmly attached to the free end of the spring. The particle is held so that the spring is compressed a distance of 0·1 m from its uncompressed state. The particle is then released. Show that subsequently

$$v^2 = 3 + 20y - 500y^2$$

where v ms^{-1} is the speed of the particle and y m is the compression of the spring. Find

(a) v^2 when the *extension* of the spring is 0·01 m,

(b) the value of y when the speed is a maximum,

(c) the maximum extension of the spring. (*AEB 1983*)

21 A bead of mass m is on a smooth straight wire joining the point, A, whose position vector is $8a\mathbf{i} + 6a\mathbf{j} + 3a\mathbf{k}$ to the point, B, whose position vector is $2a\mathbf{i} - 6a\mathbf{j} - a\mathbf{k}$ where $\mathbf{i}$, $\mathbf{j}$ are perpendicular unit vectors in the horizontal plane and $\mathbf{k}$ is the unit vector vertically upwards. The bead is released from rest at A and is acted on by the force of gravity $-mg\mathbf{k}$ together with a constant force $\mathbf{F}$ acting in the direction of the vector $\mathbf{i} - 2\mathbf{j} + 2\mathbf{k}$. If the particle arrives at B with speed $\sqrt{(34ga/3)}$ ms^{-1}, calculate the magnitude of the force $\mathbf{F}$. (*JMB*)

22 A constant force $\mathbf{F}$ has magnitude 10 units and its direction is the same as that of the vector $2\mathbf{i}+3\mathbf{j}+6\mathbf{k}$. The force acts on a particle of unit mass which moves on a smooth rail connecting the points A and B whose position vectors are $\mathbf{i}+\mathbf{j}+2\mathbf{k}$ and $3\mathbf{i}-\mathbf{j}+3\mathbf{k}$ respectively. If the particle is initially at rest at A, find its speed when it reaches B, given that $\mathbf{F}$ is the only force acting on the particle. (*JMB*)

23 A mass of 5 kg starts from rest at A and is moved to B in two seconds by a constant force F N. The position vectors of A and B are $(-12\mathbf{j}-18\mathbf{k})$ and $(8\mathbf{i}+12\mathbf{j}+18\mathbf{k})$ respectively, where $\mathbf{i}$, $\mathbf{j}$ and $\mathbf{k}$ are mutually perpendicular vectors of magnitude 1 m and $\mathbf{k}$ is vertically upwards. Find the direction vector and the magnitude of F. Find also

(i) the work done (a) against gravity, (b) by F, as the mass moves from A to B,

(ii) the speed and the position vector of the mass $\frac{1}{2}$ second after leaving A. (*AEB 1973*)

24 A particle of mass 3 kg takes 2 seconds to move from A to B under the action of gravity and a constant force $\mathbf{F} = (12\mathbf{i}-3\mathbf{j}+12\mathbf{k})$ N. The position vector of B is $(15\mathbf{i}+7\mathbf{j}-6\mathbf{k})$ m and the particle arrives at B with a velocity of $(12\mathbf{i}+\mathbf{j}-4\mathbf{k})$ ms^{-1}. Find the position vector of A and the velocity with which the particle leaves A.

Find also the work done by $\mathbf{F}$ as the particle moves from A to B and the changes in the potential and kinetic energies of the particle in this period. (*L*)

25 A particle of mass 2 kg starts with velocity vector $(\mathbf{i}+2\mathbf{j}+3\mathbf{k})$ ms^{-1} from a point A which has position vector $(4\mathbf{i}+3\mathbf{j}+2\mathbf{k})$ m. The particle moves under the action of its own weight and a constant force $\mathbf{F} = (3\mathbf{i}+4\mathbf{j}+8\mathbf{k})$ N and travels from A to a point B in 4 seconds. Find the position vector of B and the speed of the particle when it reaches B. Find also the work done by the force $\mathbf{F}$ as the particle moves from A to B. (*L*)

26 The forces acting on a particle of unit mass reduce to a single force $\mathbf{F} = -12t\mathbf{i}+2\mathbf{j}+6t\mathbf{k}$, where t is the time. Find the velocity vector and position vector at time t if the particle starts (at time $t = 0$) with velocity $\mathbf{i}+2\mathbf{j}$ from the point with position vector $\mathbf{r} = \mathbf{j}+\mathbf{k}$. Find the work done by $\mathbf{F}$ during the time interval $t = 0$ to $t = 2$.

Show that the particle is at the point $-10\mathbf{i}+9\mathbf{j}+9\mathbf{k}$ at $t = 2$. At this instant the particle collides and coalesces with a stationary particle B whose mass is twice that of A. Find the velocity vector of the combined particles

(i) immediately after impact,

(ii) when $t = 3$,

assuming that the force $\mathbf{F}$ is the only force which acts on the combined particles. (*AEB 1977*)

12 Centres of Mass

The centre of mass of a set of particles or of an extended body is an important concept both in statics and dynamics. In this chapter the concept is defined and the positions of the centres of mass of a number of different types of bodies is determined. Use is made of the centre of mass in some simple problems of balancing.

Definition

The centre of mass of two particles of masses m_1 and m_2, which are situated at the points A_1 and A_2 with position vectors $\mathbf{r}_1$ and $\mathbf{r}_2$ respectively relative to an origin O, is the point G_{12} with position vector given by

$$\overrightarrow{OG}_{12} = \frac{m_1\mathbf{r}_1 + m_2\mathbf{r}_2}{m_1 + m_2}. \qquad 12.1$$

If a third particle of mass m_3, with its centre of mass at a point with position vector $\mathbf{r}_3$ relative to O, is introduced, we may combine the mass $(m_1 + m_2)$ situated at G_{12} with this third mass m_3 by equation 12.1 to obtain

$$\overrightarrow{OG}_{123} = \frac{(m_1 + m_2)\dfrac{m_1\mathbf{r}_1 + m_2\mathbf{r}_2}{m_1 + m_2} + m_3\mathbf{r}_3}{(m_1 + m_2) + m_3}$$

$$= \frac{m_1\mathbf{r}_1 + m_2\mathbf{r}_2 + m_3\mathbf{r}_3}{m_1 + m_2 + m_3}.$$

This result may be extended to n particles with masses m_i and position vectors $\mathbf{r}_i$, $i = 1, 2, \ldots, n$. The position vector of the centre of mass of these n particles is given by

$$\bar{\mathbf{r}} = \frac{\sum_{i=1}^{n} m_i\mathbf{r}_i}{\sum_{i=1}^{n} m_i}. \qquad 12.2$$

The separate coordinates of the centre of mass referred to rectangular axes through the origin are, therefore,

$$\bar{x} = \frac{\sum_{i=1}^{n} m_i x_i}{\sum_{i=1}^{n} m_i}, \qquad \bar{y} = \frac{\sum_{i=1}^{n} m_i y_i}{\sum_{i=1}^{n} m_i}, \qquad \bar{z} = \frac{\sum_{i=1}^{n} m_i z_i}{\sum_{i=1}^{n} m_i} \qquad 12.3$$

where (x_i, y_i, z_i) are the coordinates of the particle of mass m_i.

Note that when the centre of mass of the n particles is chosen as the origin of reference, then $\sum_{i=1}^{n} m_i \mathbf{r}_i = 0$.

12.1 Uniform laminae and solids

A uniform lamina is typified by a thin rectangular sheet of paper. It is a sheet or plate, the thickness of which may be neglected. If the lamina is in the form of a plane geometrical figure with a centre – that is, a point about which the lamina has half-turn symmetry – then the centre is the centre of mass of the lamina. This can be shown as follows. By the definition of the centre, for each one of the elements into which the lamina may be divided, with mass m_i and centre of mass at the point with position vector $\mathbf{r}_i$ referred to the centre of the lamina, there is a corresponding element of mass m_i, the centre of mass of which has position vector $-\mathbf{r}_i$, as shown in Fig. 12.1. Hence, summing over the whole lamina,

$$\sum m_i \mathbf{r}_i = 0,$$

that is, the centre of mass is at the centre of the lamina.

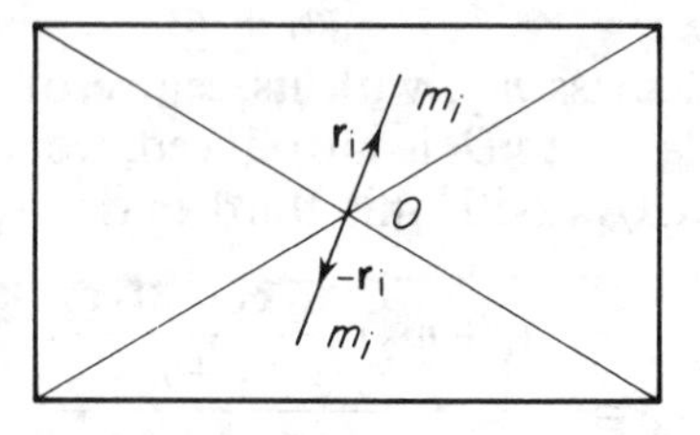

Fig. 12.1

Similarly, if any uniform solid body has a centre, this centre will be the centre of mass. Examples of such solids are uniform spheres, cylinders and rods of uniform cross-section.

EXAMPLE 1 *A uniform lamina in the form of a letter* 'E' *has dimensions as shown in Fig.* 12.2. *Find the position of the centre of mass of the lamina.*

Taking the origin at the lower left-hand corner of the lamina with the axes of x and y as shown and dividing the lamina into five strips, we have the following masses together with their associated coordinates of centres of mass: $4k$, $(2, \frac{1}{2})$; $2k$, $(\frac{1}{2}, 2)$; $3k$, $(1\frac{1}{2}, 3\frac{1}{2})$; $2k$, $(\frac{1}{2}, 5)$; $4k$, $(2, 6\frac{1}{2})$; where $15k$ is the total mass of the lamina. Thus,

$$\bar{x} = \frac{4k.2 + 2k.\frac{1}{2} + 3k.1\frac{1}{2} + 2k.\frac{1}{2} + 4k.2}{15k}\text{ cm} = \frac{22\frac{1}{2}}{15}\text{ cm} = \mathbf{1\frac{1}{2}\text{ cm}},$$

$$\bar{y} = \frac{4k.\frac{1}{2} + 2k.2 + 3k.3\frac{1}{2} + 2k.5 + 4k.6\frac{1}{2}}{15k}\text{ cm} = \frac{52\frac{1}{2}}{15}\text{ cm} = \mathbf{3\frac{1}{2}\text{ cm}}.$$

and by symmetry $\bar{y} = 3\frac{1}{2}$ cm.

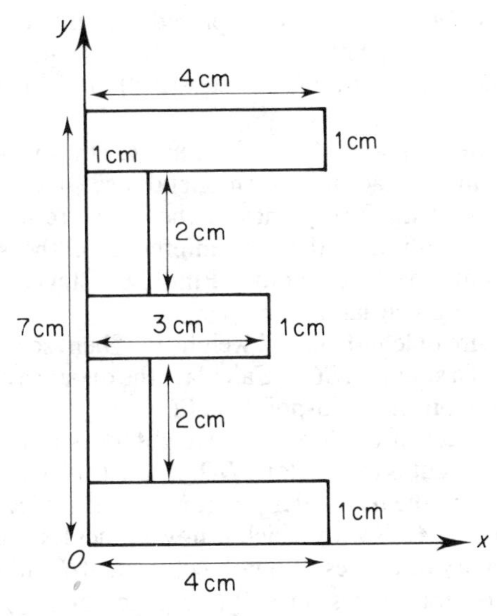

Fig. 12.2

EXERCISE 12.1

1 A lamina consists of a rectangle $ABCD$ with the lengths of AB and BC equal to 8 cm and 6 cm respectively, and a second rectangle $PQRS$ outside $ABCD$. P and Q lie on AB with the lengths of AP and BQ being respectively 3 cm and 2 cm and the length of QR being 4 cm. Calculate the distances of the centre of mass of the lamina from the sides AD and CD.

2 Particles of masses $3k$, $4k$ and $5k$ are attached to the vertices A, B and C respectively of a light triangular lamina ABC, in which angle $ABC = 90°$, $AC = 10$ m and $BC = 6$ m. Find the distance of the centre of mass of these three particles from (a) AB, (b) BC. (*L*)

3 A uniform rectangular lamina $ABCD$ is of mass $3M$; $AB = DC = 4$ cm and $BC = AD = 6$ cm. Particles of mass M are attached to the lamina at B, C and D. Calculate the distance of the centre of mass of the loaded lamina (a) from AB, (b) from BC. (*L*)

4 The figure represents a uniform L-shaped lamina. All the angles are right angles; $DE = EF = 6$ cm, $AB = 7$ cm and $BC = 8$ cm. Calculate the distances of the centre of mass of the lamina from AB and BC. (*L*)

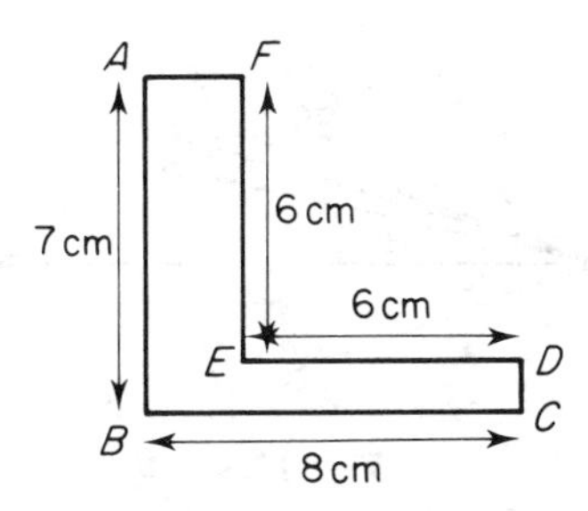

5 Particles of mass $2M$, xM and yM are placed at points whose coordinates are (2, 5), (1, 3) and (3, 1) respectively. Given that the centre of mass of the three particles is at the point with coordinates (2, 4), find the values of x and y. *(L)*

6 A uniform wooden circular disc, of radius 12 cm and of mass 0·3 kg, has attached to its plane surface a uniform metal disc which is circular, of radius 6 cm and of mass 0·9 kg. The planes of the discs are in contact so that the circles touch internally and the circumference of the smaller disc passes through O, the centre of the larger disc. Find the distance from O of the centre of mass of the combined lamina. *(L)*

7 A thin uniform wire of length $6a$ and weight W is bent so as to form three sides AB, BC and CD of a square $ABCD$. Calculate the distance of the centre of mass of the bent wire from the mid-point of BC. *(L)*

8 A uniform thin sheet of cardboard of weight W is in the form of a square $ABCD$ of side $6a$. A cut is made along DO, where O is the point of intersection of the diagonals, and the triangular portion AOD is folded over along AO and stuck to the portion AOB with which it now coincides. Find the distances of the centre of gravity of the resulting object from AB and from BC and show that the centre of gravity lies on BD. (see page 250.) *(L)*

9 A uniform lamina $ABCD$ is in the form of a trapezium in which the angles A and B are right angles, $BC = 2AB = 2a$ and $AD = x$. Find the distance of the centre of gravity of the lamina from AB. *(L)*

12.2 Use of calculus

It is possible to determine the position of the centre of mass of some bodies by using methods of integration.

A wire in the form of a circular arc.

Fig. 12.3 illustrates a wire bent in the shape of a circular arc of radius r, which subtends an angle of 2α at the centre O of the circle of which the arc is a part. We take the radius to the mid-point of the arc as the x-axis, and divide the arc into a number of elements, a typical one subtending an angle $\delta\theta$ at O, where θ is the angle between the radius OP and Ox.

Consider the element bounded by OP, OP' and the arc PP', where angle $POP' = \delta\theta$. The contribution to Σmx of this element is $kr\,\delta\theta\,.\,r\cos\theta$ to the

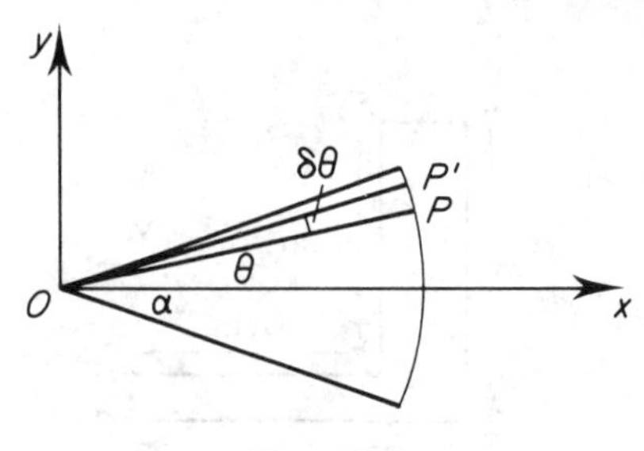

Fig. 12.3

first order of $\delta\theta$, where k is the mass of the wire per unit length. Thus,

$$\bar{x} = \frac{\sum m_i x_i}{\sum m_i} \approx \frac{\sum_{\theta=-\alpha}^{\theta=\alpha} kr^2 \cos\theta\,\delta\theta}{k.2r}$$

The limiting value of the sum $\sum_{\theta=-\alpha}^{\theta=\alpha} kr^2 \cos\theta\,\delta\theta$ is

$$k\int_{-\alpha}^{\alpha} r^2 \cos\theta\,\mathrm{d}\theta = 2kr^2 \sin\alpha.$$

Hence

$$\bar{x} = \frac{r\sin\alpha}{\alpha}.$$

Clearly $\bar{y} = 0$, as Ox is an axis of symmetry of the arc.

It is interesting to note the following particular cases.
As $\alpha \to 0$, $x \to r$. If $\alpha = \pi/2$, $x = 2r/\pi$ (a semicircular arc).
If $\alpha = \pi$, $x = 0$ (a complete circle).
The consideration of particular cases can test the plausibility of a result obtained. If an error had been made, a particular case might well highlight the error.

A plane lamina with a curved boundary.

Suppose that a uniform plane lamina is bounded by the curve $y = x^2 + 4$, the ordinates $x = 1$, $x = 3$ and the x-axis, as illustrated in Fig. 12.4.

Let k be the mass per unit area of the lamina. The lamina may be divided into a number of strips of width δx, a typical strip being shown in Fig. 12.4. For this strip

$$\text{mass} \approx ky\,\delta x,$$

$$\text{contribution to } \Sigma mx \approx kxy\,\delta x,$$

$$\text{total sum } \Sigma m_i x_i \text{ in the limit as } \delta x \to 0 = \int_1^3 kxy\,\mathrm{d}x,$$

$$\text{total mass} = \int_1^3 y\,\mathrm{d}x.$$

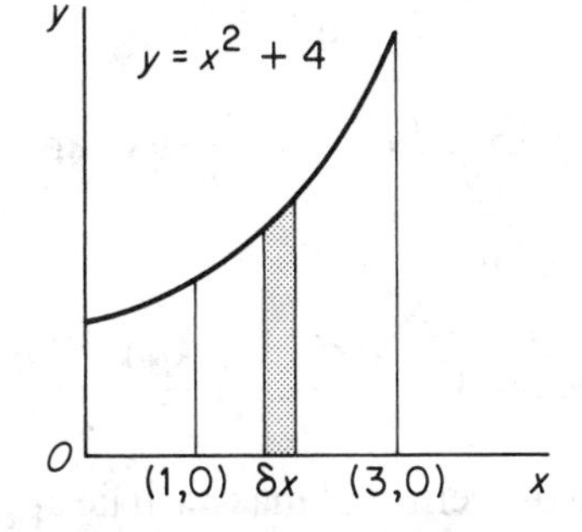

Fig. 12.4

Hence,

$$\bar{x} = \frac{k\int_1^3 (x^3+4x)\,dx}{k\int_1^3 (x^2+4)\,dx} = \frac{\left[\frac{x^4}{4}+2x^2\right]_1^3}{\left[\frac{x^3}{3}+4x\right]_1^3} = \frac{36}{16\frac{2}{3}} = 2{\cdot}16.$$

Note that the result $\bar{x} = 2{\cdot}16$ is reasonable, as clearly $\bar{x} > 2$, for the lamina is 'taller' to the right of the central value of x than to the left. For the y-coordinate

contribution to Σmy of the elementary strip $\approx ky\,\delta x\,.(\frac{1}{2}y)$,

sum $\Sigma my \approx \frac{1}{2}ky^2\,\delta x$ and the limit of this sum is $k\int_1^3 \frac{1}{2}y^2\,dx$.

Hence,

$$\bar{y} = \frac{k\int_1^3 \frac{1}{2}(x^2+4)^2\,dx}{16\frac{2}{3}k} = \frac{3}{100}\int_1^3 (x^4+8x^2+16)\,dx$$

$$= \frac{3}{100}\left[\frac{x^5}{5}+8\frac{x^3}{3}+16x\right]_1^3$$

$$= \frac{3}{100}\left[\left(\frac{243}{5}+72+48\right)-\left(\frac{1}{5}+\frac{8}{3}+16\right)\right]$$

$$\approx 4{\cdot}5.$$

Again referring to Fig. 12.4, it can be seen that this value of $\bar{y}$ is also reasonable.

A plane triangular lamina

ABC is a uniform plane triangular lamina with mass k per unit area. Let $BC = a$ and the altitude $AD = h$. Consider the strip of thickness δx with PQ as its upper edge, where PQ is parallel to BC and at a distance x from A. By similar triangles

$$\frac{y}{a} = \frac{x}{h}.$$

The mass of the strip $\approx ky\,\delta x = \frac{kax}{h}\,\delta x$ and the contribution to $\Sigma mx \approx \frac{kax^2\,\delta x}{h}$. Therefore,

$$\bar{x} = \int_0^h \frac{kax^2}{h}\,dx \bigg/ \frac{kah}{2} = \frac{2h}{3}.$$

For a very thin strip the centre of mass will lie approximately at the mid-point of PQ and, thus, it may be seen that the centre of mass of the lamina

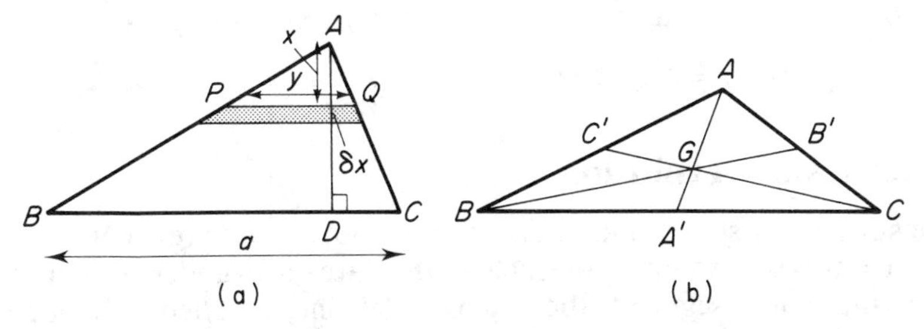

Fig. 12.5

lies on the median through A and similarly the centre of mass will lie on each of the medians. As shown in Fig. 5, the medians AA', BB' and CC' of a triangle meet at a point G where $AG = \frac{2}{3}AA'$, $BG = \frac{2}{3}BB'$ and $CG = \frac{2}{3}CC'$. The centre of mass of a triangular lamina is at the point G where the medians meet.

A circular sector

In Fig. 12.6, OAB represents a uniform lamina in the form of a sector of a circle with the arc AB subtending an angle 2α at the centre O of the circle. Let k be the mass per unit area and Ox the axis of symmetry.

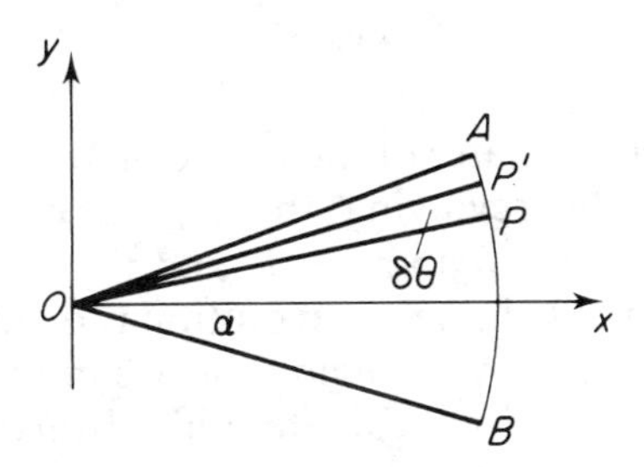

Fig. 12.6

The elementary sector OPP' has mass $\dfrac{kr^2\,\delta\theta}{2}$. Considering the triangle OPP' as an approximation to this sector, the contribution to

$$\Sigma mx \approx \frac{kr^2\,\delta\theta}{2}\,.\,\tfrac{2}{3}r\cos(\theta + \delta\theta/2)$$

and to the first order of $\delta\theta$, this is equal to $\frac{1}{3}kr^3\cos\theta\,\delta\theta$. Hence,

$$\bar{x} = \frac{k\displaystyle\int \tfrac{1}{3}r^3\cos\theta\,\mathrm{d}\theta}{\tfrac{1}{2}kr^2\,.\,2\alpha} = \frac{2r\sin\alpha}{3\alpha}.$$

The y-coordinate is zero as Ox is an axis of symmetry.

Note that for a semicircle, $\bar{x} = 4r/(3\pi)$; for a quadrant $\bar{x} = \frac{2}{3}r\frac{1}{\sqrt{2}} \Big/ (\pi/4) = 4r\sqrt{2}/(3\pi)$.

Volumes of revolution

The surfaces of some solid bodies may be obtained by revolving a curve about an axis. Examples of such bodies are – a cylinder, obtained by revolving a line segment about a parallel line; a sphere, obtained by revolving a semicircle about a diameter; and a cone, obtained by revolving a line segment AB about an axis passing through the end A (see Fig. 12.7)

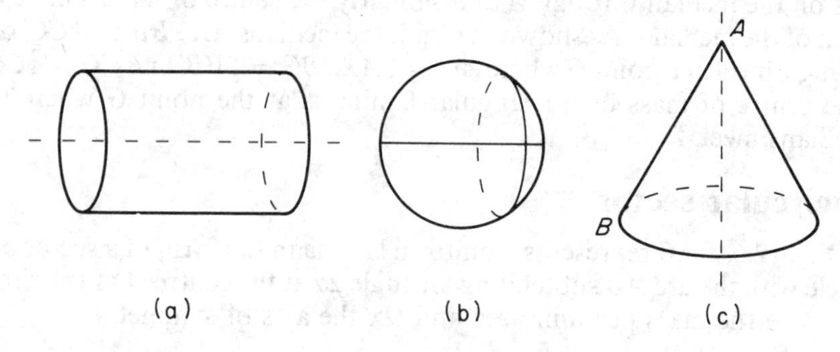

Fig. 12.7

We now find the coordinates of the centre of mass of a uniform solid of revolution obtained by revolving the region bounded by the curve $y = f(x)$, the ordinates $x = a$ and $x = b$, and the x-axis, about the x-axis (see Fig. 12.8). First, the volume of revolution is divided into a number of disc-like elements formed by revolving the arc of the curve between two points on the curve whose x-coordinates differ by δx. Let ρ be the density.

$$\text{Mass of element} \approx \rho \pi y^2 \,\delta x.$$

$$\text{Contribution to } \Sigma mx \approx \rho \pi x y^2 \,\delta x.$$

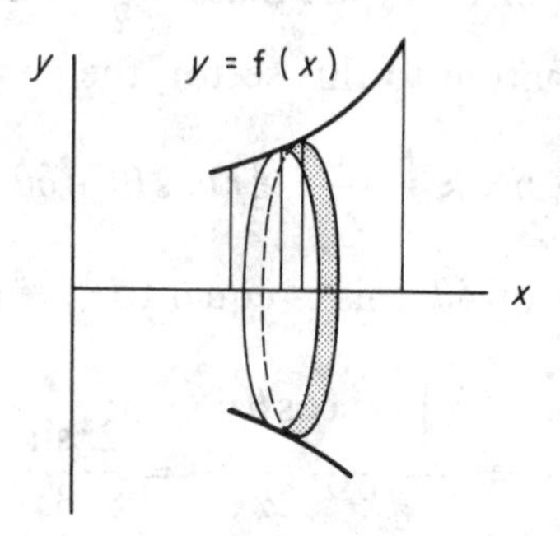

Fig. 12.8

Thus,

$$\bar{x} = \frac{\int_a^b xy^2\,\mathrm{d}x}{\int_a^b y^2\,\mathrm{d}x}.$$

If the curve is a quadrant of a circle with equation $x^2+y^2=r^2$, a hemisphere is generated (Fig. 12.9)

$$\text{Mass of element} \approx \rho\pi(r^2-x^2)\,\delta x.$$

$$\text{Contribution to } \Sigma mx \approx \rho\pi x(r^2-x^2)\,\delta x.$$

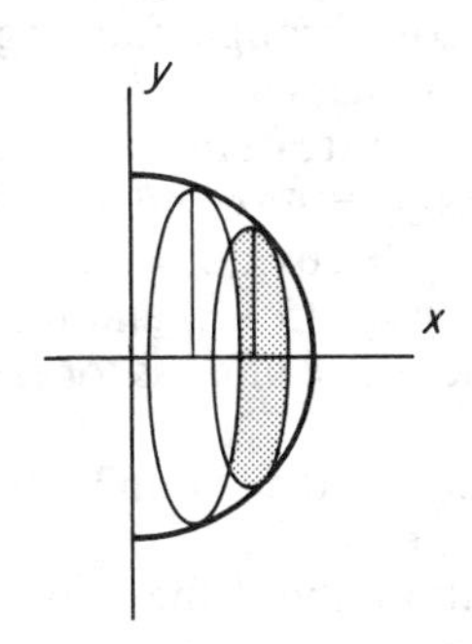

Fig. 12.9

Hence

$$\bar{x} = \frac{\rho\pi\int_0^r x(r^2-x^2)\,\mathrm{d}x}{\frac{2}{3}\pi r^3\rho} = \left[\frac{r^2x^2}{2}-\frac{x^4}{4}\right]_0^4 \Big/ \frac{2}{3}r^3$$

$$= 3r/8.$$

Again note that the centre of mass is expected to be nearer to the plane base of the hemisphere than to the top of the hemispherical dome, that is, we expect $\bar{x} < r/2$ so our result is reasonable.

EXERCISE 12.2

1 The region in the first quadrant bounded by the curve $y = x^4$, the y-axis and the line $y = 16$ is rotated through four right angles about the y-axis to form a solid of uniform density. Calculate the coordinates of the position of the centre of mass of the solid. (*L*)

2 A lamina is in the shape of the region bounded by the curve $y^2 = 4ax$ and the line $x = 9a$. Find the coordinates of the centre of mass of this uniform lamina.

If this lamina is now cut in half by the line $y = 0$, find the coordinates of the centre of mass of the half that lies above the x-axis.

3 A solid is formed by the rotation through four right angles about the x-axis of the region bounded by the curve $y = x^2/a$, $x = 2a$ and $y = 0$. Find the coordinates of the centre of mass of the uniform solid of revolution.

4 Prove by rotating the area bounded by the curve $y^2 = x(2a-x)$, $x = a$ and the x-axis about the x-axis, that the position of the centre of mass of a uniform hemisphere is at a distance of $3a/8$ from the centre of the sphere, where a is the radius.

5 A lamina is formed by the region defined by $0 \leqslant x \leqslant \pi$, $y \leqslant 6\sin x$, $y \geqslant 3$. Find the position of the centre of mass of this lamina.

6 A solid of revolution is formed by revolving the region bounded by $y^2 = 4ax$, $x = 0$ and $y = 2a$ about the y-axis. Find the coordinates of the centre of mass.

7 A cone is obtained by revolving the region defined by $y \leqslant x\tan\theta$, $y \geqslant 0$, $0 \leqslant x \leqslant h$, about the x-axis through four right angles. Prove that the coordinates of the centre of mass are $(\frac{3}{4}h, 0)$.

8 Find the coordinates of the centre of mass for each of the following laminae
(a) bounded by the curves $y^2 = 4ax$ and $x^2 = 4ay$;
(b) bounded by the curve $y = \cos x$, $x = 0$, $x = \pi/2$, $y = 0$;
(c) bounded by the curve $y = x(5-x)$ and the line $y = x$.

9 Find the coordinates of the centre of mass for each of the following solids of revolution
(a) formed by the revolution of the region bounded by $x^2 + y^2 = a^2$, $x = 0$, $x = \frac{1}{2}a$ and $y > 0$, about the x-axis;
(b) formed by the revolution about the x-axis of the region bounded by $y = rx/h$, $x = \frac{1}{2}h$, $x = h$ and $y = 0$;
(c) formed by the revolution of the region bounded by $x^2 = 4ay$, $x = 0$ and $y = a$, $(x > 0)$ about the y-axis.

10 A uniform solid is formed by the rotation about the y-axis through two right angles of the area cut off between the parabola $x^2 = a(y+a)$ and the lines $y = 0$ and $y = 3a$. Prove that the volume of the solid is $(15\pi a^3)/2$ and find the height of its centre of mass above its base. (*O&C*)

11 Show by integration that the centre of mass of a uniform solid right circular cone of height h is at a distance $h/4$ from its plane base. (*L*)

12 Show by integration that the centre of gravity of a uniform solid hemisphere of radius r is situated at a distance of $3r/8$ from the centre of its plane face.

12.3 Composite bodies

Some bodies may be regarded as a composite of separate parts, the masses and the positions of the centres of mass of these separate parts being known. The centre of mass of the composite body may be obtained using equations 12.1, 12.2 or 12.3, with a suitable choice of origin.

EXAMPLE 1 *A uniform cone, of height h and base radius r, is mounted on a hemisphere, of radius r and made of the same material, so that the circular faces of both bodies coincide. Find the position of the centre of mass of the composite body.*

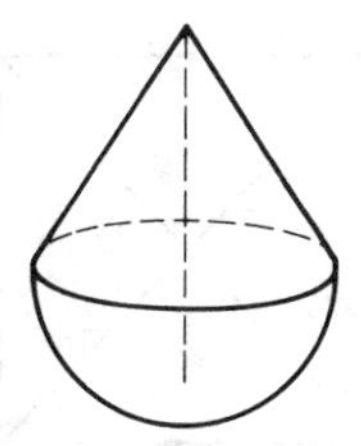

Fig. 12.10

Suppose that the body stands vertically. The centre of mass of the composite body will lie on the common axis of symmetry. If the density is ρ, the masses of the cone and hemisphere are $\frac{1}{3}\pi r^2 h\rho$, and $\frac{2}{3}\pi r^3 \rho$, respectively. The height above the lowest point of the centres of mass of the cone and hemisphere are $r + h/4$ and $5r/8$, respectively. (The position of the centre of mass of a cone is given in question 7 of Exercise 12.2). In this example, it is useful to tabulate the results.

part	mass	height of centre of mass (above the lowest point)
cone	$\frac{1}{3}\pi r^2 h\rho$	$r + h/4$
hemisphere	$\frac{2}{3}\pi r^3 \rho$	$5r/8$
whole body	$\frac{1}{3}\pi r^2 \rho (h + 2r)$	x

Hence

$$\bar{x} = \frac{\frac{1}{3}\pi r^2 \rho (h(r + h/4) + 2r.5r/8)}{\frac{1}{3}\pi r^2 \rho (h + 2r)} = \frac{\mathbf{h^2 + 4rh + 5r^2}}{\mathbf{4(h + 2r)}}.$$

We can check this result by considering special cases. If $h = 0$, x should be $5r/8$ by consideration of the hemisphere alone. As $r \to 0$, x should tend to $h/4$. Thus, again our result is reasonable.

12.4 Remainders

In some problems, part of a body is removed and one is required to find the position of the centre of mass of the remainder. Such problems can be solved by regarding the original body as the composite body, with the part remaining and the part cut out as the constituent parts.

EXAMPLE 1 *From a uniform square lamina ABCD of mass M, the diagonals of which meet at X, the triangular part BXC is removed. Find the position of the centre of mass of the remainder. (See Fig.* 12.11.)

part	mass	distance of centre of mass from AD
BCX	$\frac{1}{4}M$	$a/2 + \frac{2}{3}(a/2) = 5a/6$
remainder	$\frac{3}{4}M$	$\bar{x}$
whole square	M	$a/2$

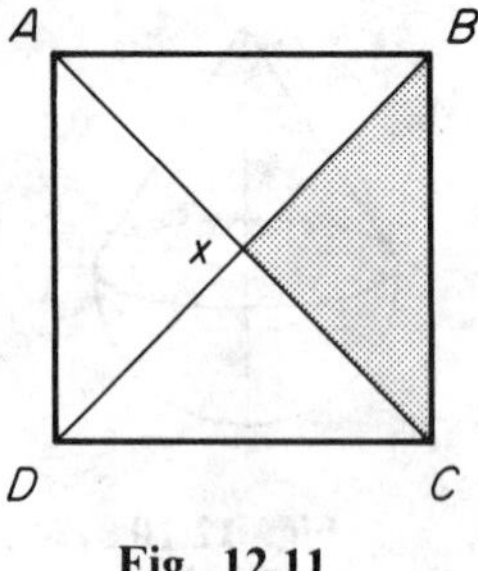

Fig. 12.11

The centre of mass is clearly on the axis of symmetry. Here we use

$$(m_1+m_2)x = m_1x_1+m_2x_2.$$

$$M\frac{a}{2} = \frac{1}{4}M\frac{5a}{6}+\frac{3}{4}Mx$$

$$\bar{x} = \mathbf{7a/18}.$$

Note that this result is less than $a/2$, as we would expect.

12.5 The centre of gravity

The centre of gravity of a body, or of a set of particles, is the point through which the total weight of the body, or of the particles, may be considered to act. If we assume that **g**, the acceleration due to gravity, is constant over the body or the particles in question, the centre of gravity will coincide with the centre of mass. We prove this as follows.

Consider the centre of gravity of two particles of masses m_1 and m_2 situated at A_1 and A_2, the position vectors of which with respect to an origin O are $\mathbf{r}_1$ and $\mathbf{r}_2$, respectively. If A_1A_2 is horizontal, (Fig. 12.12(a)) then taking moments about G, the centre of gravity, we obtain

$$m_1g\,.\,A_1G = m_2g\,.\,A_2G \Rightarrow A_1G/A_2G = m_2/m_1.$$

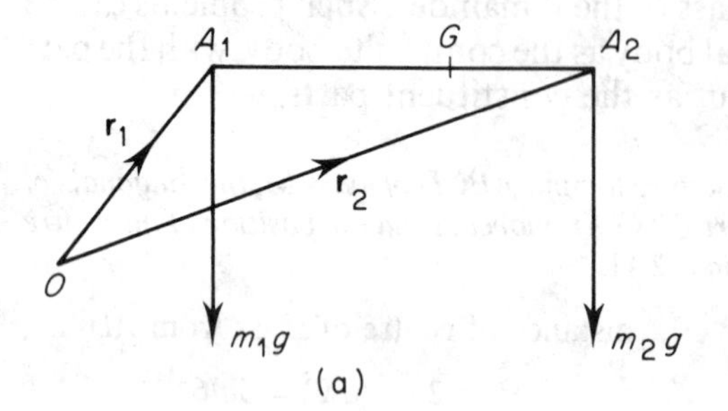

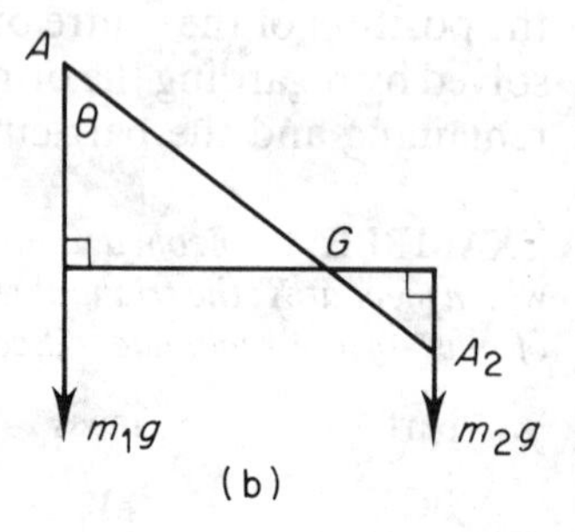

Fig. 12.12

If A_1A_2 is inclined at θ to the vertical (Fig. 12.14(b)), taking moments about G we obtain

$$m_1 g \,.\, A_1G \sin\theta = m_2 g \,.\, A_2G \sin\theta \Rightarrow A_1G/A_2G = m_2/m_1 .$$

Thus

$$\mathbf{OG} = \frac{m_1\mathbf{r}_1 + m_2\mathbf{r}_2}{m_1 + m_2},$$

that is, the centres of mass and gravity coincide for these two particles.

For almost all practical purposes, the centre of mass and the centre of gravity may be considered to be coincident but it should be remembered that they are not, by definition, the same point. The centre of mass is the point defined by equations 11.1, 11.2 and 11.3, whereas the centre of gravity is the point through which the total weight of a body, or of a set of particles, may be considered to act. Associated with the centre of gravity there are a number of simple problems studying the balancing of an extended body. Although so far in this book only the equilibrium of forces acting on a particle have been studied, it is considered appropriate to include these problems in this chapter. The theory of equilibrium of forces acting on a rigid body is to be found in chapter 13.

EXAMPLE 1 *ABCD is a uniform rectangular lamina from which the triangle BEC has been removed. AB* = 20 cm, *BC* = 10 cm *and E is the mid-point of AB. The lamina AECD rests in equilibrium when suspended by a vertical string attached to D. Find the angle between DC and the vertical when the lamina is in equilibrium.*

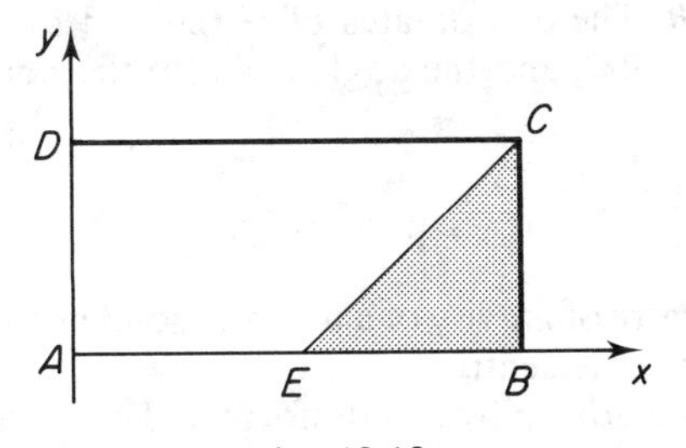

Fig. 12.13

First we find the position of the centre of mass of *AECD* (Fig. 12.13). Choose *A* as origin, and the axes of *x* and *y* along *AB* and *AD* respectively. Let *k* be the mass per unit area of the lamina. The table of the parts is as follows

part	mass	coordinates of centre of mass
triangle *BEC*	$50k$	$(10 + 6\frac{2}{3}, 3\frac{1}{3})$
AECD	$150k$	$(\bar{x}, \bar{y})$
ABCD	$200k$	$(10, 5)$

$\bar{x}$-coordinate

$$200k \,.\, 10 = 150k \,.\, x + 50k \,.\, 16\tfrac{2}{3}$$
$$40 = 3x + 16\tfrac{2}{3}$$
$$\bar{x} = 7\tfrac{7}{9}.$$

$\bar{y}$-coordinate

$$200k\,.\,5 = 150k\,.\,y + 50k\,.\,3\tfrac{1}{3}$$
$$\bar{y} = 5\tfrac{5}{9}.$$

When the lamina is suspended by a vertical string attached to D, the only external forces acting on the lamina are the weight, W and the tension T in the string (Fig. 12.14). Referring to the conditions of equilibrium in Chapter 3, these forces can only be in equilibrium if they are equal in magnitude, act in opposite directions, and have the same line of action. Thus, the centre of gravity G of the lamina will be vertically below D and we assume that G coincides with the centre of mass.

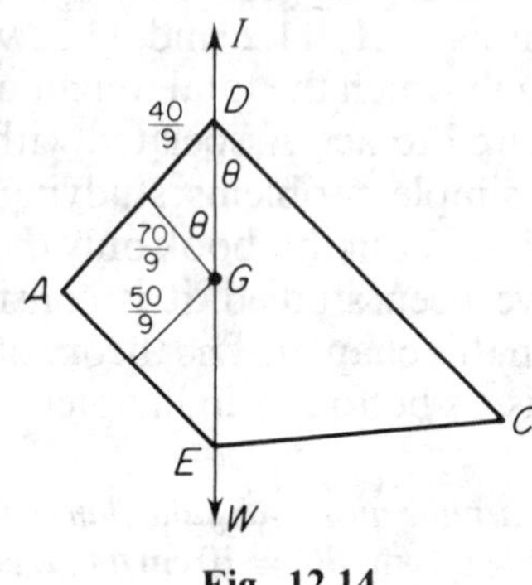

Fig. 12.14

Let angle $GDC = \theta$. The coordinates of G referred to axes AE and AD are $(\frac{70}{9}, \frac{50}{9})$. Hence, $\tan\theta = 4/7$, and the angle between the side DC and the vertical $\approx$ **30°**.

EXERCISE 12.5

1 Prove that the centre of gravity of a uniform solid hemisphere of radius r is a distance $3r/8$ from the centre.

A child's toy is made up from a uniform and solid right circular cone and hemisphere. The radius of the cone is r, and its height $3r$. The radius of the hemisphere is r. The base of the cone and hemisphere are sealed together. The material from which the hemisphere is made is three times as heavy per unit volume as the cone material. Find the distance of the centre of gravity of the toy from the vertex of the cone. (The centre of gravity of a cone is $\frac{1}{4}$ of the way up the central axis from the base.) (*SUJB*)

2 Prove by integration that the centre of mass of a uniform solid right circular cone of height h and base radius r is at a distance $\frac{3}{4}h$ from the vertex.

Such a cone is joined to a uniform solid right circular cylinder, of the same material, with base radius r and height l, so that the plane base of the cone coincides with a plane face of the cylinder. Find the centre of mass of the solid thus formed. (*JMB*)

3 A cylindrical can, made of thin material and open at the top, is of height $2a$ and the radius of the plane base is a. The mass per unit area of the uniform material making up the base of the can is twice the mass per unit area of the uniform material making up the curved surface of the can.

(a) Find the distance of the centre of mass of the empty can from the centre of the base.

(b) The can is suspended by a string attached to a point on the rim of its open end and hangs freely under gravity. Calculate, correct to the nearest degree, the angle which the plane base of the can makes with the horizontal. (L)

4 A uniform solid cylinder of radius 4 cm and height 6 cm has a cylindrical hole of radius 2 cm and depth 3 cm bored centrally at one end. Find the position of the centre of mass of the remainder. (L)

5 A uniform circular lamina, centre C and radius 8 cm, has ACB as a diameter. A circular lamina of radius 4 cm whose centre D lies on AB, where $AD = 11$ cm, is removed from the original lamina. Find the distance of the centre of mass of the remaining lamina from A. (L)

6 A uniform wire of length πa is bent to form a semicircle of radius a. Prove that the distance of the centre of mass of the wire from the centre of the circle is $2a/\pi$.

Another uniform wire of length $(2+\pi)a$ is bent into the form of a semicircular arc together with the diameter AB joining the ends of the arc, and hangs in equilibrium from the point A. Show that AB makes an angle θ with the vertical where $\tan\theta = 2/(2+\pi)$. (JMB)

7 $ABCD$ is a thin uniform rectangular metal sheet with $AB = 16$ cm and $AD = 12$ cm. The rectangle $FECH$ is cut out as shown in the diagram and stuck on the original sheet in the position $AXFY$. Find the distance of the centre of gravity of the resulting sheet from AB and from AD.

If the sheet is now freely suspended from H and hangs in equilibrium, find the angle which HF makes with the vertical. (L)

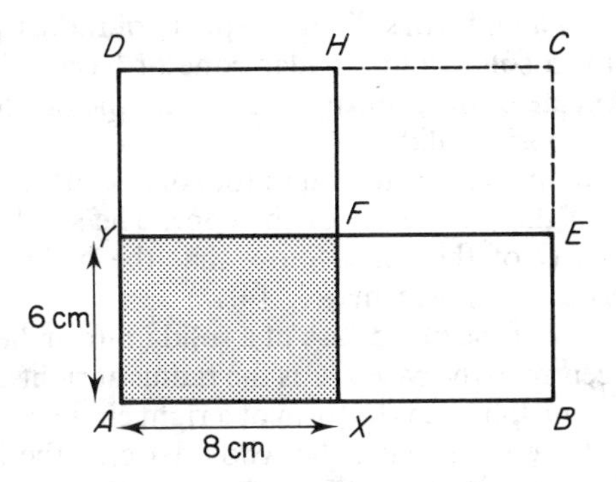

8 Find by integration the distance of the centroid of a uniform solid hemisphere of radius a from its plane face.

A hollow sphere of external radius R and internal radius r is cut in half by a plane through its centre. Show that the distance of the centroid of each half from the centre of the sphere is

$$\frac{3(R+r)(R^2+r^2)}{8(R^2+Rr+r^2)}. \qquad (L)$$

9 A uniform rectangular metal plate $ABCD$ has sides $AB = 40$ cm and $BC = 30$ cm. A right-angled isosceles triangle BCX is removed from the plate, where X is the point in AB such that $XB = 30$ cm. Find the distances of the centre of gravity of the remaining part of the plate from the sides AB and AD.

This remaining part of the plate has weight W and is freely suspended from the corner D. Show that if a weight of $\frac{2}{9}W$ is hung at A then AC is horizontal. *(MEI)*

10

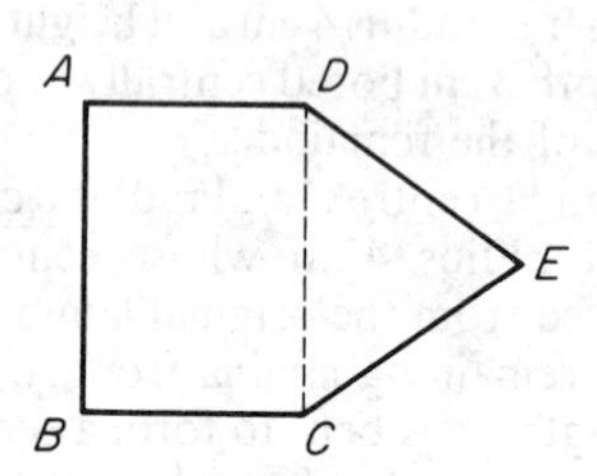

The diagram shows a uniform lamina $ABCED$ consisting of a rectangle $ABCD$ and an isosceles triangle DCE; $AB = DC = 8$ cm, $BC = AD = 6$ cm and $DE = EC = 2\sqrt{13}$ cm.

(a) Show that E is distant 6 cm from DC and that the mass of the rectangle $ABCD$ is twice the mass of the triangle DCE.

(b) Find the distance of the centre of mass of the lamina $ABCED$ from AB.

The lamina $ABCED$, which is of mass $3M$, is suspended freely from D and hangs in equilibrium with DC vertical when a particle of mass m is attached to the lamina at E.

(c) Show that $2M = 3m$. *(L)*

11 The plane base of a uniform solid hemisphere of radius a coincides with the base of a uniform solid right circular cone of base radius a, semi-vertical angle $\pi/6$ and made of the same material. Find the position of the centre of mass of the composite solid.

A particle P, of mass equal to that of the solid, is attached to a point on the circumference of the plane base of the cone. The solid is freely suspended from the vertex O of the cone. Show that the inclination of OP to the downward vertical is approximately 14°. *(L)*

12 AOB is a diameter of the plane base of a solid uniform hemisphere of radius a, O being the centre of the base. OC is the radius at right angles to the base. A portion of the hemisphere in the form of a right circular cone is cut away, the circular base of the cone being the whole base of the hemisphere and its vertex the mid-point of OC. Find the distance of the centre of gravity G of the remaining body from O, showing that G lies outside the material of the body. *(L)*

Reference list of the positions of centres of mass.

1. A triangular lamina.
   G is the point where the medians meet. $AG/AA' = BG/BB' = CG/CC' = 2/3$.
2. A uniform wire in the form of an arc of a circle subtending an angle 2α at the centre of the circle. $OG = a \sin \alpha/\alpha$.

3. A uniform lamina in the form of a sector of a circle. $OG = \dfrac{2r\sin\alpha}{3\alpha}$.
4. A uniform solid right circular cone. The distance of G from the base $= \frac{1}{4}h$, where h is the height of the cone.
5. A uniform hemisphere of radius r. Distance of G from the circular face $= 3r/8$.

MISCELLANEOUS EXERCISE 12

1 A uniform plane lamina OAC is bounded by the arc OA of the curve $y^2 = x$ for which y is positive, the x-axis from O to C and the line $x = 9$ from C to A (see figure). State the coordinates of A and calculate
(a) the area of the lamina OAC,
(b) the coordinates of the centre of gravity of the lamina.

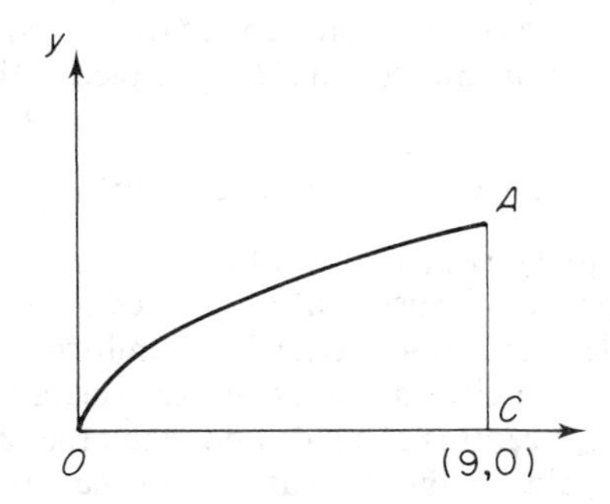

The lamina is now suspended from the point C. Find the tangent of the angle which AC makes with the vertical. (L)

2 In the uniform triangular lamina ABC, the mid-point of BC is D; $AB = AC$, $BC = 4a$ and $AD = 6a$. Prove, by integration, that the centre of gravity of this lamina is at a distance $2a$ from BC. The corner B is smoothly hinged to a fixed point so that the lamina can turn freely in its own plane which is vertical.
(a) When the lamina hangs freely in equilibrium, find the angle, to the nearest degree, made by AB with the vertical.
(b) When a couple of moment G is applied to the lamina in its vertical plane, the lamina remains at rest with AD horizontal. If the lamina is of weight W, find the value of G in terms of W and a. (L)

3 Show that the distance of the centroid of a uniform circular sector AOB from the centre O is $(2a\sin\theta)/3\theta$, where 2θ is the angle AOB and a is the radius.
Find the distance from O of the centroid of the segment of which AB is the chord, given that $\theta = \pi/6$. If a uniform lamina in the shape of this segment hangs at rest freely suspended from A, show that the tangent of the angle which AB makes with the downward vertical equals

$$(11 - 2\pi\sqrt{3})/(2\pi - 3\sqrt{3}).$$ (L)

4 Prove that the centre of mass of a uniform semicircular disc of radius a is distant $4a/3\pi$ from the bounding diameter.

A uniform circular cone of base radius a and height h is divided into equal halves by a plane through its axis of symmetry. Find the distance of the centre of mass of one of these halves from its triangular face. *(O&C)*

5 Prove that the centroid of a uniform solid hemisphere of radius a is at a distance $3a/8$ from O, the centre of its plane face.

The hemisphere is suspended by two vertical strings, one fastened at O and the other at a point P on the rim of the plane face. Given that the tension in one string is three times the tension in the other string, find the two possible values of the tangent of the angle made by OP with the horizontal. *(L)*

6 Prove that the centre of mass of a uniform lamina in the shape of a sector of a circle and subtending an angle 2θ at its centre is at a distance $(2a \sin \theta)/(3\theta)$ from the centre of the circle, where a is the radius of the circle.

A square lamina $ABCD$ of side $2a$ is made of uniform thin metal. When a semi-circular piece with CD as diameter is removed from the square, show that the centre of mass of the remainder of the lamina is at a distance $20a/(24-3\pi)$ from the line CD.

The remainder of the lamina is suspended from a light string attached at A and hangs in equilibrium. Show that AB is inclined to the downward vertical at an angle θ, where

$$\tan \theta = (28-6\pi)/(24-3\pi). \qquad (L)$$

7 A uniform right circular solid cone is of height h and the radius of its base is r. Prove that its centre of mass is at a distance $3h/4$ from the vertex.

The cone is made of the same material as a uniform solid cylinder of radius r and height $2h/3$. The circular bases of the cone and cylinder are joined together so that their centres coincide and when the resulting solid is freely suspended from a point on the circular edge of the join, the uppermost slant edge of the cone is horizontal. Show that $5h^2 = 36r^2$. *(L)*

8 A thin uniform wire of length $6a$ and weight W is bent so as to form three sides AB, BC, CD of a square $ABCD$. Calculate the distance of the centre of mass of the bent wire from the midpoint of BC.

The bent wire is suspended from A and hangs freely. Prove that, in equilibrium, AB makes an angle θ with the vertical, where $\tan \theta = \frac{3}{4}$.

If a horizontal force is now applied at C so as to maintain the wire in equilibrium with BC horizontal and below A, calculate the magnitude of this force and the magnitude and direction of the force exerted on the support at A. *(L)*

9 A uniform square lamina is divided into n^2 equal squares, and one of the corner squares is removed. If the board is suspended from the mid-point of one of the sides opposite this corner, prove that the angle which this side makes with the horizontal in the position of equilibrium is $\cot^{-1}(n^2+n-1)$. *(O&C)*

10 The mass per unit length of a ladder increases uniformly from the top to the bottom of the ladder, and is twice as great at the bottom as at the top. Find the position of its centre of mass. *(O&C)*

11 Prove that the centroid of a uniform triangular lamina ABC is at the centre of mass of three equal masses placed at the vertices A, B, C.

Prove that the centroid of a uniform quadrilateral lamina $ABCD$ is at the centre of mass of four equal masses m placed at the vertices A, B, C, D and one mass $-m$ placed at the intersection O of the diagonals AC, BD. *(O&C)*

13 Further Statics

In this chapter, the basic principles established in Chapters 2 and 3 are applied to problems involving the equilibrium of rigid bodies. The solutions of some problems for one rigid body are considered in §13.1, and the methods extended in §13.2 to problems involving two or more connected rigid bodies. In §13.3, some of the problems associated with the ways in which equilibrium can be broken are considered. The particular problems associated with finding forces in the members of a rigid framework are described in §13.4, and the Chapter concludes with a description of the methods of reducing one system of forces to another one.

13.1 Equilibrium of single bodies

We consider equilibrium problems for
(a) plane bodies (that is, laminae),
(b) three-dimensional configurations in which all the forces act in a vertical plane through the centre of gravity of the body.

The latter problems come within the ambit of the general theory of Chapter 3, since all the forces act in one plane. We shall, in fact, assume that problems concerning bodies symmetric about a vertical plane through the centre of mass can be treated as two-dimensional, provided that all the applied forces are in this plane. It is possible to justify this assumption.

A body will, as explained in §3.3, be in equilibrium provided
(a) the vector sum of the forces acting is zero,
(b) the total moment about any one point of all the forces is also zero.
Statical equilibrium problems effectively reduce to determining some quantities – for example, forces – such that these conditions hold.

The first step is to decide what forces act and at what points, and to mark them clearly on a diagram. The definitions of §2.3 must be used to decide the nature of the forces, and it should always be remembered that the force exerted by gravity on a body acts vertically downwards through its centre of mass. In problems involving rough contact it is again better, unless limiting friction is to be explicitly assumed, to treat the two components of the frictional reaction as unknowns and, if necessary, to apply the condition, $F \leqslant \mu R$, at the end of a problem.

Once the existence and nature of the forces has been established, the next step is to equate to zero separately the sum of the components in two non-parallel directions. In practice, this is most easily carried out, as described in Chapter 2, by equating the sum of the upwards (leftwards) components to the sum of the downwards (rightwards) components. Usually the most convenient directions for equating components are horizontal and vertical but occasionally, for example, in problems involving inclined planes, other directions are more convenient.

After equating components in two separate directions, the sum of the moments about a point are equated to zero. Again, this is usually done by equating the clockwise to the counter-clockwise moments. It does not matter which point is chosen, but it is usually best to take moments about a point through which most unknown forces act because they will then not occur in the moment equation. It is possible, as mentioned in §3.3, to replace one (or both) component equations by a moment equation. This is generally not advisable, as moment equations are slightly harder to obtain in the correct form than component equations. In any case, *it should be remembered that, for a single body, only three independent equations can be found.*

For the particular case of only three forces acting on a body in equilibrium, it is possible to use an apparently simpler method which does not involve taking moments. If the three forces are not all parallel, then the line of action of some pair of them must intersect at some point O. The sum of the moments of all three forces about O must be zero. The lines of action of two of them pass through O and so these two forces have zero moment about O. Therefore, the third also has zero moment about O and its line of action must pass through O. Hence, when any three non-parallel forces act on a body in equilibrium, their lines of action intersect at a point. Such problems can, therefore, be solved by finding the point of intersection and considering equilibrium at a point, as in §2.4. Though solutions obtained in this way can be rather short, they can also be tricky as some geometrical ingenuity or insight is often needed to find the point of intersection.

In determining the forces acting on a rigid body, the nature of the contact with some other body (possibly fixed) must be taken into account. We assume that this contact will be one of three types

(a) *simple point contact*, where one body exerts a force on another;

(b) *smooth hinging* which is conventionally taken to mean that the action of one body on another is a single force, but that the hinge permits rotation about the point of contact (if the hinge is not stated to be smooth, then a couple should also be assumed to be exerted at the point of contact);

(c) *contact along a line.*

In case (c), if the contact is smooth, there will be forces acting at various points in the region of contact. These forces will all be parallel and in the same sense and will, therefore, have a resultant which will act at some point in the region of contact. It may well be that both the position and magnitude of the resultant will be unknown. For rough contact, there will also be a force acting along the line of contact. We assume that, for bodies in contact with a plane and symmetric about a vertical plane through the centre of mass, the reaction of the plane will also be in the vertical plane through the centre of mass, provided that the applied forces are also in this plane. It is possible to justify this assumption.

EXAMPLE 1 *A uniform rod AB of mass 2m and length 2a is smoothly hinged at A to a fixed point. A particle of mass 3m is attached to B, and the rod and particle are maintained in equilibrium, with AB at an angle θ to the downward vertical, by a horizontal force of magnitude P applied at B. Find P, and the magnitude and direction of the reaction at the hinge.*

The forces acting on the rod are shown in Fig. 13.1, with X and Y denoting the vertical and horizontal components of the force exerted by the hinge on the rod at A. Equating the vertical components

$$Y = 5mg.$$

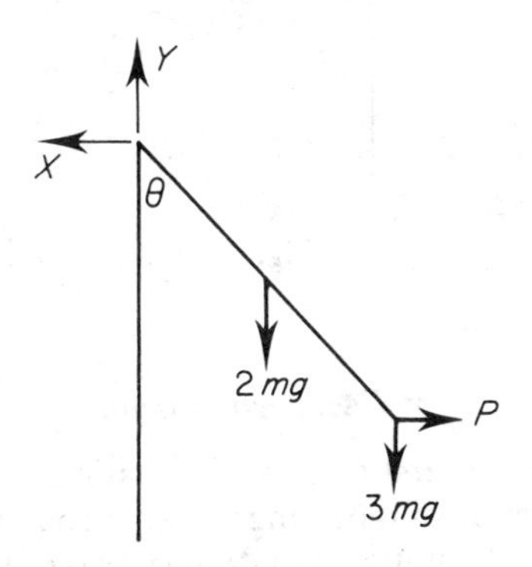

Fig. 13.1

Equating the horizontal components

$$X = P.$$

Since there are two unknown forces acting at A, it seems advisable to take moments about it, and this gives

$$2Pa \cos\theta = 2mg.a \sin\theta + 3mg.2a \sin\theta,$$

so that $P = \mathbf{4\,mg \tan\theta}$.

The magnitude of the resultant at the hinge is

$$(X^2 + Y^2)^{\frac{1}{2}} = \mathbf{mg(25 + 16 \tan^2\theta)^{\frac{1}{2}}}$$

and this resultant acts in the direction $\tan^{-1}(Y/X) = \mathbf{\tan^{-1}(5\cot\theta/4)}$ to the horizontal.

EXAMPLE 2 *A smooth sphere, of centre O, mass m and radius a, rests against a smooth vertical wall, being supported by a string of length 1·6a. One end of the string is attached to a fixed point A on the wall and the other to a point on the surface of the sphere. Find the force exerted by the sphere on the wall and the tension in the string.*

Fig. 13.2 shows the forces acting, with R denoting the magnitude of the horizontal force exerted by the wall on the sphere at the point of contact B. It is assumed that the string is inclined at an angle θ to the wall. Resolving vertically and horizontally

$$T\cos\theta = mg,$$
$$T\sin\theta = R.$$

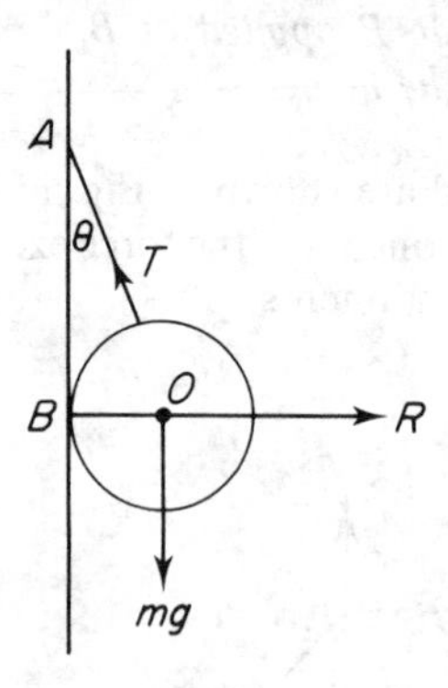

Fig. 13.2

Taking moments about B

$$TAB\sin\theta = mga.$$

Eliminating T between the first and third equations gives $\tan\theta = a/AB$, which shows that the line of action of the string passes through O. This result could have been found immediately by noticing that, since two of the forces pass through O, the line of action of the third must also.

The straight line OA is, therefore, of length $2{\cdot}6a$ and $\sin\theta = 5/13$, $\cos\theta = 12/13$. Therefore

$$R = \frac{5mg}{12}, \quad T = \frac{13mg}{12}.$$

EXAMPLE 3 *A uniform rod of mass m and length 2a rests in equilibrium, inclined at an angle* $\tan^{-1}(3/4)$ *to the horizontal, with one end resting on a smooth vertical wall and the other end on a rough horizontal floor. The vertical plane containing the rod is perpendicular to the wall. Find the least value of the coefficient of friction.*

The forces acting on the rod are shown in Fig. 13.3. The reaction of the wall is normal to it (that is, horizontal) and of magnitude R and the horizontal and vertical components of the force exerted by the ground on the rod are denoted by F and S respectively. Resolving horizontally and vertically gives

$$S = mg,$$
$$R = F.$$

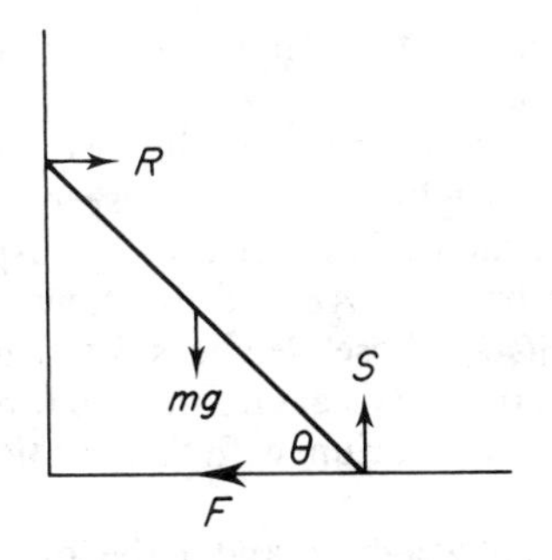

Fig. 13.3

Taking moments about the lower end of the rod avoids introducing F and S, and we get

$$2Ra \sin\theta = mga \cos\theta$$

$$R = \tfrac{1}{2}mg \cot\theta.$$

Therefore

$$\frac{F}{S} = \tfrac{1}{2}\cot\theta = \frac{2}{3}.$$

For equilibrium to be possible, this ratio must be less than or equal to μ, so $\mu \geqslant 2/3$. The least possible value for μ is, therefore, $\mathbf{\frac{2}{3}}$.

EXERCISE 13.1

1 A uniform rod can turn freely about a smooth fixed hinge attached to one of its ends. It is pulled aside from the vertical by a horizontal force, acting at the other end and equal to a quarter of the weight of the rod. Find the tangent of the angle between the rod and the downward vertical.

2 Find, for the configuration of the previous question but with the horizontal force equal to three-quarters of the weight of the rod, the reaction at the hinge, given that the rod is of weight W.

3 Given that the rod in the previous two questions is of length $2a$ and weight W, and that the joint exerts a couple of magnitude aW in the sense opposite to that through which the rod has been rotated, find the horizontal force required to turn it through an angle of $\tan^{-1}(3/4)$.

4 A uniform rod AB of weight W and length $2a$ is in equilibrium in a vertical plane, with one end A against a smooth vertical wall and the other end attached to one end of a light inextensible string of length l. The other end of the string is fixed at a point C at a distance h vertically above A. In the equilibrium position, the rod is inclined at an acute angle to the downward vertical. Find the tension in the string, the cosine of the angle between the string and the downward vertical, and the cosine of the angle between the rod and the downward vertical.

5 A heavy uniform rod of weight W and length $2a$ is suspended from a fixed point O by two strings, one of which is of length $2b$ and the other string is such that the angle between the strings at O is a right angle. Find the tensions in the strings.

6 An L-shaped structure is formed by joining two uniform rods AB and BC, of lengths $2a$ and $2b$ and weights W and $5W$ respectively, at the point B, so that angle ABC is a right angle. The structure is suspended from A. Find the tangent of the angle between AB and the downward vertical.

7 A square lamina $ABCD$, whose weight is $4W$, can turn in a vertical plane about a smooth horizontal axis at A. Find the force acting along BC which will keep the lamina in equilibrium with this side horizontal. Find also, in this case, the reaction at A.

8 A uniform plate ABC, of weight W and in the form of an equilateral triangle, can turn in a vertical plane about a smooth horizontal axis at A. It is kept in equilibrium with AB vertical by a string acting along BC produced. Find the tension in the string and the tangent of the angle between the vertical and the reaction at A.

9 A uniform circular disc, of weight W and centre O, can rotate freely about a horizontal axis through a point C on its rim. It is supported by a smooth peg at the point B on its rim, where $BCO = \theta$, so that CO is inclined at an angle α to the horizontal. Find the reaction of the peg and, for fixed α, determine θ such that the reaction has the least value.

10 A uniform rod AB rests in a vertical plane with A in contact with a rough wall and B supported by a light inextensible string BC, where C is vertically above A and $AB = BC$. Given that the coefficient of friction is μ and that the string is inclined at an angle θ to the downward vertical, find the minimum value of $\tan\theta$.

11 A uniform ladder rests at an angle α to the horizontal, with one end on a smooth vertical wall and the other on rough horizontal ground, the coefficient of friction being μ. Show that for equilibrium to be possible $\mu > \frac{1}{2}\cot\alpha$. When $\mu = 2\cot\alpha$, a horizontal force applied at the lower end of magnitude P_1 is just sufficient to move the foot of the ladder away from the wall, and a horizontal force of magnitude P_2 at the same point is just sufficient to move the foot of the ladder towards the wall. Find P_2/P_1.

12 When no force is applied at the foot of the ladder in question 11 and $\mu = 0{\cdot}8\cot\alpha$, find the maximum weight that can be placed at the top of the ladder without disturbing equilibrium, taking the weight of the ladder as W.

13 A uniform ladder rests with one end on horizontal ground and the other against a vertical wall, the coefficients of friction being 3/5 and 1/3 respectively. Find the tangent of the angle between the horizontal and the ladder when the ladder is on the point of slipping.

14 A light ladder of length 12 m stands in a vertical plane on rough ground and leans against an equally rough wall. The ladder is inclined at an angle $\tan^{-1}(1{\cdot}5)$ to the horizontal and the coefficient of friction is 0·5. Find the greatest distance along the ladder at which a weight can be placed on the ladder without causing it to slip.

15 A uniform cube of side $4a$ is at rest on a rough horizontal plane, the coefficient of friction being 0·5. A horizontal force is applied perpendicular to a vertical face and at a height a above the mid-point O of the lowest edge of that face, so that the cube is just on the point of sliding. Find the point at which the reaction of the plane acts.

16 A uniform circular cylinder rests between two inclined planes, as shown in the diagram. The plane AB is smooth whilst the plane AC is rough, the coefficient of friction being μ. A horizontal force equal to the weight of the cylinder acts at D, the mid-point of the highest generator of the cylinder. Show that equilibrium is not possible unless $\alpha > 45^\circ$ and find, when $\tan\alpha = 4/3$, the least value of μ for which equilibrium is possible.

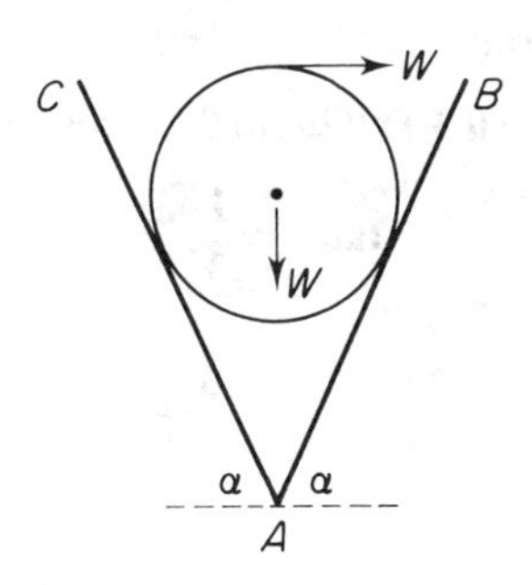

13.2 Equilibrium of connected bodies

Problems involving equilibrium of several bodies are only more complicated than those for a single body in that it may be necessary to consider the equilibrium of each body separately. The same approach as in §13.1 is still appropriate, though it should be remembered, in working out the forces at a point of contact, that the force exerted by body A on body B is equal and opposite to that exerted by body B on body A. Care is needed, when marking forces on a diagram, to ensure that forces acting at points of contact satisfy the law that 'action and reaction are equal and opposite'.

However, it is not always necessary to consider the equilibrium of each body separately. Indeed, much information can be found from considering the bodies as one large body. Using this approach avoids introducing the forces between bodies but, if these are actually the forces required, there is no alternative to considering the equilibrium of the separate bodies.

If two bodies A and B are connected by a light inextensible string, then the tension in the string makes no contribution to either the component equations or the moment equation *for the whole system* since the action of the string on A is equal and opposite to that on B, and, therefore, cancels when both bodies are considered.

EXAMPLE 1 *Two uniform rods AB and BC, each of length 2a and mass m, are smoothly hinged at B and suspended from A. The middle points of the rods are joined by a light inextensible string so that angle ABC is a right angle. Find*

(*i*) *the inclination of AB to the vertical,*
(*ii*) *the tension in the string,*
(*iii*) *the reaction at B.*

(i) Fig. 13.4(a) shows the configuration in equilibrium with AB inclined at an angle θ to the downward vertical. First we consider the two rods as one system. Resolving horizontally and vertically only shows that the force exerted on the rod at A is vertical and of magnitude $2mg$. Taking moments about A gives (since the string makes no contribution)

$$mga\sin\theta + mg(2a\sin\theta - a\cos\theta) = 0$$

$$\Rightarrow \tan\theta = \frac{1}{3}.$$

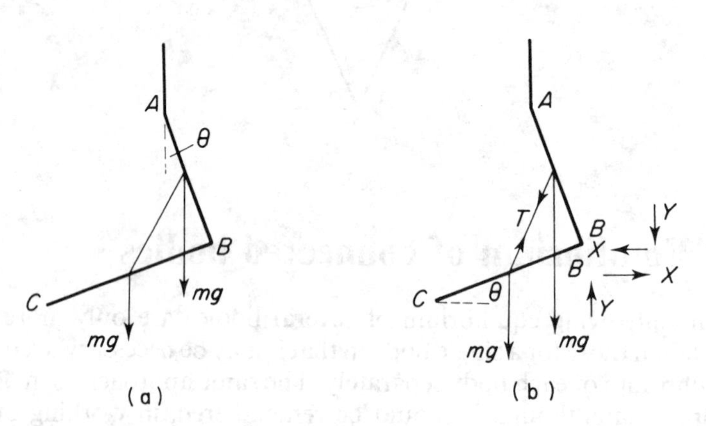

Fig. 13.4

(ii) In order to find the tension in the string, it is necessary to consider the bodies separately. The forces acting on the separate bodies are as shown in Fig. 13.4(b), with the law of 'action and reaction' applied at B. The string is at a perpendicular distance $a/\sqrt{2}$ from B. Taking moments about B for rod BC gives

$$\frac{Ta}{\sqrt{2}} = mga\cos\theta$$

$$\Rightarrow T = 3mg/\sqrt{5}.$$

(iii) The next step should be to consider the vertical and horizontal components of the forces acting on BC. However, here this step can be avoided by noticing that, since two of the three forces acting on the rod BC pass through the mid-point of BC, then so must the third. The reaction at B, therefore, acts along BC and is of magnitude

$$T\cos\pi/4 - mg\sin\theta = 2mg/\sqrt{10}.$$

EXAMPLE 2 *Fig. 13.5(a) shows two smooth cylinders, each of radius 2a and mass 3m, which rest in contact with both a smooth floor and a smooth vertical wall. The axes of the cylinders are at a distance 6a apart. A third smooth cylinder, of mass 4m and radius 3a, rests on the cylinders. Find the force between the cylinders and between the cylinders and the wall and floor.*

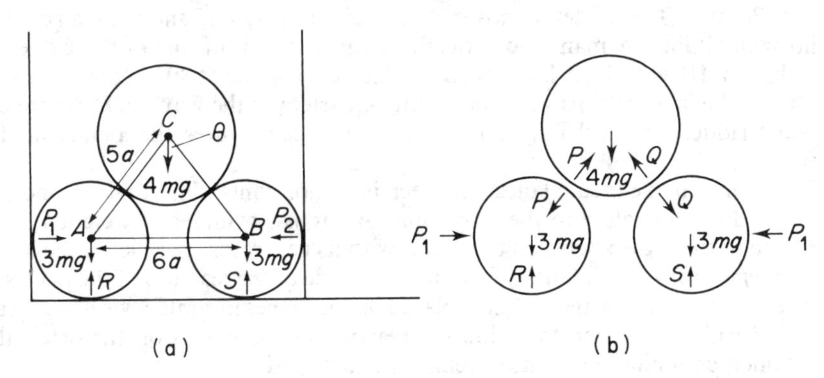

Fig. 13.5

The forces exerted by the wall and floor on the cylinders are shown in Fig. 13.5(a). Resolving horizontally and vertically gives $P_1 = P_2$ and

$$R+S = 10mg.$$

Taking moments about C gives $R = S$, so that $R = S =$ **$5mg$**. Equality of R and S also follows from consideration of symmetry.

Fig. 13.5(b) shows the forces acting between the cylinders. Resolving horizontally for the upper cylinder gives $P = Q$, and resolving vertically gives

$$2P\cos\theta = 4mg,$$

where θ is the angle between AC and the vertical.

Since the triangle ABC is isosceles, the vertical through C bisects AB, so that $\sin\theta = \frac{3}{5}$ and hence $\cos\theta = 4/5$. This gives $P =$ **$2{\cdot}5mg$**. Resolving horizontally for the cylinder centre A gives

$$P_1 = P\sin\theta = \mathbf{1{\cdot}5mg}.$$

EXERCISE 13.2

1 Two uniform rods AB and BC of equal length are smoothly hinged at B, and are of weights W and $2W$ respectively. They stand in a vertical plane at right angles to each other with A and C on a smooth horizontal plane. Determine the horizontal forces that have to be applied at A and C to maintain equilibrium.

2 A stepladder of weight $2W$ consists of two equal uniform parts, smoothly hinged at the top and held together by a rope halfway between the top and bottom so that the angle between the parts is 2θ. A man of weight $5W$ mounts the ladder and stops a quarter of the way up. Find, neglecting friction at the ground, the tension in the rope, and the horizontal and vertical components of the reaction at the hinge.

3 Two equal uniform rods of mass W are smoothly hinged at B, and attached to light strings at A and C. The strings pass over small smooth pegs in the same horizontal line and carry equal weights of $2W$ at their other ends. The rods hang in equilibrium symmetrically between the two pegs. Find, correct to the nearest degree, the inclinations of the strings and the rods to the horizontal.

4 A stepladder is formed by smoothly hinging two identical uniform rods AB and BC at B. The ladder stands in a vertical plane with A and C on a rough horizontal floor. A man whose weight is equal to that of one of the ladders walks up AB when both ladders are inclined at an angle of 30° to the vertical. One of the ladders slips when he is three-quarters of the way up. Determine which ladder slips, and find, correct to two significant figures, the coefficient of friction.

5 Two smooth inclined planes intersect in a horizontal line and are both inclined at an angle α to the horizontal. A circular cylinder of weight W is placed in the region between the plane, with its generators parallel to the line of intersection of the planes. Find the magnitude of the reaction of the planes. A second identical cylinder is now placed on the planes in contact with the first cylinder along a generator. Find the reaction of the planes on the original cylinder, assuming its position remains unchanged.

6 Two rough equal spheres, each of weight W, rest in contact with each other on a rough horizontal plane. The coefficient of friction at each of the three points of contact is μ. Forces $5W$ and $4W$ are applied to the spheres inwards along the line of centres. Find the range of μ such that equilibrium is possible.

7 Find, in the previous question, the range of μ when the forces are $4W/5$ and $3W/5$.

8 Two smooth spheres, each of radius a and weight W, lie symmetrically in contact in a smooth spherical bowl of radius na. Find the reaction between the smaller spheres.

13.3 Problems involving slipping and toppling

Fig. 13.6 shows the vertical section through the centre of mass of a uniform cube of mass m and side $2a$, which rests in equilibrium on a rough horizontal plane. A force of magnitude P is applied at the mid-point of an upper edge as shown. Instinct suggests that for relatively small values of θ, the cube will tend, as P increases, to tilt about A, whereas for values of θ near 90°, it is more likely to slide. This instinctive approach gives no method of finding out what will happen for a particular value of θ as P increases, and so a more accurate analysis is clearly necessary.

The reaction of the plane is assumed to have components F and R as shown, and to act at a distance h from A. Resolving horizontally and vertically gives

$$F = P\sin\theta,$$

$$R = mg + P\cos\theta.$$

Taking moments about A gives

$$Rh = mga - 2Pa\sin\theta.$$

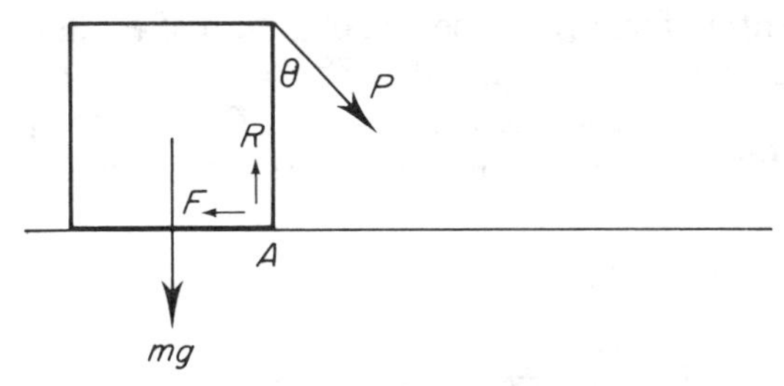

Fig. 13.6

From these equations we see that when $P = 0$, the point at which the reaction acts is vertically below the centre of mass, but as P is increased, it moves gradually to the right, and passes through A when $P = \frac{1}{2}mg \operatorname{cosec} \theta$. If P increases beyond this value, equilibrium is not possible because the reaction would be outside the cube and, therefore, the cube tilts.

So far we have assumed that the cube does not slide but equilibrium can also be broken by sliding. This occurs when

$$\frac{F}{R} = \mu = \frac{P \sin \theta}{mg + P \cos \theta},$$

and so
$$P = \frac{\mu mg}{(\sin \theta - \mu \cos \theta)}$$

provided that $\tan \theta > \mu$.

Toppling will occur before sliding if the value of P which produces toppling is less than that which gives sliding, that is,

$$\frac{mg}{2 \sin \theta} < \frac{\mu mg}{(\sin \theta - \mu \cos \theta)}$$

$$\Rightarrow \sin \theta < \mu(\cos \theta + 2 \sin \theta).$$

For $\sin \theta = 0{\cdot}1$ this will hold provided that $\mu > 0{\cdot}084$ (that is, even for relatively smooth surfaces) whereas for $\theta = \frac{1}{2}\pi$ it only holds for $\mu > 0{\cdot}5$. This confirms the instinctive view that for small values of θ tilting is more likely than sliding.

In problems where equilibrium can be broken by tilting or sliding, the two possibilities have to be considered separately and the condition for each to occur found. Tilting occurs when equilibrium cannot be maintained by a reaction in the region of contact, whereas sliding occurs when the friction exceeds limiting friction. Once the condition for each has been found it has to be decided, as above, which condition is first satisfied.

Since there are effectively only three forces acting on the cube, this problem can be solved (possibly more directly) by using the result that the lines of action of the forces acting must intersect at a point. The vertical through the centre of mass and the line of action of P intersect at the point O at a distance $a \cot \theta$ above the cube. The line OR in Fig. 13.7, therefore, represents the line of action of the resultant force of the plane. The cube slips before it tilts if friction is limiting (that is, if the angle ϕ is equal to λ)

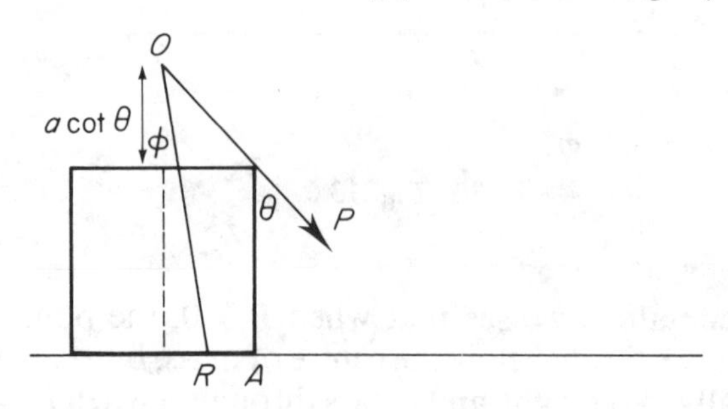

Fig. 13.7

with AR positive. We have that

$$\tan \phi = \frac{a - AR}{2a + a \cot \theta}$$

and, therefore, $\phi = \lambda$ with $AR > 0$ requires that

$$\mu < \frac{1}{2 + \cot \theta}.$$

Lami's theorem gives
$$\frac{mg}{\sin(\theta - \phi)} = \frac{P}{\sin \phi},$$

so that
$$P = mg \frac{\sin \phi}{\sin(\theta - \phi)}.$$

For toppling to occur
$$\tan \phi = \frac{a}{2a + a \cot \theta}$$

which, when substituted into the above expression for P, gives $P = \frac{1}{2}mg \operatorname{cosec} \theta$. Slipping occurs when $\phi = \lambda$, and substituting ϕ equal to λ in the expression for P gives the value previously obtained.

EXAMPLE 1 *A uniform square lamina, of mass m and side 2a, is placed on a line of greatest slope of a rough plane, inclined at an angle α to the horizontal. The plane of the lamina is vertical. Assuming the plane to be sufficiently rough for the lamina not to slip, find the greatest value of* tan α *for which equilibrium is possible. When α exceeds this value, find the greatest and least forces parallel to a line of greatest slope that can be applied at the highest point of the lamina without disturbing equilibrium.*

Fig. 13.8(a) shows the configuration without the applied force. The reaction of the plane acts at a distance h above the lowest point A and has components R and F. For equilibrium, the resultant of R and F must be vertical and pass through the centre of mass of the lamina. Equivalently, the vertical from the centre of mass must intersect the base of the square. This vertical, in fact, intersects the base at a distance $a(1-\tan\alpha)$ from the lowest point A, and hence the required condition is $\mathbf{\tan\alpha \leqslant 1}$.

Fig. 13.8(b) shows the configuration when a force of magnitude P acts parallel

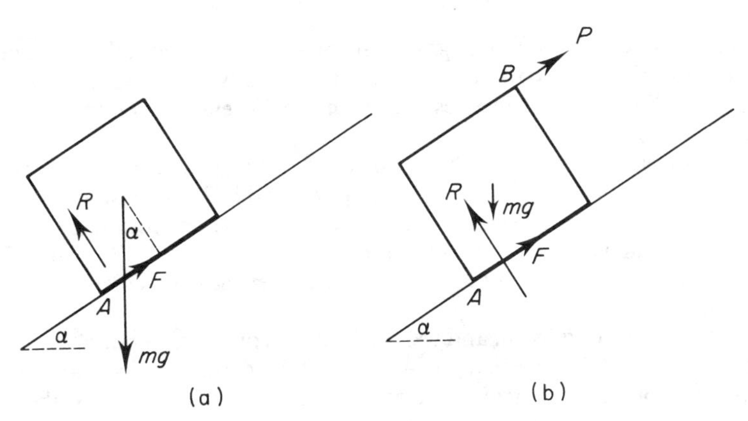

Fig. 13.8

to the plane at the highest point B. Resolving along and perpendicular to the plane gives

$$P+F = mg\sin\alpha$$
$$R = mg\cos\alpha.$$

Taking moments about A

$$2Pa + mga\cos\alpha = mga\sin\alpha + Rh.$$

For $P = 0$, h has to be positive, which gives $\tan\alpha \leqslant 1$ as obtained previously.

For $\tan\alpha > 1$, equilibrium will not be possible unless $h \geqslant 0$ which requires

$$2P \geqslant mga\cos\alpha(\tan\alpha - 1),$$
$$\mathbf{P \geqslant \tfrac{1}{2}mga\cos\alpha(\tan\alpha - 1).}$$

This gives the lowest possible value of P. As P increases, h will increase but cannot exceed $2a$ because otherwise the reaction would be outside the lamina. Therefore,

$$2Pa + mga\cos\alpha \leqslant mga\sin\alpha + 2mga\cos\alpha$$

or

$$\mathbf{P \leqslant \tfrac{1}{2}mg(\cos\alpha + \sin\alpha).}$$

When P has its maximum value, the lamina is about to turn about its highest point.

The problem can also be solved fairly easily by considering the point of intersection of the three forces acting.

In the above problems, the location of the centre of mass was obvious but problems are often set where this is not the case and the centre of mass has first to be located. The bodies in this case are often composite ones (for example, a hemisphere attached to the plane base of a cylinder) and the method described in Chapter 12 has to be used to find the centre of mass. Once this has been done the problem can be solved as described above.

EXERCISE 13.3

1 A cone of base radius r and height h rests on a rough horizontal plane, and the inclination of the plane to the horizontal is gradually increased. Show that the cone will slide before it topples over if the coefficient of friction is less than $4r/h$.

2 A cone of base radius r, height h and weight W rests on a rough horizontal plane, the coefficient of friction being μ. A gradually increasing horizontal force is applied at the vertex. Find the values of the magnitude of the force when the cone is just about to tilt and when it is just about to slide. Hence show that it will slide or tilt according to whether the coefficient of friction is less, or greater, than r/h.

3 A uniform cube of edge $4a$ and weight W stands on a rough horizontal plane, the coefficient of friction being μ. A gradually increasing horizontal force P is applied to one of its vertical face at a point a distance a directly above the centre of the face. Find the values of P when the cube is just about to tilt and when it is just about to slide. Hence determine the condition for the equilibrium being broken by tilting.

4 A square lamina of weight W stands upright on a rough horizontal plane, the coefficient of friction being μ. A string attached to one of the upper corners of the lamina is pulled with gradually increasing force at right angles to the diagonal passing through that corner, the angle between the upward vertical and the direction of the pull being 45°. Find the force P in the string when the lamina is about to tilt and when it is about to topple. Prove that it tilts before it topples if $\mu > \frac{1}{3}$.

5 Obtain the values of P, in the previous question, when the angle between the direction of pull and the downward vertical is 45°. Find, for this case, the condition that tilting occurs before sliding.

13.4 Light frameworks

One particularly important class of problems involving the equilibrium of several connected bodies is that involving finding the forces in a framework of rods, such as the one shown in Fig. 13.9. Rigid frameworks form the basis of many structures, and it is important for structural engineers to be able to estimate the forces in the members of a frame. In order to make such estimates, two assumptions are made
(a) the weights of the rods may be neglected,
(b) the rods are freely jointed (hinged, pinned) at their points of contact so that no couples are exerted at the joints.

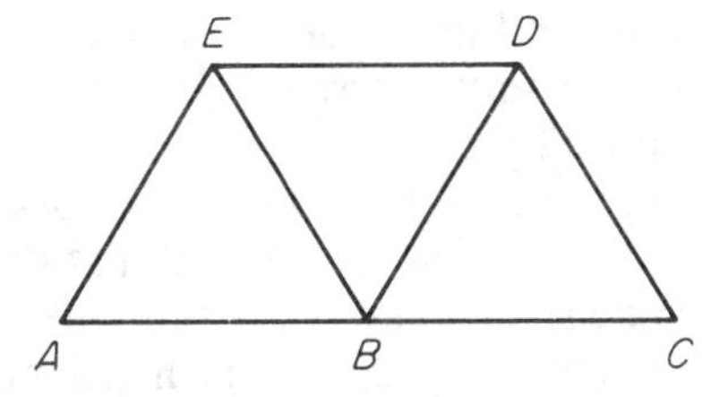

Fig. 13.9

Although such idealised frames do not correspond exactly to those used in building construction, their study can give useful results since the forces ignored (the weights) are usually small when compared to the loads applied. For example, a comparatively light steel girder may be subjected to loads many times its own weight.

For any particular rod, such as AB in Fig. 13.9, there will be forces exerted on the rod by the joints at A and B, and the rod will be in equilibrium under the action of these forces. Therefore, since the weight of the rod is neglected, the force at B exerted on AB is equal and opposite to that at A exerted on AB. The possible alternatives are shown in Fig. 13.10(a) and (b), with the forces acting towards AB and away from AB respectively. The rods will also exert equal and opposite forces on the joints, and Fig. 13.10(c) and (d) show, respectively, the forces exerted on the joints corresponding to those shown in (a) and (b). In (c) the rod is said to be in compression (thrust) whilst (d) corresponds to the rod being in tension.

Each member of the frame will be either in tension or thrust and the joints will be in equilibrium under the action of these forces. Therefore, the determination of the forces in members of a framework reduces to

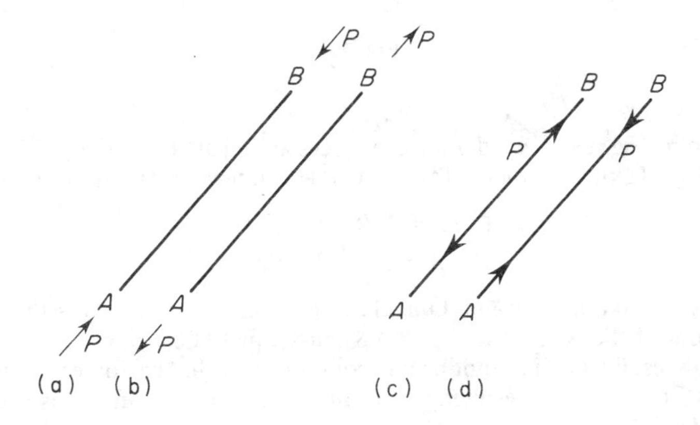

Fig. 13.10

establishing equilibrium conditions for forces acting at a point as in §2.4. Equations involving the various forces will hence be obtained and can be solved to give any required force.

It should also be remembered that the framework is itself fastened to (or supported by) some external structure and that the force of support is generally unknown. Whenever possible, it simplifies matters to consider the equilibrium of the framework as a whole in order to find the unknown forces of support. Once this is done, the next step is to systematically establish the equations of equilibrium at each joint. In doing so, one should start with joints at which there are fewest unknown forces. If there are only two unknown forces at a joint at which other known forces act, the former can be found immediately by resolving in two separate directions. Often, work can be avoided by careful choice of the directions in which components are resolved, since resolving in a direction perpendicular to an unknown force means that this force will not enter the equation. It is also important to take advantage of any symmetries occurring.

EXAMPLE 1 *Find the forces in the rods of the framework of Fig.* 13.11 *assuming that the rods are of equal length, the framework is supported by vertical forces at A and C, and vertical loads of* 100 N *and* 500 N *respectively are applied at E and D as shown.*

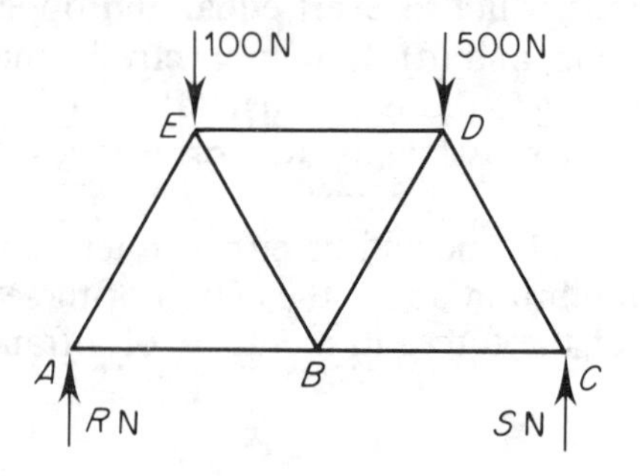

Fig. 13.11

As shown in Fig. 13.11, we denote the forces acting at A and C by R N and S N respectively. Taking moments for the whole framework about A and C

$$4S = 100 + 3 \times 500$$
$$4R = 500 + 3 \times 100,$$

so that $R = 200$ and $S = 400$. One check on these calculations is that, for the equilibrium of the whole system, $R + S$ must equal 600.

We now establish the equilibrium conditions at A, the forces acting being shown in Fig. 13.12(a). Resolving vertically, the force at A in AE is a thrust of magnitude T given by

$$T \sin 60° = 200 \text{ N}$$

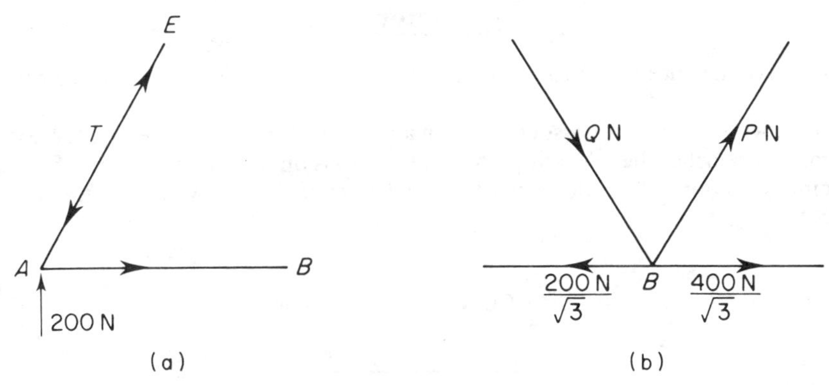

Fig. 13.12

so that $T = (\mathbf{400/\sqrt{3}})\,\mathbf{N}$. Resolving horizontally shows that the force at A in AB is a tension of magnitude $T\cos 60°$ and is, therefore, $(\mathbf{200/\sqrt{3}})\,\mathbf{N}$. The configuration of rods at C and A are geometrically identical but the applied force at C is twice that at A. The forces in BC and CD are, therefore, twice those in AB and AE respectively, so that BC sustains a tension of magnitude $(\mathbf{400/\sqrt{3}})\,\mathbf{N}$ and CD a thrust of magnitude $(\mathbf{800/\sqrt{3}})\,\mathbf{N}$.

There are now two unknown forces at each of the joints B, E and D, and it is relatively immaterial which joint is considered next. It is possibly slightly easier to look at B first, since here two of the rods act along the same line. The forces acting are shown in Fig. 13.12(b). The forces acting at B in BE and BD are assumed to be of magnitude Q N and P N in the senses shown. Resolving vertically gives $Q = P$ and resolving horizontally gives

$$Q\cos 60° + P\cos 60° + \frac{400}{\sqrt{3}} = \frac{200}{\sqrt{3}},$$

giving $Q = P = -\mathbf{200/\sqrt{3}}$. Therefore BD is in thrust and BA in tension.

The force in ED can now be found by considering equilibrium at E. The forces acting are shown in Fig. 13.13. Resolving horizontally shows that the force of magnitude K N at E in ED (shown as a thrust) is given by

$$K = \frac{400}{\sqrt{3}}\cos 60° + \frac{200}{\sqrt{3}}\cos 60° = \mathbf{100\sqrt{3}}.$$

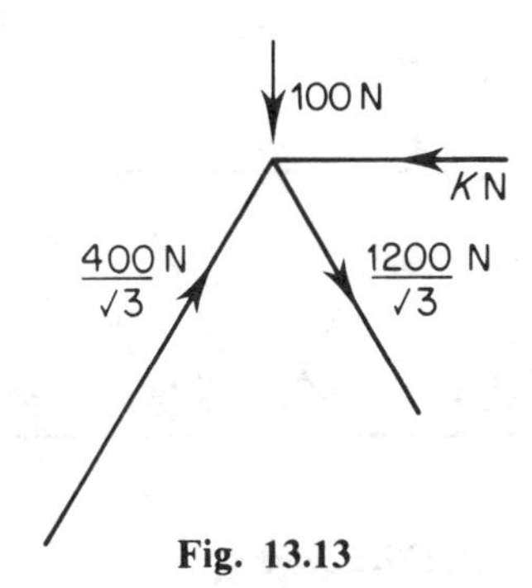

Fig. 13.13

EXERCISE 13.4

In each of the questions 1 to 4, find the unknown external forces and the forces in the bars named.

Arrows which are not associated with a numerical value indicate that there is a simple support at the relevant point if the arrow is directed upwards and that the point is anchored by a downward vertical force if the arrow is downwards.

1

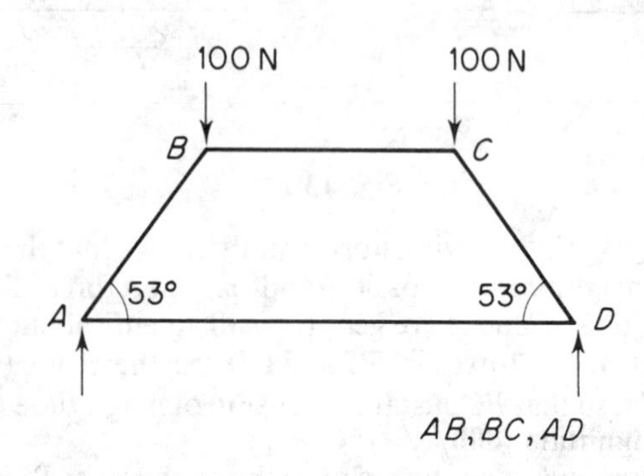

2

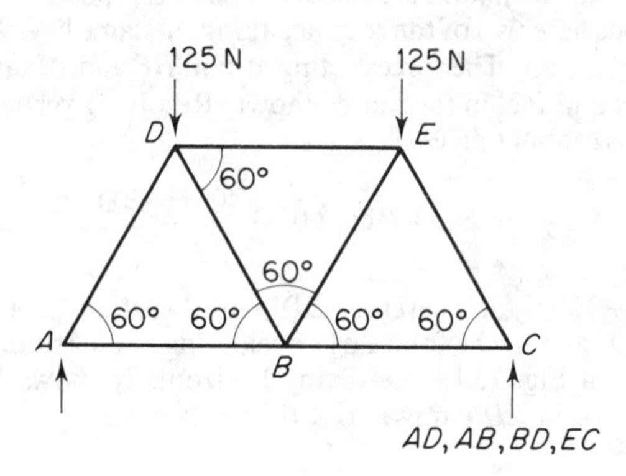

3

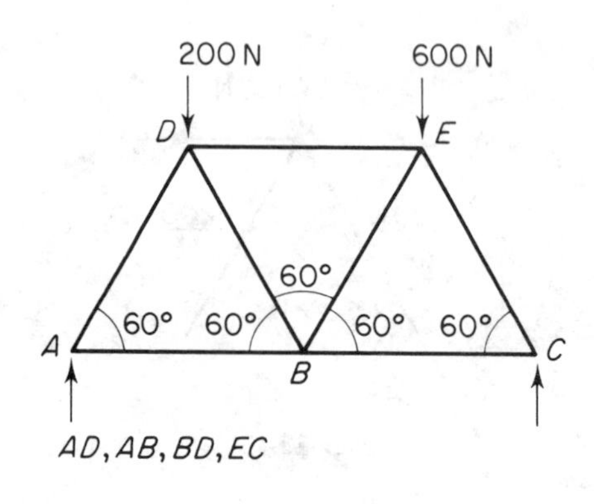

4

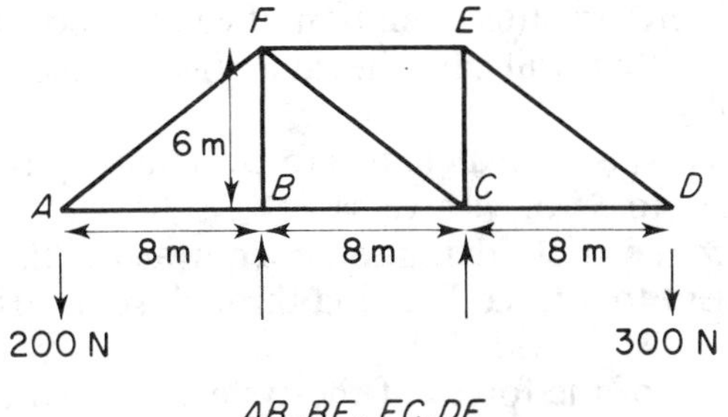

AB, *BF*, *FC*, *DE*

5 The framework in the diagram consists of four light bars, *AB*, *BC*, *CD*, *DB* freely jointed at *B*, *C*, *D* and attached to a vertical wall at *A* and *D*. A weight of 100 N is suspended from *C*. Find the forces in all the bars and the reactions at *A* and *D*.

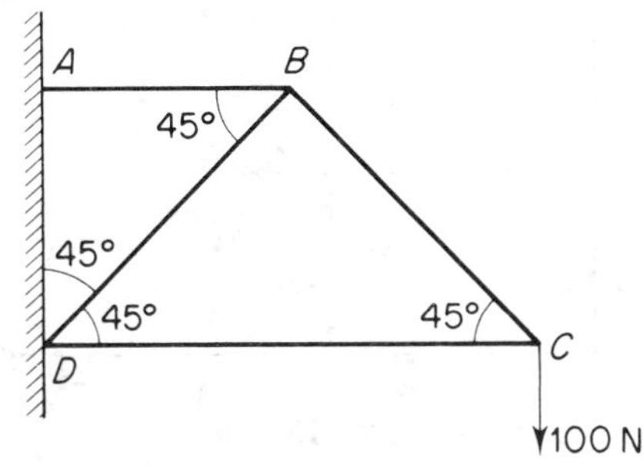

6 In the diagram, the frame is freely hinged to a vertical wall. Find the forces in the bars and the reactions at the wall.

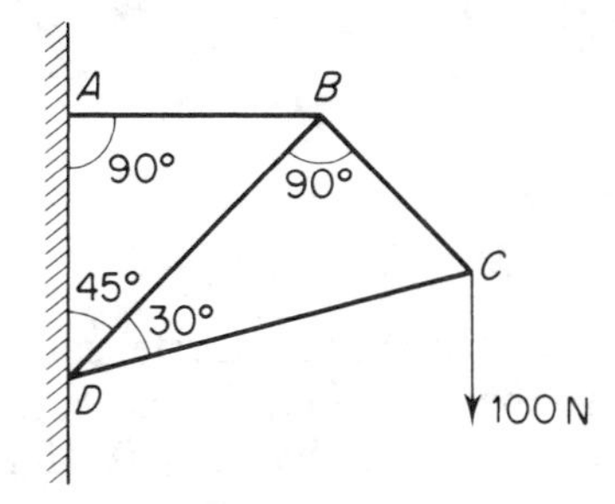

13.5 Equivalent force systems

Two force systems are defined to be equivalent if and only if
(a) the vector sum of the forces in both systems is the same, and
(b) the sum of the moments about one point of the forces of both systems is the same.
In fact, if (b) holds for one point and (a) holds also then (b) will hold for all points. This result is proved at the end of this section. (It can be shown that the motion of a rigid body is completely determined by the vector sum of the forces acting on it and by the sum of the moments of those forces

about a point. Therefore, this definition of equivalence ensures that two equivalent force systems will have the same effect on the motion of a rigid body.)

If a force system S_1 is equivalent to a force system S_2, then it is said that S_1 can be reduced to S_2 or vice versa.

It follows from the above definition of equivalence that any system of forces is equivalent to a force $\mathbf{F}$ acting through some arbitrarily chosen point P and a couple provided that

(i) the vector sum of the forces of the system is equal to $\mathbf{F}$,

(ii) the moment of the whole system about any point O is equal to the sum of the moment of the couple and the moment about O of the force $\mathbf{F}$ through P.

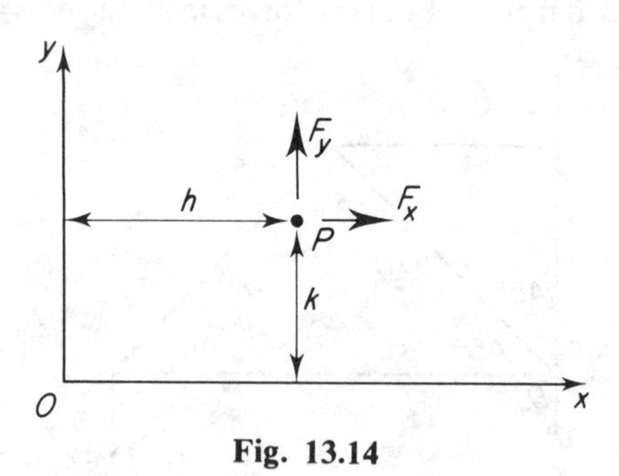

Fig. 13.14

If the counter-clockwise moment of the force system about O is M, whilst that of the couple is G, and

$$\mathbf{F} = F_x\mathbf{i} + F_y\mathbf{j},$$

then (ii) gives (Fig. 13.14)

$$M = hF_y - kF_x + G. \qquad 13.1$$

M, F_x and F_y are determined by the force system and, therefore, by varying h and k we can vary G so that, provided F_x and F_y are not both identically zero, we can find points P such that $G = 0$. Hence, there are points, satisfying

$$M - (hF_y - kF_x) = 0, \qquad 13.2$$

such that the original system is equivalent to a single force acting at such points. Equation 13.2 shows that all these points lie on a line parallel to $\mathbf{F}$. We refer to the single force acting through points on this line as the resultant of the system. If $\mathbf{F} = 0$, then the system is equivalent to a couple and, if the couple is also zero, then the system is in equilibrium.

In problems on reduction of force systems, it is generally required either to reduce the system to a force through a given point and a couple, or to find the resultant completely (–that is, to find the resultant force and the line along which it acts, generally referred to as the line of action of the resultant). In both cases the first step is to find the magnitude and direction of the resultant force. This is done by adding the components, in two perpendicular directions, of the forces of the system. In the first type of problems, the couple is the moment, about the given point, of the forces of the system. In the second type of problem, it is sufficient to find the point of intersection of the resultant with some given line (for example, a coordinate axis) and the point is determined by the condition that the moment about it (that is, the couple) is zero.

EXAMPLE 1 *Forces of magnitude P, $2P$, $3P$ and $4P$ act along the sides AB, BC, CD, DA respectively, in the senses indicated by the order of the letters, of the square $ABCD$ of side $2a$. Reduce the system to a force acting through the mid-point of AB and a couple.*

The system is shown in Fig. 13.15. The vector sum of the forces has components $2P$ in the directions BA and DA, so the resultant has magnitude $2\sqrt{2}P$ and is parallel to CA. The moment of the system about E, the mid point of AB, is $2Pa+3P.2a+4Pa = 12Pa$ in the counter-clockwise direction. Therefore, the system is equivalent to a force of magnitude $\mathbf{2\sqrt{2}P}$ acting through E parallel to CA and a counter-clockwise couple whose moment is of magnitude $\mathbf{12Pa}$.

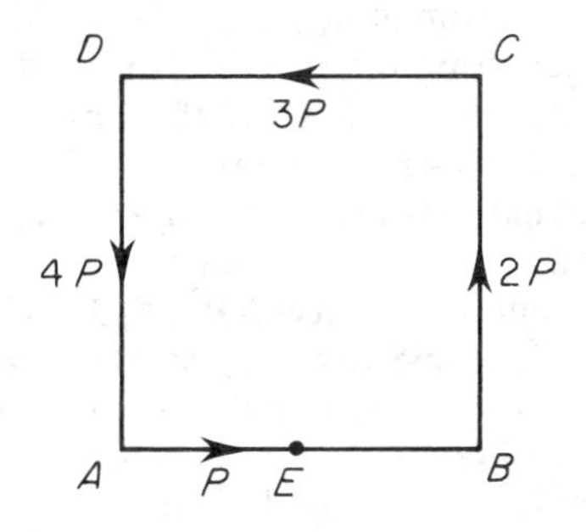

Fig. 13.15

EXAMPLE 2 *Determine the resultant of the force system in the previous example and, taking x and y axes along AB and AD respectively, find the equation of the line of action of this resultant.*

We assume the resultant intersects AB at the point F, where $AF = b$. The moment about F has to be zero and this gives

$$4Pb+2P(2a-b)+6Pa = 0$$

Hence, $b = -5a$ and so F is on **BA produced at a distance $5a$ from A.**

The resultant passes through the point $(-5a,0)$ and is parallel to AC. Therefore, the equation of its line of action is

$$\mathbf{y = x+5a.}$$

EXERCISE 13.5

1 $ABCD$ is a square of side 1 m whose diagonals intersect at O. Forces 1 N, 2 N, 3 N, 4 N, $6\sqrt{2}$ N act along AB, BC, CD, DA, AC. Find the magnitude and direction of the single force at O and the magnitude of the couple which together are equivalent to the given forces. Also find at what distance from A the resultant of the five forces cuts the line AB.

2 Forces of 1 N, 2 N, 3 N and 4 N act along the sides AB, BC, CD and DA respectively of a square $ABCD$ of side 1 m in the directions indicated by the letters. Reduce the system to a single force and find where its line of action cuts AB produced.

3 Forces $18P$, $32P$, $24P$ and $40P$ act along the sides $\overline{AB}$, $\overline{CB}$, $\overline{CD}$ and $\overline{AD}$ of a square $ABCD$ of side 1 m. Reduce the system to a single force and find where its line of action cuts BA produced.

4 For the system of the previous question, calculate the Cartesian equation of the line of action of the resultant, taking AB and AD as axes of x and y respectively.

5 Forces 2 N, 6 N, 10 N, 14 N and $16\sqrt{2}$ N act along the sides AB, BC, CD, DA and the diagonal BD of a square of side. Taking AB and AD as axes of x and y respectively, find the magnitudes of the resultant force and the equation of its line of action.

6 ABC is an equilateral triangle of side 5 m and forces 4 N, 2 N and 2 N act along AB, AC, BC in the directions indicated by the letters. Find the resultant force and the distances from A at which it cuts AB and AC.

7 Forces each of 5 N act along the sides BC and DA of a square $ABCD$ and forces 10 N each act along CD and XC in the senses indicated by the order of the letters, where X divides AB in the ratio 1:3. Find the equation of the line of action of the resultant referred to AB and AD as axes, taking the side of the square to be one unit.

8 $ABCD$ is a square of side 1 m. Forces $7P$, $8P$, $9P$, $10P$ and $2\sqrt{2}P$ act along AB, CB, CD, AD and DB respectively in the senses indicated by the order of the letters. Show that the system reduces to a couple and find its moment.

9 $ABCDEF$ is a regular hexagon. Forces each of one unit act along AB, BC, CF, FE and ED in the senses indicated by the order of the letters. Reduce the system to a single force and state its line of action.

We now present a proof of the result stated earlier that if the vector sum of the moments for two systems about one point are equal and the sum of the components in two perpendicular directions are equal, then the sum of the moments about any point are equal. We consider two force systems one with forces with components (X_i, Y_i) acting at points $[(x_i, y_i), i = 1, 2, \ldots, n]$ and the other involving forces with components (X'_j, Y'_j) acting at $[(x'_j, y'_j)\ j = 1, 2, \ldots, m]$.

For the force systems to be equivalent we require that

$$\sum_{i=1}^{n} X_i = \sum_{j=1}^{m} X'_j \qquad 13.3$$

$$\sum_{i=1}^{n} Y_i = \sum_{j=1}^{m} Y'_j. \qquad 13.4$$

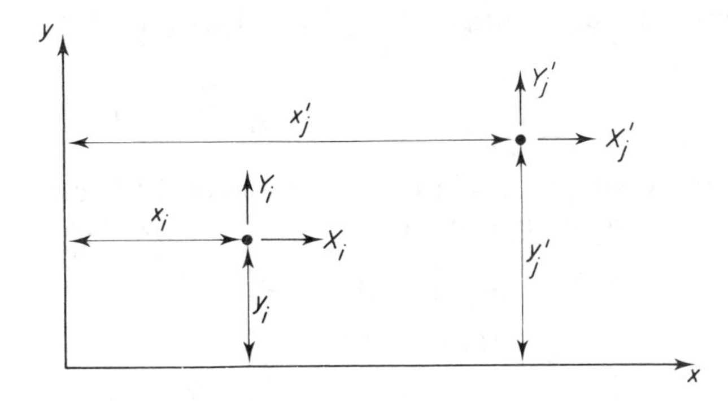

Fig. 13.16

We also assume that the moments of the two force systems about the origin are the same, that is,

$$\sum_{i=1}^{n} (Y_i x_i - X_i y_i) = \sum_{j=1}^{m} (Y'_j x'_j - X'_j y'_j). \qquad 13.5$$

The moments of the force systems about the point (h, k) will be equal if

$$\sum_{i=1}^{n} \{Y_i(x_i - h) - X_i(y_i - k)\} = \sum_{j=1}^{m} \{Y'_j(x'_j - h) - X'_j(y'_j - k)\} \quad 13.6$$

that is,

$$\sum_{i=1}^{n} (Y_i x_i - X_i y_i) - h \sum_{i=1}^{n} Y_i + k \sum_{j=1}^{n} X_i = \sum_{j=1}^{m} (Y'_j x'_j - X'_j y'_j) - h \sum_{j=1}^{m} Y'_j$$
$$+ k \sum_{j=1}^{m} X'_i.$$

It follows immediately from equations 13.3 to 13.5 that this last equation is true.

It is relatively easy using the above formulae to prove the result, referred to in §3.3, that each component equation in the equilibrium could be replaced by a further moment equation. It follows from the left-hand side of equation 13.6 that the moments M_2 and M_3 about the points (h, k) and (h', k') are given by

$$M_2 = M_1 - h \sum_{i=1}^{n} Y_i + k \sum_{i=1}^{n} X_i$$

$$M_3 = M_1 - h' \sum_{i=1}^{n} Y_i + k' \sum_{i=1}^{n} X_i,$$

where $$M_1 = \sum_{i=1}^{n} (Y_i x_i - y_i X_i)$$

is the moment about the origin. Also, R_1 and R_2 defined by

$$R_1 = \sum_{i=1}^{n} X_i, \quad R_2 = \sum_{i=1}^{n} Y_i$$

are the x- and y-components of the vector sum of the forces. Therefore, assuming one equilibrium condition to be $M_1 = 0$, we have

$$M_2 = -hR_2 + kR_1$$
$$M_3 = -h'R_2 + k'R_1$$

Setting R_1 (or R_2) equal to zero and equating M_2 (or M_3) to zero gives R_2 (or R_1) to be zero, so that all the components vanish. If both M_2 and M_3 are set equal to zero, then R_1 and R_2 will both be zero unless

$$\frac{h}{k} = \frac{h'}{k'},$$

which means that the three points about which moments are taken are collinear. Therefore, equating one component of the total force to zero and equating to zero the moment of the forces about two separate points gives equilibrium. Also, equating to zero the moment about three non-collinear points also yields equilibrium.

MISCELLANEOUS EXERCISE 13

1 A uniform ladder of length $2a$ rests in limiting equilibrium in a vertical plane with its lower end on rough horizontal ground and its upper end against a smooth vertical wall. The ladder makes an angle 60° with the ground. Show that the coefficient of friction is $\sqrt{3}/6$.

The ladder is lowered in its vertical plane whilst still resting against the smooth wall and the ground to make an angle 30° with the ground. The coefficient of friction between the ladder and the ground remains at $\sqrt{3}/6$. A man whose weight is four times that of the ladder starts climbing up the ladder. Find how far he can climb up the ladder before it slips. (*L*)

2 A uniform rod AB, of length $2a$ and weight W, is hinged to a vertical post at A and is supported in a horizontal position by a string attached to B and to a point C vertically above A, where $\angle ABC = \theta$. A load of weight $2W$ is hung from B. Find the tension in the string and the horizontal and vertical resolved parts of the force exerted by the hinge on the rod. Show that, if the reaction of the hinge at A is at right angles to BC, then

$$AC = 2a\sqrt{5}. \qquad (L)$$

3 A uniform rod AB, of mass $4m$ and length l, is smoothly jointed at the end A to a fixed point of a fixed straight horizontal rough wire AC. The end B is attached by means of a light inextensible thread of length l to a bead of mass m, which can slide along the wire. Given that the system is in equilibrium,

with the rod and thread both inclined at an angle θ to the horizontal, show that

(a) the tension in the thread is $mg \operatorname{cosec} \theta$,

(b) the normal reaction of the bead on the wire has magnitude $2mg$.

Given that the coefficient of friction between the bead and the wire is $\frac{1}{2}$, show that, for equilibrium to be possible, θ must lie in the interval $\pi/4 \leqslant \theta \leqslant \pi/2$. (*L*)

4 A light structure ABC consists of a vertical post AB of height h, fixed in the ground at A, joined rigidly to a horizontal crossbar BC of length b. A force of magnitude F is applied at C, in the plane ABC, at an angle θ to the downward vertical, as shown in the diagram. Express the system of forces exerted by the ground on the post as a force through A together with a couple, giving the magnitude of the force and the moment M_A of the couple in terms of h, b, θ and F. Show on a diagram the direction of the force and the sense of the couple. Express likewise the system of forces exerted by the post on the crossbar as a force through B and a couple M_B.

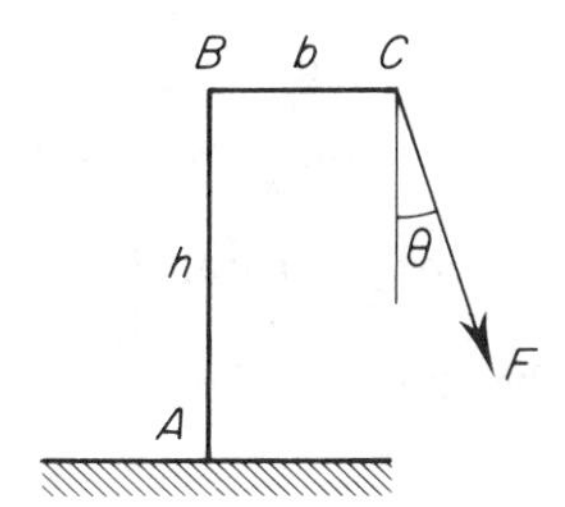

The angle θ is varied while the magnitude F of the force applied at C remains constant. State the directions of the force applied at C for which M_A and M_B, respectively, are maximum. (*JMB*)

5 A light rod AB of length $5a$ has its end A on level ground and its end B against a smooth vertical wall. A weight W is attached to the rod at C, where $AC = 2a$. The rod is in equilibrium in a vertical plane perpendicular to the wall and makes an angle arctan (4/3) with the ground.

(i) If the ground is rough, find the least possible value of the coefficient of friction between the rod and the ground.

(ii) When the ground is smooth, equilibrium is maintained by an elastic string, one end of which is attached to the rod at C and the other end to a point D on the wall. Given that D is vertically below B and CD is horizontal, find

(a) the tension in the string,

(b) the natural length of the string given that its modulus is $5W/2$. (*L*)

6 The diagram shows two uniform rods AB and AC each of weight $\frac{1}{2}W$ and length $4a$ rigidly connected at A so that they are at right angles to each other. The rods are free to turn in a vertical plane about a horizontal axis at A. Two small rough rings P and Q, each of weight W, are threaded on AB and AC respectively and joined by a light inextensible string of length $2a$ and passing through a small smooth ring at A. The coefficient of friction between the rings and the rods is μ, where $\mu < \frac{1}{2}$. Given that AB is inclined at an angle θ to the vertical and P is just about to slip down AB, show that the tension T in the string is given by

$$T = W(\cos\theta - \mu\sin\theta).$$

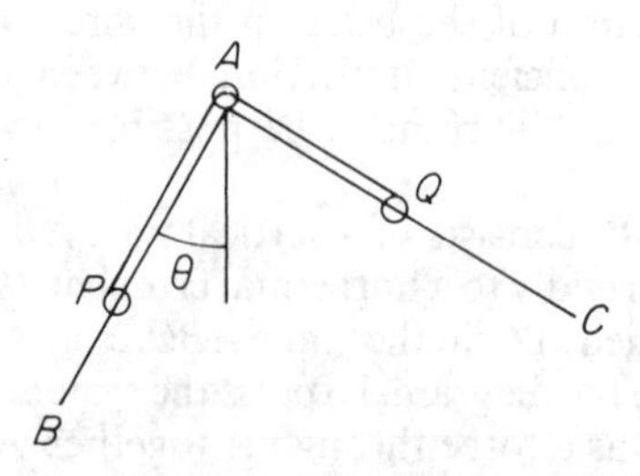

By considering the equilibrium of Q when P is just about to slip down AB, obtain a second expression for T in terms of W, μ and θ. Hence express $\tan\theta$ in terms of μ.

Given that x denotes the distance of P from A when it is about to slip down AB show, by considering the equilibrium of the whole system, that

$$x = a + 2a\mu.$$

Find also the distance of P from A when it is on the point of moving up AB. *(AEB 1982)*

7 A small rough horizontal peg is parallel to a smooth vertical wall at a horizontal distance c from it. A uniform heavy rod AB of length $2a$ ($> c$) rests on the peg with the end A resting against the wall.

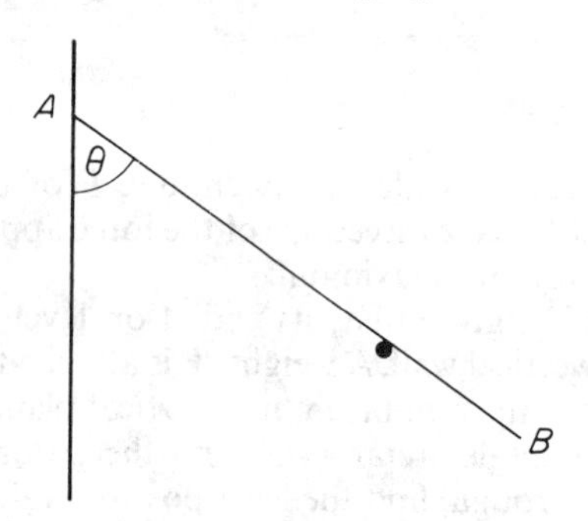

If the rod is in limiting equilibrium, prove that

$$c = a\sin^2\theta(\sin\theta + \mu\cos\theta),$$

where μ is the coefficient of friction between the rod and the peg and θ is the angle between the rod and the vertical. *(O)*

8 A heavy uniform rod of length $2a$ rests in equilibrium against a small smooth horizontal peg with one end of the rod on a rough horizontal floor. If the height of the peg above the floor is b and friction is limiting when the rod makes an angle $\pi/6$ with the floor, show that the coefficient of friction between the floor and the rod is $\dfrac{\sqrt{3}a}{8b - 3a}$. *(O)*

9 State the conditions which must be satisfied when a body is in equilibrium under the action of three non-parallel forces.

A uniform rod AB of weight W and length $2a$ has its end A resting on a smooth inclined plane of slope 30°. A light elastic string BC of natural length $\frac{12}{5}a$ joins the end B to a fixed peg C. The peg is at a perpendicular distance $3a$ from the inclined plane. The system rests in equilibrium with the rod horizontal. Prove that the angle between the string and the rod is 60°.

Find the tension and the extension of the string, and show that its modulus is $\frac{1}{2}W\sqrt{3}$. (*C*)

10 A uniform rod of length $2l$ and weight w hangs freely from its end which is attached to a fixed point A. A light inextensible string of length a has one end attached to A and the other to a point on the surface of a smooth uniform sphere of radius a and weight W. The system hangs in equilibrium with the rod tangential to the sphere. Prove that, if $2aW = lw$, the string and the rod are equally inclined to the vertical and show that the tension in the string is

$$W \cos 15°/\cos 30°. \quad (C)$$

11 A uniform circular hoop of weight W hangs over a rough horizontal peg A. The hoop is pulled with a gradually increasing horizontal force P which is applied at the other end B of the diameter through A and acts in the vertical plane of the hoop. Given that the system is in equilibrium and that the hoop has not slipped when AB is inclined at an angle θ to the downward vertical, find the value of P in terms of W and θ. Show also that the ratio of the frictional force to the normal reaction at the peg is

$$(\tan \theta)/(2 + \tan^2\theta).$$

Show that, when the coefficient of friction is $\frac{1}{2}$, the hoop never slips, however hard it is pulled. (*L*)

12 A uniform hexagonal lamina $ABCDEF$ has weight W. The lamina is smoothly pivoted at B. It is kept in equilibrium in a vertical plane with AB horizontal and the vertices C, D, E and F above AB by a force of magnitude P acting at D in the direction AD. Find P and the magnitude and direction of the force exerted on the lamina by the pivot.

The force at D is reversed in direction whilst retaining its magnitude. Find the weight which must be attached at C to keep the lamina in this position of equilibrium. (*L*)

13 A uniform rod AB, of length $a\sqrt{3}$ and weight W, has a light ring attached to the end A as shown in the diagram. The ring is free to slide along a long horizontal wire. One end of a light inextensible string of length a is attached to the end B of the rod and the other end is attached to a fixed point O on the wire. The rod, string and wire are in a vertical plane.

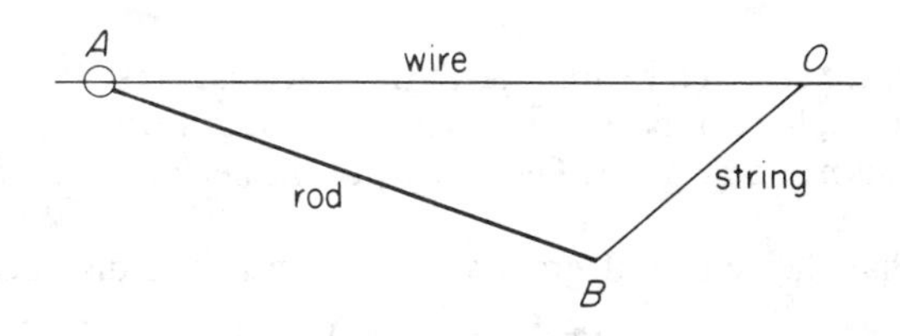

(a) Given that the wire is smooth, find the angle between the rod and the wire when the rod is in equilibrium.

(b) Given that the coefficient of friction between the ring and the wire is μ, and that the rod is in equilibrium when inclined at an angle of 30° to the horizontal, show that $\mu \geqslant (\sqrt{3})/5$. (*L*)

14 A uniform lamina $ABCD$ is in the form of a trapezium in which $DC = AD = a$, $AB = 2a$ and $\angle BAD = \angle CDA = 90°$. When the lamina is freely suspended from a point E in BC the edge AB is vertical. Show that $BE{:}EC = 4{:}5$.

Find the tangent of the angle which AB makes with the vertical when the lamina is freely suspended from B.

The lamina is suspended by two vertical strings attached at C and D. Find the ratio of the tensions in the strings if CD is horizontal. (*L*)

15 A toy tower is constructed of thin uniform sheet metal and consists of a right circular cylinder, with open ends, of height h and radius r, where $h > r$, covered by a hemispherical cap of radius r. Show that, for such a tower, the centre of gravity is at a distance $\bar{x}$ from the ground where

$$\bar{x} = \frac{1}{2}(h + r)$$

The tower is suspended from a string which is attached to a point on the rim of the hemispherical cap. Show that, when the tower hangs in equilibrium, the axis of symmetry of the tower is inclined at an angle α to the horizontal, where

$$\tan \alpha = \frac{h - r}{2r}.$$ (*L*)

16 $ABCD$ is a square lamina, of side 2 units and weight W, and E is the point on DC such that $DE = \frac{2}{3}$ unit. EF is drawn perpendicular to DE meeting BD at F. The piece $BFEC$ is cut away. Show that the distance from DA of the centre of gravity of the remainder is $\frac{29}{45}$ unit.

The lamina stands in a vertical plane with DE resting on a horizontal table. Find the least force which applied to the lamina will make it topple in its own plane. (*L*)

17 A heavy uniform circular disc centre X and radius R has a circular hole centre Y and radius r cut from it, where $r < R$. If $XY = R - r$, and the centre of gravity of the crescent-shaped lamina is at a distance $4r/9$ from X, show that $R = 5r/4$.

The lamina is now suspended from a point on its outer rim lying on the perpendicular to XY through X. Find the angle which XY makes with the vertical. If the weight of the crescent is W, find the smallest weight which must be attached to the lamina to maintain XY in a horizontal position. (*L*)

18 A toy top is constructed by joining, at their circular rims, the bases of a solid uniform hemisphere of base radius a and a solid uniform right circular cone of base radius a and height h. The density of the hemisphere is b times that of the cone.

(i) Show that the centre of gravity of the top is at a distance

$$\frac{3h^2 + b(3a^2 + 8ah)}{4(h + 2ab)}$$

from the vertex of the cone.

(ii) The top, when suspended from a point on the rim of the base of the cone, rests in equilibrium with the axis of the cone inclined at an acute angle, α, to the downward vertical. Find $\tan\alpha$.
(iii) For the case $h = 4a$, determine the range of values of b such that the top cannot rest in equilibrium with the slant surface of the cone in contact with a smooth horizontal plane. (*AEB 1982*)

19 Two uniform rods AB and AC, each of weight w and of length $2a$, are smoothly jointed at A. An additional weight kw is also placed on AB at a distance pa from A. The rods stand in equilibrium in a verticle plane, each inclined at an angle θ to the downward vertical, with B and C in contact with a horizontal plane.
(i) Given that B and C are joined by a light inextensible string and that the plane is smooth find the reactions at B and C and show that the tension in the string is

$$\tfrac{1}{2}w\tan\theta\,[k(1-\tfrac{1}{2}p)+1].$$

(ii) Given that B and C are no longer joined by a string but that the plane is uniformly rough with coefficient of friction μ,
(a) determine whether, as k increases, equilibrium is broken by slipping at B or at C.
(b) find, when $p = 0$, the least value of k, in terms of θ and μ, which will cause slipping.
Find the range of values for $\tan\theta$ for which your expression in (b) for k is valid. State what happens in each of the two cases when $\tan\theta$ is outside this range. (*AEB 1982*)

20 Two equal uniform rods AB, BC, each of weight W, are smoothly jointed at B, and AB is smoothly hinged to a fixed point A. The rods are kept at right angles to each other by a light string AC. Prove that
(i) AB is inclined at an angle $\tan^{-1}(\frac{1}{3})$ to the vertical,
(ii) the tension in the string is $3W/2\sqrt{5}$.
By resolving horizontally and vertically for the rod BC, or otherwise, prove that the resultant reaction at B is $\frac{1}{2}W$. (*O*)

21 The diagram shows two parallel vertical walls at a distance $3a$ apart, with two uniform smooth cylinders, each of radius a and weight W. The axes of the cylinders are horizontal and parallel to the walls. Each cylinder is in contact with the other and with one wall; the lower cylinder rests on horizontal ground. Find the magnitudes of the reactions between the two cylinders and between each cylinder and the corresponding wall. (*C*)

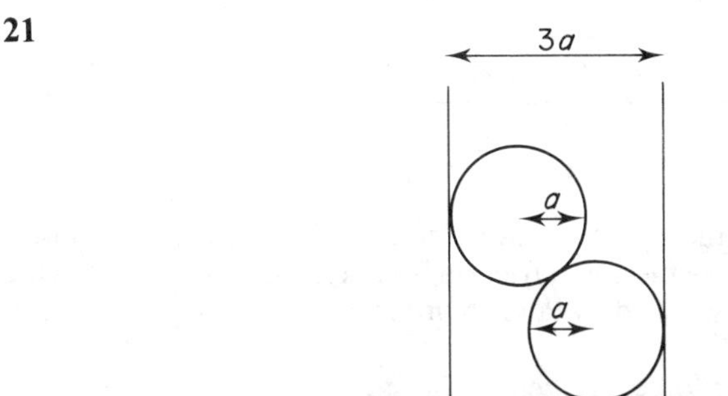

The diagram shows a uniform rectangular lamina $ABCD$ of weight W, with edges $AB = CD = 2a$, $BC = DA = a$. The lamina rests in a vertical plane, with two small studs at A and B making contact with a horizontal plane. The coefficients of friction at A and B are μ and μ' respectively. A horizontal force of slowly increasing magnitude P acts at C, in the plane of the lamina.

22

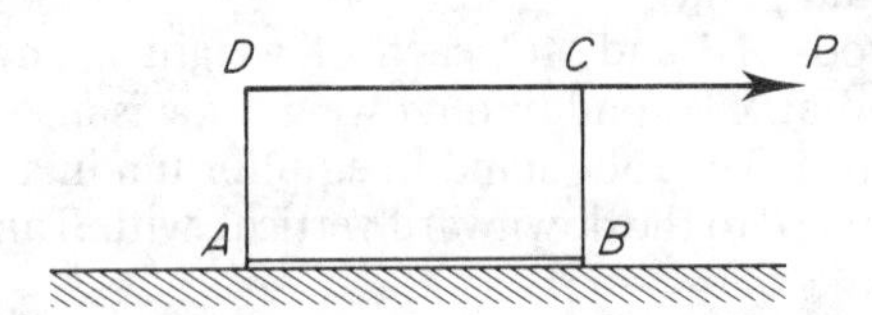

Assuming that equilibrium is broken by sliding, show that this occurs when

$$P = \frac{(\mu + \mu')\,W}{2 + \mu - \mu'}.$$

Given that this value for P is sufficient to cause the lamina to turn about B, deduce that $\mu' < 1$. (*C*)

23 A square lamina $ABCD$ rests in a vertical plane with AB in contact with a rough horizontal table. The coefficient of friction between the lamina and the table is μ. A gradually increasing force is applied at C in the plane of the lamina in an upwards direction and making an angle α with DC (so that part of the line of action of the force lies inside the lamina). Prove that equilibrium is broken as follows:

(i) if $\tan\alpha > 2$ and $\mu > 1/(\tan\alpha - 2)$, by the lamina tilting about A,
(ii) if $\tan\alpha < 2$ and $\mu > 1/(2 - \tan\alpha)$, by the lamina tilting about B,
(iii) if $\mu < 1/|\tan\alpha - 2|$, by the lamina sliding. (*O*)

24 The figure represents a solid formed by the removal of a sphere of radius $a/2$ from a uniform solid hemisphere of base-radius a. The point O denotes the centre of the plane base of the hemisphere and C is that point on the hemisphere which is farthest from the base. Show that the centre of gravity of the solid is on OC at a distance of $a/3$ from O.

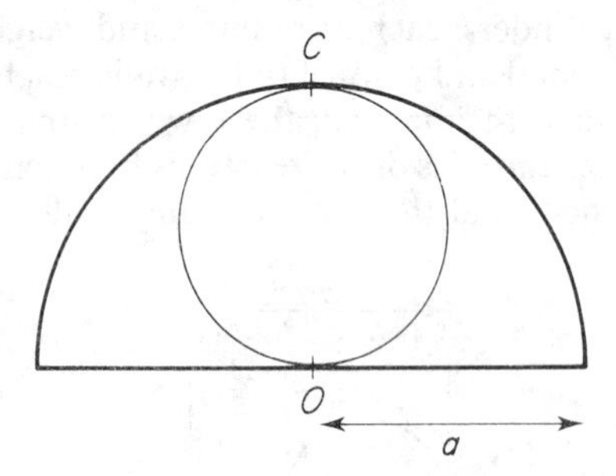

The solid is placed with its plane face on a rough plane inclined at an angle α to the horizontal. The coefficient of friction between the solid and the plane is $\frac{2}{3}$. Find the condition satisfied by $\tan\alpha$ in order that the solid does not slip immediately.

A gradually increasing force P is then applied to the solid. The force acts at the point C on the solid in a direction parallel to the line of greatest slope and up the plane. Given that the solid is of weight W show that, when it is just about to tilt

$$P = W\cos\alpha + \tfrac{1}{3}W\sin\alpha.$$

Obtain the condition that has to be satisfied by $\tan\alpha$ in order that the body tilts before it slips. *(AEB 1983)*

25 (i) Forces $4P$, $4P$, $6P$ act respectively along AB, BC, CA, where ABC is an equilateral triangle of side d. The line of action of the resultant, R, of these forces meets BC produced at D. Calculate
(a) the magnitude and direction of R,
(b) the distance BD.
(ii) A system of forces acts in the plane of an equilateral triangle LMN of side a. This system has moments $+K$, $+K$, 0 about L, M, N respectively ($+$ indicates the sense LMN). Show that the resultant of this system is parallel to ML and find its magnitude. *(L)*

26 A rigid square lamina $ABCD$ of side a is subject to forces of magnitude 1, 2, 3, 1, $3\sqrt{2}$ and $\lambda\sqrt{2}$ units acting along AB, BC, CD, AD, AC and DB respectively in the directions indicated by the order of the letters. Given that the direction of the resultant force is parallel to AC, find λ.

With this value of λ, find the total moment about A of the forces acting on the lamina. Hence, or otherwise, find AE in terms of a, where E is the intersection of AB with the line of action of the resultant force. *(JMB)*

27 The midpoints of the sides BC, CA and AB of a triangle ABC are L, M and N respectively. The centroid of this triangle is G, and O is any point in the plane of the triangle. Show that
(i) the resultant of the forces $\overrightarrow{OA}$, $\overrightarrow{OB}$ and $\overrightarrow{OC}$ is $3\overrightarrow{OG}$,
(ii) the resultant of the forces $\overrightarrow{AB}$, $\overrightarrow{BC}$ and $\overrightarrow{CA}$ is a couple whose magnitude can be represented by twice the area of the triangle ABC.

Find the resultant of each of the following systems of forces
(iii) $\overrightarrow{OL}$, $\overrightarrow{OM}$ and $\overrightarrow{ON}$,
(iv) $\overrightarrow{AL}$, $\overrightarrow{BM}$ and $\overrightarrow{CN}$. *(AEB 1970)*

28 Forces of magnitudes $5P$, $7P$, mP and nP act along $\overrightarrow{AB}$, $\overrightarrow{BC}$, $\overrightarrow{CD}$ and $\overrightarrow{DA}$, where $ABCD$ is a square of side a. With A as origin and AB and AD as x-axis and y-axis respectively, the equation of the line of action of the resultant of this system of forces is $3x - 4y - 8a = 0$. Find the values of m and n.

A couple is now added to the system and the line of action of the resultant of the enlarged system acts through B. Find the moment and state the sense of this couple. *(AEB 1976)*

29 Forces P, $2P$, $3P$, $4P$ act along the sides $\overrightarrow{AB}$, $\overrightarrow{BC}$, $\overrightarrow{CD}$, $\overrightarrow{DA}$ of a square of side a. Prove that the system is equivalent to a single force acting through D together with a couple. Find the magnitudes of the force and the couple.

Prove further that the system is equivalent to a single force and find the distance of the line of action of this force from D. *(O)*

30 In a rectangle $ABCD$, $AB = DC = 4$ m and $AD = BC = 3$ m. Forces of magnitude 3, 5, 4, 6, P, Q newtons act along AB, BC, CD, DA, BD, AC respectively in the directions indicated by the order of the letters. Show that this system of forces cannot be in equilibrium.
(i) If the system reduces to a couple, show that $Q = 7P$.
(ii) If the system reduces to a single resultant force acting through B, show that $Q = 15$. Given $P = 10$, find the magnitude of the resultant force. *(L)*

31 The diagram shows a symmetrical framework consisting of nine smoothly jointed light rods, with

$$AB = BC = BD = AD\sqrt{2} = CD/\sqrt{2}.$$

If AD is vertical and the framework rests on smooth supports at C and E, determine the stresses in the rods when the framework carries weights P at A, Q at B and Q at F. (O)

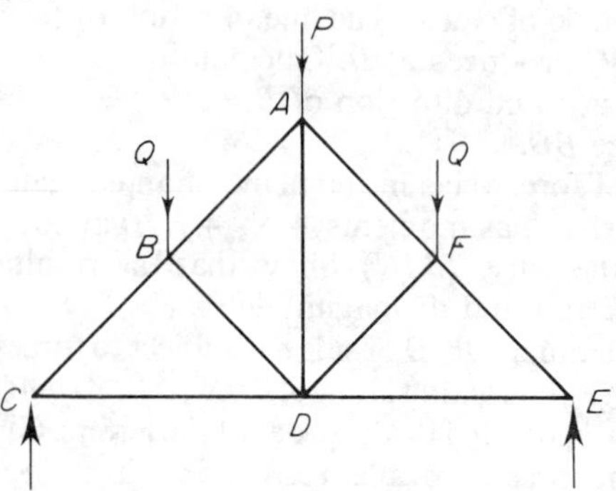

32 The figure represents a framework consisting of nine smoothly jointed light rods. AD is vertical. $CD = DE$ and the acute angles in the figure are either 30° or 60°. The framework carries weights $2W$ at A, W at B and W at F and rests on smooth supports at C and E. Determine the stresses in the rods, specifying which are tensions and which are thrusts. (O)

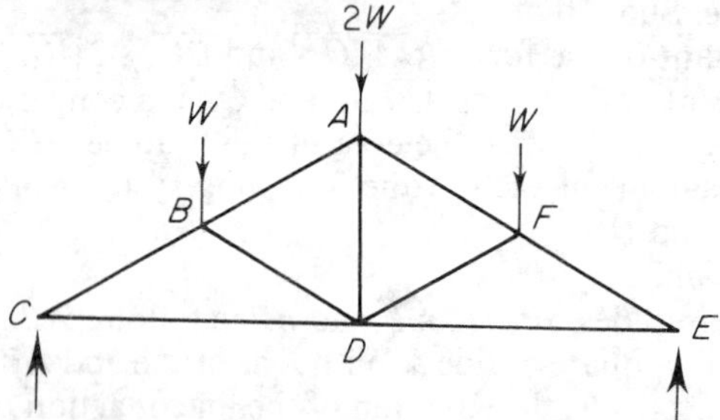

33 The figure represents a framework of five smoothly jointed light rods with $AB = BC = CD = DA = DB$. The framework is freely hinged to a support at A. A weight W is hung from D and the framework is supported with AD horizontal by a vertical force at C. Find the stresses in the rods, specifying whether they are tensions or thrusts. (O)

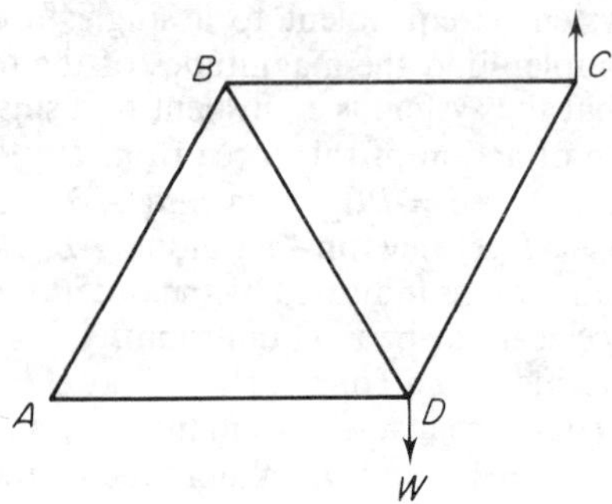

14 Simple Harmonic Motion

There is one particular type of rectilinear motion – simple harmonic motion (often abbreviated to S.H.M.) – which occurs so frequently in practice that it needs to be examined separately in detail. The motion is essentially an oscillatory one, and the motion of a particle at the end of a spring (or, very approximately, the motion of a car on its suspension) is simple harmonic.

14.1 Basic results

We start by considering the motion of a particle whose displacement x m, relative to a fixed origin O, at time t seconds is given by

$$x = 3\sin\left(2t + \frac{\pi}{6}\right). \qquad 14.1$$

The graph of x against t is shown in Fig. 14.1. The particle is at the point where $x = 0$ when $t = \frac{-\pi}{12}$. When $t = \frac{\pi}{6}$, $\dot{x}$ is zero and x attains its maximum value of 3; $\dot{x}$ is again zero when $t = \frac{2\pi}{3}$ and then $x = -3$. The particle also passes through $x = 0$ when $t = \frac{5\pi}{12}$ and $\frac{11\pi}{12}$, and after the latter time the cycle is repeated so that the total time for a complete cycle is π seconds.

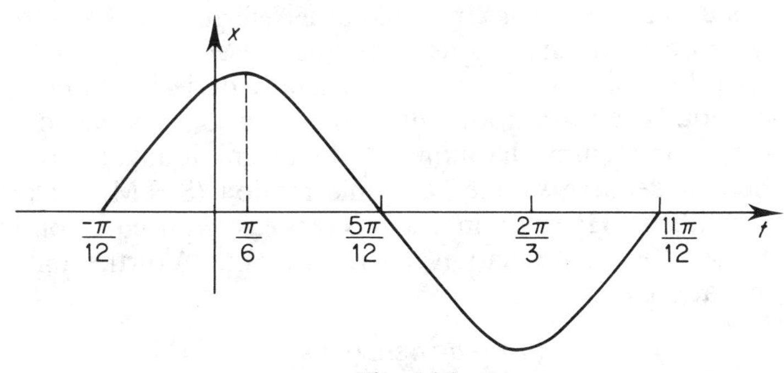

Fig. 14.1

The motion is, therefore, an oscillatory one. A particle moving to the right through O continues to move in this direction until it reaches the point where $x = 3$, where it is instantaneously at rest. It then starts moving to the left until it reaches the point where $x = -3$, when it stops instantaneously and then moves back to the right towards its original position and the cycle is repeated.

The motion defined by equation 14.1 is a particular example of simple harmonic motion with centre O. The most general form of simple harmonic motion with centre O is defined by

$$x = a \sin(\omega t + \varepsilon), \qquad 14.2$$

where a, ω and ε are independent constants.

The graph of x, defined in equation 14.2, against time will have a similar form to that of Fig. 14.1. Differentiating equation 14.2 with respect to t gives

$$\dot{x} = \omega a \cos(\omega t + \varepsilon). \qquad 14.3$$

$|x|$ attains the value a when $\sin(\omega t + \varepsilon) = \pm 1$; for these values of t we see that $\dot{x} = 0$, so the particle is at rest when at its furthest distances from the origin. The time interval between the points where the particle is at rest is π/ω. The maximum distance, a, of the particle from the centre is referred to as the *amplitude* of the motion.

The maximum value of the speed is ωa and occurs when $\cos(\omega t + \varepsilon) = \pm 1$. At these times the particle is at the origin, so that the centre is effectively the point at which the speed is a maximum. The particle passes through the origin at intervals π/ω. Hence, the particle passes through the origin, in the direction of x increasing, with speed ωa and comes to instantaneous rest when $x = a$; it then starts moving in the direction of decreasing x, passing through the origin with speed ωa, and comes to rest instantaneously when $x = -a$. It then starts moving in the direction of increasing x again and again passes through the origin with speed ωa, and the cycle repeats itself. The time taken between two successive passages in the same direction is $2\pi/\omega$ and this is called the *period*, T, of the motion. Its inverse, $1/T$ $(\omega/2\pi)$ is equal to the number of complete oscillations per unit time and is called the *frequency*, while ω is referred to as the angular frequency and has the dimensions of angular speed.

In solving dynamical problems it is generally desirable to be able to recognise immediately from an equation of motion whether or not a particular motion is simple harmonic. At first sight, it might appear that our method of defining simple harmonic motion (S.H.M.) makes this recognition difficult. However, in fact, it is very easy from equation 14.2 to determine the differential equation governing S.H.M. A further differentiation of equation 14.3 gives

$$\ddot{x} = -\omega^2 a \sin(\omega t + \varepsilon)$$

and from equation 14.2 this becomes

$$\ddot{x} = -\omega^2 x. \qquad 14.4$$

Equation 14.4 states that the acceleration is proportional to the displacement but opposite in direction, and this is the basic equation governing S.H.M. Indeed, simple harmonic motion is often defined by equation 14.4, and the methods described in §16.1 have then to be used to derive equation 14.2. We can easily show here, without using the methods described in §16.1, that equation 14.2 represents the general solution of equation 14.4. Direct substitution shows that the right-hand side of equation 14.2 represents a solution. To show that it is a general solution we need to show that a and ε can be found so that the displacement and velocity can be prescribed independently at any one value of t. In order to simplify the algebra we assume conditions are given at $t = 0$ (this involves no loss of generality), and we re-write equation 14.2 as

$$x = A\cos\omega t + B\sin\omega t, \qquad 14.5$$

where $A = a\sin\varepsilon$ and $B = a\cos\varepsilon$. If at $t = 0$, $x = x_0$ and $\dot{x} = v$, then substituting in the expression for x gives $A = x_0$, whilst substituting in the corresponding expression for $\dot{x}$ gives $\omega B = v$. Thus, the solution is sufficiently general to cope with any prescribed displacement and velocity. Equation 14.3 shows that

$$v^2 = \dot{x}^2 = \omega^2 a^2 \cos^2(\omega t + \varepsilon) = \omega^2\ [(a^2 - a^2\sin^2(\omega t + \varepsilon)\ ],$$
$$\Rightarrow v^2 = \omega^2(a^2 - x^2), \qquad (14.6)$$

which relates speed and displacement.

The basic formulae that need to be remembered for simple harmonic motion are equations 14.2, 14.4, 14.5, 14.6 and the expressions for the period and the maximum speed, that is,

$$T = 2\pi/\omega, \qquad 14.7$$

$$v_{\max} = a\omega. \qquad 14.8$$

It is also useful to remember the particular solutions

$$x = a\sin\omega t \qquad 14.9$$

which corresponds to a particle at the centre at time $t = 0$, and

$$x = a\cos\omega t, \qquad 14.10$$

which corresponds to the particle being at an extreme position at time $t = 0$.

Simple harmonic motion with centre x_0 is defined by

$$x = x_0 + a\sin(\omega t + \varepsilon), \qquad 14.11$$

that is,

$$\ddot{x} = -\omega^2(x - x_0). \qquad 14.12$$

Equation 14.12 serves as an equivalent definition of simple harmonic motion with centre x_0 and we can also deduce from it that

$$\ddot{x}+\omega^2 x = b,$$

defines simple harmonic motion with centre b/ω^2.

14.2 Kinematic problems

We consider first purely kinematic problems – that is, ones where an equation of motion does not have to be found. Most such problems in S.H.M. require some or all of A, B, ε, a and x_0 to be found from given information. For problems where time (other than the period) is not considered, most of the relevant information can usually be obtained by using equations 14.7 to 14.10. Sometimes the centre and the amplitude have to be found, and in such cases it is useful to use the results

$$\text{extreme displacement} = x_0 \pm a, \qquad 14.13$$

and

$$|\text{difference between extreme displacements}| = 2a.$$

We have that

$$v^2 = \omega^2 a^2 \sin^2(\omega t+\varepsilon)$$

for all centres, but if the centre is not at $x = x_0$ the expression for v in terms of x becomes

$$v^2 = \omega^2 [a^2 - (x-x_0)^2] \qquad 14.14$$

Summary of results

For convenience, the basic results are collected together here.

Motion centre O

$$\ddot{x} = -\omega^2 x$$

General solution

$$x = a\sin(\omega t+\varepsilon)$$

or

$$x = A\cos\omega t + B\sin\omega t$$

$$v^2 = \omega^2(a^2-x^2)$$

$$|v_{max}| = a\omega$$

Particular solutions

$$x = a\sin\omega t \ (x = 0 \text{ at } t = 0)$$

$$x = a\cos\omega t \ (x = a, \text{ or } \dot{x} = 0, t = 0)$$

Period

$$2\pi/\omega \text{ (all centres)}$$

Motion centre $\mathbf{x_0}$

$$\ddot{x} = -\omega^2 (x - x_0)$$

General solution

$$x = x_0 + a \sin(\omega t + \varepsilon)$$

or

$$x = x_0 + A \cos \omega t + B \sin \omega t.$$

$$v^2 = \omega^2 [a^2 - (x - x_0)^2]$$

The S.I. system will again be employed in numerical examples. In such examples, it will be assumed that x m and x_0 m denote the displacement at time t s of a moving point P and the centre of the oscillation from a fixed origin, and that a m denotes the amplitude of the motion. The velocity at time t s will be denoted by v ms^{-1}, and the angular frequency by ω rad s^{-1}. The quantities x, x_0, a, t, v and ω are, therefore, dimensionless numbers satisfying equations 14.3 to 14.14.

EXAMPLE 1 *In an S.H.M. the greatest value of the acceleration is* 4 ms^{-2} *and the greatest value of the speed is* 2 ms^{-1}. *Find the period, the amplitude, and the speed at a distance of* 0·5 m *from the extremities of the path.*

The acceleration is proportional to displacement, so its maximum numerical value occurs when $x = \pm a$

$$4 = \omega^2 a.$$

Equation 14.8 gives

$$2 = \omega a$$

and, therefore, $\omega = 2$, the period is π **s** and the amplitude is **1 m**. We require the speed at a distance of (1–0·5) m from the centre, and from equation 14·6,

$$v^2 = \tfrac{3}{4} \times 4$$

so the speed is $\sqrt{3}$ **ms^{-1}**.

EXAMPLE 2 *A particle describing S.H.M. with the origin as centre has a speed of* 6 ms^{-1} *at a distance of* 1 m *from O and a speed of* 2 ms^{-1} *at a distance of* 3 m *from O. Find the amplitude and period of the motion.*

In this case we are given v for two values of x and, therefore, using equation 14.6

$$36 = \omega^2 (a^2 - 1)$$

$$4 = \omega^2 (a^2 - 9).$$

Dividing these equations,

$$9 = \frac{a^2 - 1}{a^2 - 9},$$

so that $a^2 = 10$ and the amplitude is $\sqrt{10}$ **m**. Hence, $\omega^2 = 4$, giving the period to be π **s**.

EXAMPLE 3 *A particle describing S.H.M. passes through three points A, B and C, in that order, and its speeds at these points are* 4 ms^{-1}, $\sqrt{13}$ ms^{-1} *and* 1 ms^{-1} *respectively. Given that* $AB = 1$ m, $AC = 3$ m, *find the amplitude and centre of the motion.*

We are not given any information about the centre of the motion, so we have to assume it is not at the origin and use equation 14.14. The choice of origin of x is arbitrary, and we choose it at A so that the conditions of the problem give

$$16 = \omega^2(a^2 - x_0^2),\ 13 = \omega^2[a^2 - (1 - x_0)^2],\ 1 = \omega^2[a^2 - (3 - x_0)^2].$$

Eliminating ω^2 and a^2 gives $x_0 = -1$, and then $a = \sqrt{17}$ and $\omega = 1$, so the period is 2π s. The amplitude is $\sqrt{17}$ **m** and the centre of motion is at $\boldsymbol{x = -1}$.

A harder type of problem involving S.H.M. is one in which the time from one point to another is required. Problems of this type are usually best solved by using equation 14.9 so that time is measured from when a particle leaves O. The time from the centre to each point is then found and the time difference found. As the particle goes through each point (other than the extreme points) twice in a period, it is important to make certain whether the time required refers to points where the velocities are in the same direction or to points where the velocities are in opposite directions. If the centre is at x_0, then equation 14.9 has to be replaced by

$$x = x_0 + a\sin\omega t.$$

EXAMPLE 4 *A particle describes simple harmonic motion between two points A and B at a distance of* 10 m *apart, and the time to travel once from A to B is* 2 s. *Find the time taken to move from A to a point C, a distance of* 7·5 m *away, and from C to D, where D is between C and B and CD* = 1·5 m.

The centre of the motion is midway between A and B, and the amplitude is 5 m. The period is twice the time taken from A to B and is 4 s, so that $\omega = \pi/2$.

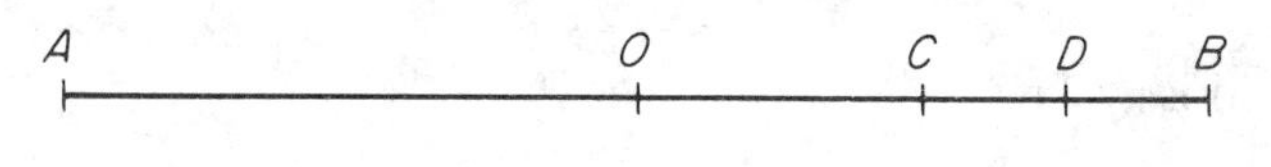

Fig. 14.2

If we refer to Fig. 14.2 we see that the time from A to C is the time from A to the mid-point O, plus the time from O to C. The time from A to O is 1 s. We can find the time from O to C by using equation 14.9, measuring x from O

$$2{\cdot}5 = 5\sin\frac{\pi t}{2}$$

giving

$$\frac{\pi t}{2} = \frac{\pi}{6}.$$

Hence the additional time is $\frac{1}{3}$ s, and the total time is $\boldsymbol{1\frac{1}{3}}$ **s.**

We also require the time from C to D and we obtain this by finding the time from O to D and subtracting from it the time from O to C. The time T s from O to D satisfies

$$4 = 5\sin(\pi T/2)$$

Therefore, $\pi T/2 = \sin^{-1}(0{\cdot}8)$. The inverse sine has to be obtained in radians and this gives

$$\pi T/2 = 0{\cdot}927$$

so that $T = 0{\cdot}59$ and hence the time from C to D is approximately **0·26 s**.

EXERCISE 14.2

Answers involving multiples of powers of π need not be simplified further.

The following questions refer to a particle P describing simple harmonic motion with centre O. A and B denote the extremities of the motion, and C denotes a point between O and B.

1 When $OP = 2$ m the acceleration is of magnitude 8 ms^{-2}. Find the period.

2 The greatest speed is 5 ms^{-1} and the period is $\pi/5$ s. Find OP when the speed is 4 ms^{-1}.

3 P makes 10 oscillations per second and its greatest acceleration is 20 ms^{-2}. Find the greatest speed.

4 P makes 3 complete oscillations per second and the maximum speed is 10 ms^{-1}. Find the amplitude.

5 When $OP = 4$ m, P has speed 6 ms^{-1} and when $OP = 3$ m, P has speed 8 ms^{-1}. Find the amplitude and period.

6 $AB = 6$ m, $OC = 1{\cdot}5$ m and the period is 4 s. Find the time to go directly from O to C and from C to B.

7 When $BP = 8$ m, the particle has speed 48 ms^{-1} and when $BP = 1$ m, the particle has speed 20 ms^{-1}. Find the amplitude of the motion.

8 $AB = 6$ m, and P passes through C at intervals of $(5\pi/3)$ s and $(7\pi/3)$ s. Find OC.

14.3 Dynamical problems

In dynamical problems, the fact that a particular motion is simple harmonic is not given but must be deduced from the equation of motion. Once it has been established that a motion is simple harmonic, then the problem becomes virtually a kinematic one. The most common dynamical situation which leads to simple harmonic motion is the motion of a particle at the end of a light spring or string. That this motion is simple harmonic will be illustrated in the following examples, which also demonstrate some of the variations that occur. It is particularly important in a problem to note whether the motion is at the end of a spring or a string, because the former can sustain a compression whilst the latter cannot. This difference can materially affect the motion.

EXAMPLE 1 *A particle of mass 3m is attached to one end of a light elastic spring, of modulus 12mg and natural length 2a. The other end of the spring is held fixed and the particle is constrained to move in a horizontal line along the line of the spring, there being no other forces acting. The particle is released from rest with the spring of length 2·25a. Find the speed of the particle when the length of the spring is 2·125a and also the time taken for the length of the spring to reduce to this value.*

The configuration is illustrated in Fig. 14.3, where x denotes the length of the spring at time t. Whilst the spring is extended, the tension is $6mg(x-2a)/a$ and acts on the particle in the direction of decreasing x, as shown. When the spring is compressed, that is, $x < 2a$, it will exert on the particle a force $6mg(2a-x)/a$ in the direction of increasing x. So, for the spring both extended and compressed, the component of force in the direction of increasing x is $-6mg(x-2a)/a$. The equation of motion is, therefore,

$$3m\ddot{x} = -\frac{6mg}{a}(x-2a)$$

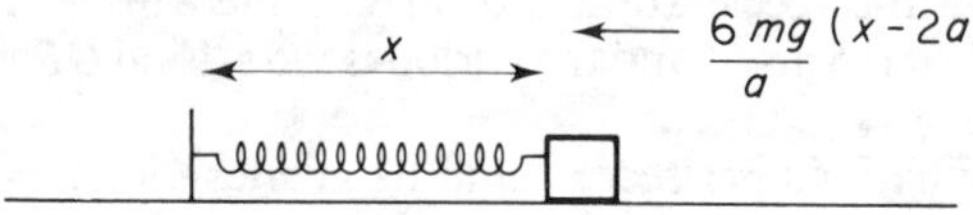

Fig. 14.3

and
$$\ddot{x} = -\frac{2g}{a}(x-2a).$$

Hence, the motion of the particle is simple harmonic, with the centre of the motion at the point where $x = 2a$. When $x = 2a$ the spring is just unstretched, so $x = 2a$ represents the equilibrium position. Letting y denote the extension, so that $y = x-2a$, we obtain

$$\ddot{y} = -\omega^2 y,$$

where $\omega^2 = 2g/a$. The particle is at rest when $y = 0{\cdot}25a$, so the amplitude of the motion is $0{\cdot}25a$ and equation 14.6 becomes

$$v^2 = \omega^2(0{\cdot}0625a^2 - y^2).$$

Substituting $y = 0{\cdot}125a$ into this equation and using the value obtained for ω gives the speed as $\mathbf{\frac{1}{8}(6ga)^{\frac{1}{2}}}$. The particle is released from rest at an extreme point, so the displacement is given by equation 14.10

$$y = 0{\cdot}25a\cos\omega t.$$

Substituting $y = 0{\cdot}125a$ gives the time as $\mathbf{(a/2g)^{\frac{1}{2}}\pi/3}$.

EXAMPLE 2 *A particle of mass 4m is attached to one end of a light elastic spring, the other end of which is held fixed so that the particle can move in a vertical line through the spring. The spring is of modulus mg/32 and natural length a, and the particle is released from rest with the spring extended a distance 0·5a. Find the time T taken for the extension to reduce to 0·25a and the speed of the particle at this time.*

Referring to Fig. 14.4, we measure x from the point of suspension, so whilst the spring is extended there is a force $mg(x-a)/32a$ acting upwards. It follows, exactly as in the previous example, that the upward component of the force, for

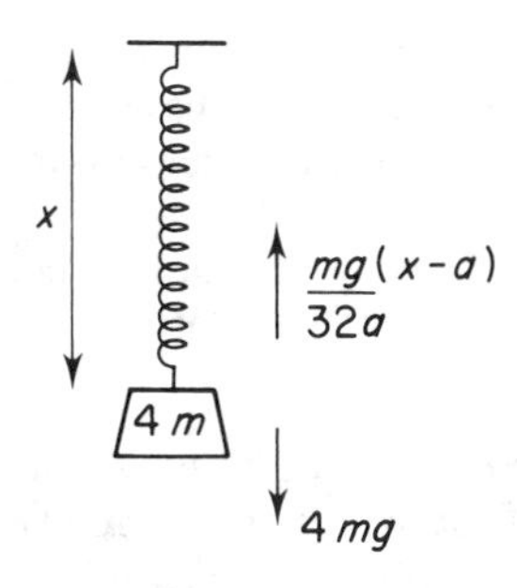

Fig. 14.4

both extension and compression, has this value. The only other force acting is the weight of the particle $4mg$ downwards, and so the equation of motion becomes

$$4m\ddot{x} = -\frac{32mg}{a}(x-a)+4mg$$

$$= -\frac{32mg}{a}(x-1{\cdot}125a).$$

The motion is, therefore, simple harmonic with its centre at the point where $x = 1{\cdot}125a$. When the particle is in equilibrium, the extension e of the spring satisfies

$$4mga = 32mge$$

so that $e = 0{\cdot}125a$. Hence, the centre of the simple harmonic motion is the equilibrium position. This result will hold for all such motions and it may save time always to choose the origin in this position. Letting y denote the displacement from equilibrium (that is, $x = 1{\cdot}125a + y$), we have

$$\ddot{y} = -\omega^2 y$$

with $\omega = (8g/a)^{\frac{1}{2}}$.

In the initial position, $x = 1{\cdot}5a$, so $y = 0{\cdot}375a$. This is, the amplitude of the motion, and the particles rises a distance $0{\cdot}75a$ above its initial position. When the extension is $0{\cdot}25a$, $y = 0{\cdot}125a$, and the time T is found from equation 14.6 with $\omega = (8g/a)^{\frac{1}{2}}$ and a replaced by $0{\cdot}375a$

$$T = (a/8g)^{\frac{1}{2}}\cos^{-1}(1/3).$$

The required speed v satisfies

$$v^2 = \frac{8g}{a}[(0{\cdot}375a)^2 - y^2]$$

and when $y = 0{\cdot}125a$, $v = (ga)^{\frac{1}{2}}$.

EXAMPLE 3 *Find the maximum height reached above the initial position by the particle in the previous example when the spring is replaced by a light string of the same modulus and natural length.*

The essential difference here is that the string cannot sustain a compression, so the equation of S.H.M. only holds for $x \geqslant a$, that is, $y \leqslant -0{\cdot}125a$. Therefore, when $y = -0{\cdot}125a$ the particle starts moving under gravity and its initial speed, as found earlier, is $(ga)^{\frac{1}{2}}$. Hence, it travels upwards a further distance $0{\cdot}5a$ under gravity and then falls freely a distance $0{\cdot}5a$, when the motion becomes simple harmonic again. The height reached above the original position is, therefore, $\boldsymbol{a}$.

EXAMPLE 4 *A particle of mass* 0·2 kg *is attached to one end of a light elastic string, of modulus* 16 N *and natural length* 0·4 m. *The other end of the string is fixed at a point O and the particle is allowed to fall freely from a point* 0·2 m *below O. Find*
(*i*) *the speed of the particle when the string is extended a distance* 0·1 m,
(*ii*) *the depth below O at which the particle comes to instantaneous rest,*
(*iii*) *the time taken before the particle returns to its original position.*

In this type of question, it is necessary to think rather carefully about the motion. The particle falls freely under gravity until the string is extended, and we know that then the motion becomes simple harmonic. This motion continues until the particle again reaches the point where the string becomes slack, where its speed is the same as it had at that point earlier, though it is now moving upwards. It will come to instantaneous rest at the initial position, and the cycle will then be repeated.

Possibly the most obvious starting point is to calculate the speed when the string first stops being slack, determine the position of the centre of the S.H.M. and calculate the ω appropriate to the motion. Substituting the appropriate values into $v^2 = \omega^2(a^2 - x^2)$ determines the amplitude, which is equal to the depth below the centre at which the particle is first instantaneously at rest.

This is a somewhat elaborate method, and a far better way of finding the relation between position and speed in any problem involving strings or springs is to use the equation of energy. Taking the zero of potential energy at the point of suspension and letting x denote the extension, the equation of energy is (approximating to g by $10\,\text{ms}^{-2}$)

$$-0{\cdot}2 \times 0{\cdot}2 \times 10 = -0{\cdot}2 \times (0{\cdot}4 + x) \times 10 + \frac{1}{2}\frac{16 \times x^2}{0{\cdot}4} + \frac{1}{2}0{\cdot}2 \times v^2.$$

The particle comes to rest when

$$20x^2 - 2x - 0{\cdot}4 = 0,$$

that is,

$$(10x - 2)(2x + 0{\cdot}2) = 0.$$

Therefore, the particle first comes to rest when $x = 0{\cdot}2$, that is, at a distance **0·6 m** below O. Substituting $x = 0{\cdot}1$ in the energy equation gives the required speed as $\mathbf{2\,ms^{-1}}$. In order to find the total time, we need to establish the equation of motion and this is

$$0{\cdot}2\ddot{x} = \frac{-16}{0{\cdot}4}x + 0{\cdot}2 \times 10.$$

$$\ddot{x} = -200x + 10.$$

The period is, therefore, $2\pi/(200)^{\frac{1}{2}}\,\text{s} = (\pi\sqrt{2})/10\,\text{s}$.

We need to add to the period twice the time taken by the particle to fall freely a distance of 0·2 m. This time is t_1 s, where t_1 is given by

$$0{\cdot}2 = \tfrac{1}{2} \times 10t_1^2$$

so that $t_1 = 0{\cdot}2$ s. The total time is $[0{\cdot}2 + \pi\sqrt{2}/10]$ s, that is, approximately **0·644 s.**

Now that the equation of motion has been obtained, it seems worth using it to show the rather longer method of finding the results. The time taken for the string just to become taut is, that found above, so it reaches that position with a speed of 2 ms^{-1}. The centre, which is the equilibrium position, is determined by the condition $\ddot{x} = 0$, that is, $x = 0{\cdot}05$. Substituting in the equation

$$v^2 = \omega^2(a^2 - x^2).$$

gives

$$4 = 200(a^2 - 0{\cdot}0025)$$

so that $a = 0{\cdot}15$ m. The total depth below O is, therefore, again found to be 0·6 m.

EXERCISE 14.3

In numerical exercises, take $g = 10$ ms^{-2}. Answers involving multiples of powers of π need not be simplified further.

1 A horizontal plane is oscillating vertically in simple harmonic motion with amplitude 0·1 m and period $\pi/4$ s. Find the maximum and minimum values of the reaction of the plane on a particle of mass 1·5 kg resting on it.

2 A particle rests on a smooth table, which is oscillating horizontally in simple harmonic motion with amplitude 0·2 m and at a frequency of 5 oscillations per second. Find the least value of the coefficient of friction so that the particle stays at rest relative to the table.

3 A horizontal platform performs vertical simple harmonic oscillations of period $4\pi/5$ s. Find the amplitude such that a particle on the platform does not lose contact with it.

4 If the amplitude in the previous question is 2 m and the particle is of mass 1 kg and is attached by adhesive to the platform, find the maximum force that the adhesive has to exert.

5 A particle P of mass $3m$ is attached to one end of a light elastic string of modulus $9mg$ and natural length $2l$. The other end of the string is fixed to a point O, and the particle can move, in the line of the string, on a smooth horizontal table. The particle is released from rest with $OP = 2{\cdot}5l$. Find its speed when the string first becomes slack and the time taken.

6 A particle P of mass $2m$ is attached to one end of a light elastic spring of modulus $4mg$ and natural length l. The other end of the spring is fixed at a point O, and the particle can move along the line of the spring, on a smooth horizontal table. At time $t = 0$, when the spring is extended a distance $0{\cdot}25l$, the particle is given a velocity of $(3gl/8)^{\frac{1}{2}}$. Find the extreme positions of the motion and an expression for the extension at time t.

Questions 7 and 8 refer to a particle P of mass m suspended from one end of a light elastic spring of modulus λ and natural length l. The other end of the spring is attached to a fixed point O, and P can move in the vertical line through O.

7 $m = 10\,\text{kg}$, $\lambda = 6000\,\text{N}$, $l = 1{\cdot}5\,\text{m}$. The particle is released from rest with $OP = 1{\cdot}85\,\text{m}$. Find the period of the motion and the maximum and minimum lengths of the spring.

8 In the equilibrium position, the extension is b. P is raised a distance $\frac{1}{2}b$ from the equilibrium position and released from rest. Find the greatest speed and acceleration in terms of g and b.

9 Two equal particles P and Q are in equilibrium suspended from a point O by a light elastic spring of modulus $16mg$ and natural length a. The particle Q is then gently removed. Find the period and amplitude of the subsequent motion.

Questions 10 to 12 refer to the configuration of questions 7 and 8 with the spring replaced by a string.

10 $m = 2\,\text{kg}$, $\lambda = 170\,\text{N}$, $l = 0{\cdot}85\,\text{m}$. The particle is released from rest with $OP = 1\,\text{m}$. Find the period of the motion and the minimum length of OP.

11 $m = 4\,\text{kg}$, $\lambda = 140\,\text{N}$, $l = 0{\cdot}7\,\text{m}$. The particle is released from rest with $OP = 1{\cdot}2\,\text{m}$. Find the minimum length of OP.

12 $m = 3\,\text{kg}$, $\lambda = 525\,\text{N}$, $l = 0{\cdot}5\,\text{m}$. P is released from rest at O. Find the maximum length of OP and, correct to two decimal places, the amplitude of that part of the motion for which the string is taut.

13 Two points A and B are at a distance $2a$ apart in a horizontal line on a smooth plane. A particle P of mass m is attached to both A and B by identical light elastic springs of modulus $2mg$ and natural length a. When P is in equilibrium it is displaced a small distance d towards B and then released. Find the time taken for it to first return to the equilibrium position.

14.4 The simple pendulum

A practical example of simple harmonic motion is provided by the small amplitude oscillations of the simple pendulum. In its basic form, this consists of a heavy particle P attached to one end of a light inextensible string. The other end of the string is fixed at a point A and the particle is allowed to oscillate, in a vertical plane, with the string staying taut. At one time many time-keeping devices were based on the simple pendulum, but this is no longer the case.

The simplest method of examining the motion of the simple pendulum is by using the formula for transverse acceleration obtained in §10.1. We give this approach at the end of this section, but first we use a method which does not rely on this formula.

The idealised form of the simple pendulum is illustrated in Fig. 14.5, where the string is assumed to be of length l with the tension in it being T. Horizontal and vertical axes are taken as shown, with O being the point at which the particle P (known as the bob of the pendulum) would rest in equilibrium. The horizontal and vertical components of the equation of motion are

$$m\ddot{x} = T\cos\theta - mg$$
$$m\ddot{y} = T\sin\theta.$$

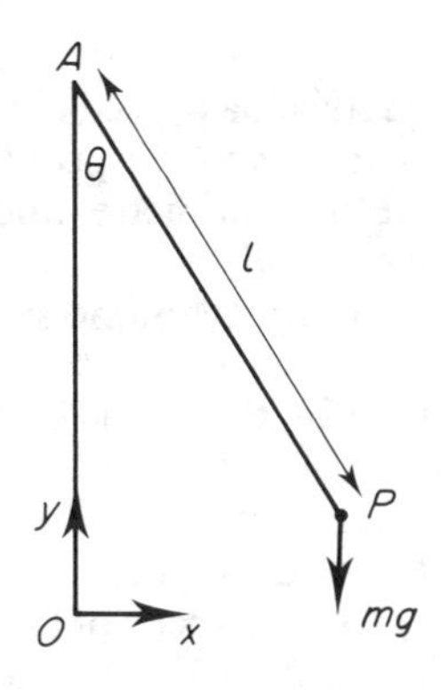

Fig. 14.5

A is the point $(l, 0)$ and

$$AP^2 = l^2 = x^2 + (y-l)^2,$$

that is,

$$2ly = x^2 + y^2. \qquad 14.15$$

It is assumed that the oscillations are sufficiently small for x^2 and y^2 to be small compared with x and y respectively. It follows from equation 14.15 that y is proportional to x^2 and so may be neglected compared to x and, similarly, $\ddot{y}$ may be neglected. Cos θ, which is equal to $(l-y)/l$ can also be set equal to unity. Therefore, to this order of accuracy, $T = mg$ and

$$\ddot{x} = \frac{gx}{l}.$$

This represents simple harmonic motion of period $2\pi\ (l/g)^{\frac{1}{2}}$.

Using the results of §10.1, the equation of motion of P normal to AP is

$$ml\ddot{\theta} = -mg \sin\theta.$$

For small values of θ we have that $\sin\theta \approx \theta$, so that

$$\ddot{\theta} + \frac{g\theta}{l} = 0,$$

again showing that the motion is simple harmonic.

MISCELLANEOUS EXERCISE 14

Unless otherwise stated, in numerical questions, g should be taken as $10\,\mathrm{m\,s^{-2}}$.

1 A particle, A, is performing simple harmonic oscillations about a point O with amplitude 2 m and period 12π s. Find the least time from the instant when A passes through O until the instant when
(i) its displacement is 1 m,
(ii) its velocity is half that at O,
(iii) its kinetic energy is half that at O. *(AEB 1978)*

2 A particle of mass 10 grams is moving along a straight line with simple harmonic motion. The particle has speeds of 9 centimetres per second and 6 centimetres per second at P and Q respectively, whose distances from the centre of oscillation are 1 centimetre and 2 centimetres respectively. Calculate the greatest speed and the greatest acceleration of the particle.

If the points P and Q are on the same side of the centre of oscillation, calculate

(i) the shortest time taken by the particle to move from P to Q,
(ii) the work done during this displacement. (*AEB 1972*)

3 A particle of mass m moves in simple harmonic motion about O in a straight line, under the action of a restoring force of magnitude proportional to the distance from O. At time $t = 0$ the speed of the particle is zero. After 1 second the speed of the particle is $2\,\text{ms}^{-1}$, after a further second the speed is $2\sqrt{3}\,\text{ms}^{-1}$ and subsequently the particle passes through O for the first time. Show that the speed of the particle when it passes through O is $4\,\text{ms}^{-1}$ and find

(i) the period and amplitude of the motion,
(ii) the time at which the speed of the particle is equal to $2\,\text{ms}^{-1}$ after it first passes through O.

A stationary particle of mass m is placed at O and the moving particle collides and coalesces with this particle. The combined particle moves under the action of the same restoring force as before. Find the period of the subsequent motion and the speed when the displacement is $12/\pi$ m from O. (*C*)

4 (a) A particle describes simple harmonic motion in a straight line. The particle is instantaneously at rest at time $t = 0$ and is next at rest at time $t = 3$ s. The speed at time $t = 2$ s is $1\,\text{ms}^{-1}$. Determine the amplitude of the motion and the maximum speed of the particle.

(b) A particle of mass m moves in a horizontal circle with constant speed $\sqrt{(3ga/2)}$ on the smooth inside surface of a fixed sphere with centre O and radius a. Find the depth below O of the plane of the circle. (*C*)

5 A particle moves in a straight line with simple harmonic motion of period 8 s about a fixed point O in the line. At a certain time the particle is at a distance 3 m from O and two seconds later it is still on the same side of O but 4 m from O. Find the amplitude of the motion.

Show that after a further two seconds the particle is on the other side of O at a distance 3 m from O. Find the speed and acceleration of the particle in this position. Find also the time which elapses before the particle next passes through this position.

6 A particle is moving in a straight line and its distance s at time t, measured from a fixed point O in the line, is given by $s = a \sin nt$, where a and n are constants. Show that the particle is executing simple harmonic motion and state

(i) the period,
(ii) the greatest speed.

In a certain tidal estuary the water level rises and falls with simple harmonic motion. On a particular day a marker indicates that the depths of water at low and high tides are 4 m and 10 m and that these occur at 1100 and 1720 respectively. Calculate

(i) the speed, in m/h, at which the water level is rising at 1235,

(ii) the time, during this tide, at which the depth of water is $8\frac{1}{2}$ m.
(*AEB 1975*)

7 Define simple harmonic motion.

A particle is executing a simple harmonic motion in a straight line of period T and amplitude a, and the particle has velocity v when at a distance x from the mean position. Show, from your definition, that

$$vT = 2\pi(a^2 - x^2)^{\frac{1}{2}}.$$

Given that the particle has velocities $3\,\text{ms}^{-1}$ and $2\,\text{ms}^{-1}$ when its distances from the mean position are 2 m and 3 m respectively, find the amplitude and the period. (*W*)

8 A particle is moving in a straight line. Its acceleration is proportional to its displacement from a fixed point O on that line and directed towards O. Write down the differential equation for the motion and prove that

$$v^2 = \omega^2(a^2 - x^2)$$

and

$$x = a\cos(\omega t + \alpha)$$

where x is the displacement of the particle from O, $v = \dot{x}$, t is the time and a, α and ω are arbitrary real numbers.

If, at time $t = 0$, $x = b$ and $\dot{x} = v_0$, show that

$$x = (v_0/\omega)\sin\omega t + b\cos\omega t.$$

A particle moving with simple harmonic motion passes successively through a fixed point P at times $t = 0, 2$ and 8 s with a speed of $3\,\text{m s}^{-1}$. Show that P is at a distance $12/\pi$ metres from the centre of motion and find the amplitude of the motion. (*WT*)

9 A particle of mass 5 kilograms executes simple harmonic motion with amplitude 2 metres and period 12 seconds. Find the maximum kinetic energy of the particle, leaving your answer in terms of π.

Initially the particle is moving with its maximum kinetic energy. Find the time that elapses until the kinetic energy is reduced to one quarter of the maximum value, and show that the distance moved in this time is $\sqrt{3}$ metres.
(*JMB*)

10 (a) The speed v of a particle moving along the x-axis is given by

$$v^2 = -4x^2 - 8x + 21.$$

Show that the motion is simple harmonic and find the centre, period and amplitude of the motion. (*W*)

11 A particle moves in a straight line with velocity v m/s, related to its distance x m from a fixed point on the line by the relation

$$v^2 = 27 + 18x - 9x^2.$$

Show that the motion is simple harmonic about the point where $x = 1$. Find the period of oscillation, the amplitude, the maximum velocity and the maximum acceleration of the particle. (*AEB 1975*)

12 A particle moves on a straight line through a fixed point O so that at time t its displacement from O is x and its equation of motion is

$$\frac{d^2x}{dt^2} = -16x.$$

Given that $x = -12$ and $dx/dt = 20$ when $t = \pi/4$, find
(i) the position and velocity of the particle when $t = \pi/2$,
(ii) the least positive value of t for which $x = 0$, giving two decimal places in your answer. (*JMB*)

13 A light elastic spring, of modulus $8mg$ and natural length l, has one end attached to a ceiling and carries a scale pan of mass m at the other end. The scale pan is given a vertical displacement from its equilibrium position and released to oscillate with period T. Prove that

$$T = 2\pi\sqrt{\left(\frac{l}{8g}\right)}.$$

A weight of mass km is placed in the scale pan and from the new equilibrium position the procedure is repeated. The period of oscillation is now $2T$. Find the value of k.

Find also the maximum amplitude of the latter oscillations if the weight and the scale pan do not separate during the motion. (*AEB 1970*)

14 The diagram shows a light rod of length l smoothly pivoted at one end O, and carrying a particle of mass m at its other end. The system hangs at rest under gravity, when at time $t = 0$ the particle is given a small initial horizontal speed u. At time t the angular displacement of the rod from the vertical is θ.

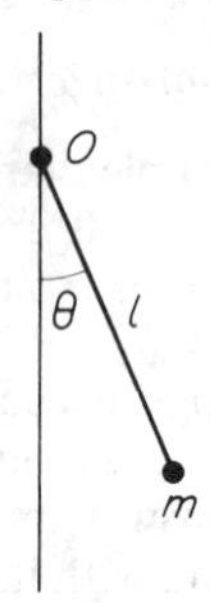

Show that if θ is small then approximately

$$\frac{d^2\theta}{dt^2} + \omega^2\theta = 0, \text{ where } \omega = \sqrt{(g/l)}$$

and deduce that the maximum value of θ is $u/\sqrt{(gl)}$. (*JMB*)

15 A light elastic string of natural length l has a particle of mass m attached at one end B and the other end A is fixed. If $AB = 3l/2$ when the particle hangs freely at rest, show that the modulus of elasticity of the string is $2mg$.

When hanging at rest the particle is suddenly given a downward vertical velocity v so that it describes simple harmonic motion of amplitude $l/2$. Find, in terms of l and g, the period of this motion and the value of v.

Find the speed of the particle and the tension in the string when $AB = 5l/4$. (*AEB 1976*)

16 Two particles of masses m, m', ($m' < m$), are hanging freely in equilibrium from the lower end of a light elastic string whose upper end is fixed. At a certain instant the particle of mass m' falls off. Show that the distance of the particle m from the upper end of the string at time t later is

$$a + b + c\cos(\sqrt{(g/b)}t),$$

where a is the natural length of the string and b, c are the distances the string would stretch when supporting the particles m, m' respectively. (*O*)

17 A light elastic string, of natural length l and modulus $6mg$, has one end attached to a fixed point O, and to the other end is attached a particle P of mass m. The particle is held at rest at a point A vertically below O, where $OA = 3l/2$, and is then released. By energy considerations, or otherwise, show that P is next at rest when it is at a distance $3l/4$ below O.

Show that if the length of the string is $l+x$ at time t after release and before the string becomes slack, then

$$\frac{d^2x}{dt^2} = g - \omega^2 x,$$

where $\omega^2 = 6g/l$.

By making the substitution $x = y + l/6$, or otherwise, show that the string first becomes slack when $t = 2\pi/(3\omega)$. (*L*)

18 A light elastic string, of natural length c and modulus of elasticity $4mg$, is attached at one end to a fixed point A. A particle of mass m is tied to the other end B of the string.
(i) If the particle hangs in equilibrium, calculate the length AB of the extended string.
(ii) If the particle is held level with A and allowed to fall vertically, use the principle of conservation of energy to find the greatest distance between A and B in the ensuing motion.
(iii) If the particle moves in a horizontal circle, centre O, where O is at a depth d vertically below A, show that the angular velocity of the particle is $\sqrt{(g/d)}$. (*AEB 1980*)

19 One end O of a light elastic string, obeying Hooke's law and of natural length l, is attached to a fixed point. To the other end P of the string is attached a particle of mass m which hangs in equilibrium with $OP = 5l/4$. The particle is pulled down vertically a further distance a and released from rest.
(i) If $a \leqslant l/4$, show that P rises a distance $2a$ before first coming to instantaneous rest after a time $\frac{1}{2}\pi\sqrt{(l/g)}$.
(ii) If $l/4 < a < 3l/4$, show that P rises a distance $(l+4a)^2/(8l)$ before first coming to instantaneous rest and find its greatest speed during this motion. (*L*)

20 One end of a light elastic string, of natural length a and modulus of elasticity kmg, is attached at a fixed point on a frictionless plane inclined at an angle θ to the horizontal. A heavy particle of mass M is attached to the other end of the string. The particle is at rest on the plane with the string along a line of greatest slope and extended by a length b. The particle is then pulled down a distance d in the line of the string and released. Show that the period of the simple harmonic motion with which the particle starts to move is independent of θ.

If $d = 2b$, find the time from release to the string going slack and find also the speed of the particle at the instant when the string goes slack. (*AEB 1973*)

21 A particle of mass m lies on a smooth horizontal table and is attached by two light elastic strings, obeying Hooke's law and each of natural length l and modulus $4mg$, to two fixed points A and B on the table, where $AB = 4l$. Show that any oscillation of the particle along the line AB for which the strings remain taut has periodic time $\frac{1}{2}\pi\sqrt{(2l/g)}$.
The particle is released from a point of the line AB at distance $5l/3$ from B. Obtain the greatest speed of the particle in the subsequent motion. (*L*)

22 A particle P of mass m lies on a smooth horizontal table and is attached by two light elastic springs, AP and BP, to the table at two fixed points A and B. Each of the springs AP and BP has natural length l and modulus of elasticity $4mg$. The distance AB is $4l$. Show that any oscillation of the particle in the line AB has a periodic time $\pi\sqrt{l/(2g)}$.

The particle is released from a point on the line AB at a distance $5l/3$ from B. Obtain, in terms of g and l,
(i) the greatest speed of the particle,
(ii) the speed at the point C at a distance $9l/5$ from B,
(iii) the shortest time to reach C. (*AEB 1975*)

23 An elastic string of natural length $2a$ and modulus λ has its ends attached to two points A, B on a smooth horizontal table. The distance AB is $4a$ and C is the mid-point of AB. A particle of mass m is attached to the mid-point of the string. The particle is released from rest at D, the mid-point of CB.

Denoting by x the displacement of the particle from C, show that the equation of motion of the particle is

$$\frac{d^2x}{dt^2} + \frac{2\lambda}{ma}x = 0.$$

Find the maximum speed of the particle and show that the time taken for the particle to move from D directly to the mid-point of CD is

$$\frac{\pi}{3}\left(\frac{ma}{2\lambda}\right)^{\frac{1}{2}}.$$ (*JMB*)

24 A small cubical block of mass $8m$ is attached to one end A of a light elastic spring AB of natural length $3a$ and modulus of elasticity $6mg$. The spring and block are at rest on a smooth horizontal table with AB equal to $3a$ and lying perpendicularly to the face to which A is attached. A second block of equal physical dimensions, but of mass m, moving with a speed $(2ga)^{1/2}$ in the direction parallel to BA impinges on the free end B of the spring.

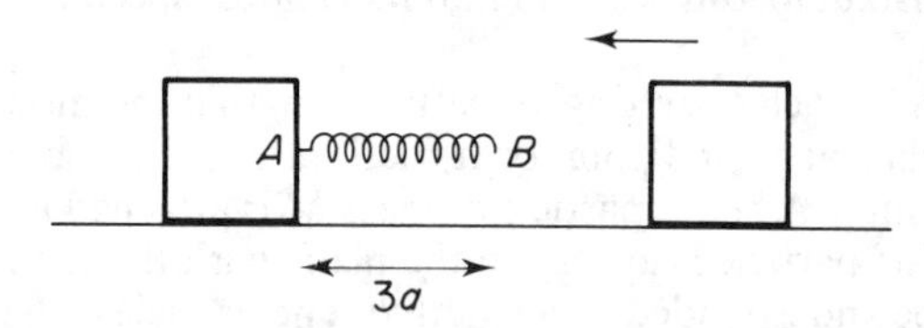

Assuming that the heavier block is held fixed and that AB remains straight and horizontal in the subsequent motion, determine
(i) the maximum compression of the spring,
(ii) the time that elapses between impact and the lighter block first coming to instantaneous rest.

Assuming now that the heavier block is also free to move determine, at the instant when the blocks are first moving instantaneously with the same velocity, the values of
(iii) the common velocity of the blocks,
(iv) the compression in the spring. (*AEB 1982*)

15 Rigid Body Dynamics

In our study of dynamics so far, the bodies in motion have either been regarded as particles, (point masses) or, equivalently, the dimensions of the bodies have been neglected. Reference has been made only to translation of the bodies. In this chapter, the rotation of extended bodies will be studied.

15.1 The motion of the centre of mass of a system of particles

We will consider first the motion of three individual particles of masses m_1, m_2 and m_3 under the action only of external forces. Suppose that at time t these particles have position vectors $\mathbf{r}_1$, $\mathbf{r}_2$ and $\mathbf{r}_3$ respectively with respect to a fixed origin O and that the forces acting on the particles are $\mathbf{F}_1$, $\mathbf{F}_2$ and $\mathbf{F}_3$ respectively (Fig. 15.1). Then

$$\mathbf{F}_1 = m_1\ddot{\mathbf{r}}_1, \quad \mathbf{F}_2 = m_2\ddot{\mathbf{r}}_2, \quad \mathbf{F}_3 = m_3\ddot{\mathbf{r}}_3.$$

Hence

$$\mathbf{F}_1 + \mathbf{F}_2 + \mathbf{F}_3 = m_1\ddot{\mathbf{r}}_1 + m_2\ddot{\mathbf{r}}_2 + m_3\ddot{\mathbf{r}}_3.$$

Now, if $\bar{\mathbf{r}}$ is the position vector of the centre of mass of the three particles, using equation 12.2

$$\bar{\mathbf{r}} = \frac{m_1\mathbf{r}_1 + m_2\mathbf{r}_2 + m_3\mathbf{r}_3}{m_1 + m_2 + m_3},$$

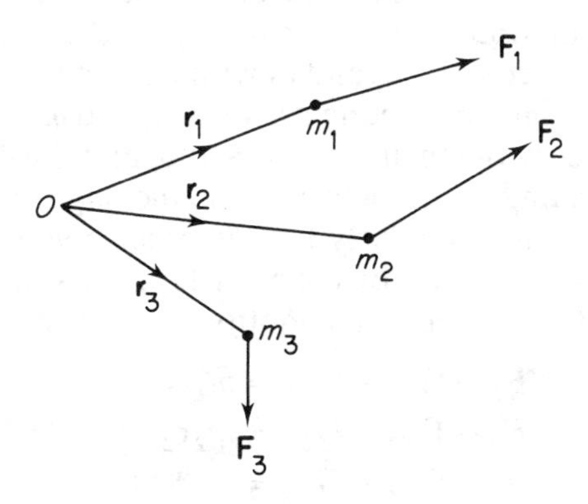

Fig. 15.1

so that if the total mass is M, where $M = m_1 + m_2 + m_3$, then

$$M\bar{\mathbf{r}} = \sum_{i=1}^{3} m_i \mathbf{r}_i.$$

Differentiating this equation twice with respect to the time gives

$$M\ddot{\bar{\mathbf{r}}} = \sum_{i=1}^{3} m_i \ddot{\mathbf{r}}_i = \sum_{i=1}^{3} \mathbf{F}_i.$$

Thus, the motion of the centre of mass of the three particles is the same as that of a single particle, of mass equal to the total mass, situated at the centre of mass and under the action of the three external forces applied at the centre of mass.

This result may easily be generalised to the case of the motion of n particles under the action of external forces only. For such a set of particles

$$\sum_{i=1}^{n} \mathbf{F}_i = M\ddot{\bar{\mathbf{r}}}, \qquad 15.1$$

where $\mathbf{F}_i$ is the external force on the particle of mass m_i, M is the total mass $\sum_{i=1}^{n} m_i$ and $\bar{\mathbf{r}}$ is the position vector of the centre of mass of the particles.

15.2 Rigid bodies

To extend the study to the consideration of the motion of solid bodies, it is necessary to explain the concept of a rigid body. Such a body maintains its shape constantly, whatever the forces that act on it and however it moves. This shape is maintained by internal forces or bonds acting between the various parts. We define a rigid body as one in which the distance between any two constituent particles remains constant.

The body is regarded as consisting of a number of separate particles bonded together. Acting on each particle there may be a number of external forces together with a number of internal forces. For simplification, consider the motion of three separate particles of masses m_1, m_2 and m_3 with position vectors $\mathbf{r}_1$, $\mathbf{r}_2$ and $\mathbf{r}_3$ with respect to a fixed origin O. Suppose that the bonding force acting between m_1 and m_2 is G_{12} acting on m_1. The bonding force between m_1 and m_2 acting on m_2 will be equal and opposite to the bonding force between m_1 and m_2 acting on m_1, as illustrated in Fig. 15.2, and similarly for the other bonding forces.

Let the resultant external force on the particle of mass m_i be $\mathbf{F}_i$, ($i = 1, 2, 3$) where $\mathbf{F}_i$ may be $\mathbf{0}$. Considering the motion of each particle separately,

$$\mathbf{F}_1 + \mathbf{G}_{12} + \mathbf{G}_{13} = m_1 \ddot{\mathbf{r}}_1,$$
$$\mathbf{F}_2 - \mathbf{G}_{12} + \mathbf{G}_{23} = m_2 \ddot{\mathbf{r}}_2,$$
$$\mathbf{F}_3 - \mathbf{G}_{13} - \mathbf{G}_{23} = m_3 \ddot{\mathbf{r}}_3.$$

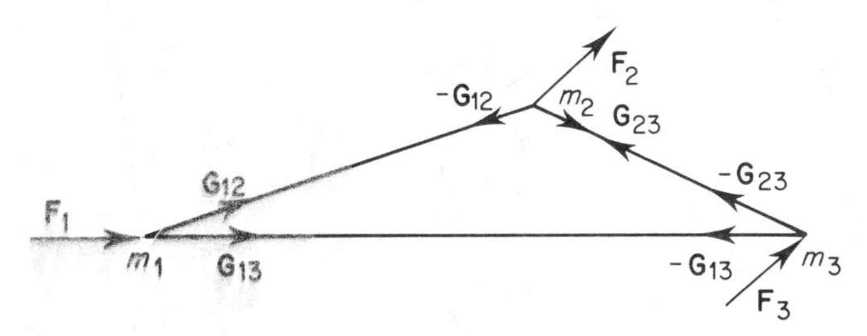

Fig. 15.2

Adding these equations

$$\sum_{i=1}^{3} \mathbf{F}_i = \sum_{i=1}^{3} m_i \ddot{\mathbf{r}}_i = M\ddot{\bar{\mathbf{r}}}, \text{ as before.}$$

Thus, again the motion of the centre of mass may be obtained by regarding all external forces as acting at the centre of mass on a particle of mass equal to the total mass of the 'body'.

This result may easily be generalised to that for any number n of particles and thus we obtain the following result for the motion of any rigid body.

The motion of the centre of mass of a rigid body is given by the equation

$$\sum_{i=1}^{n} \mathbf{F}_i = M\ddot{\bar{\mathbf{r}}}, \qquad 15.2$$

where $\sum_{i=1}^{n} \mathbf{F}_i$ is the vector sum of all the external forces, M is the mass of the body, and $\bar{\mathbf{r}}$ is the position vector of the centre of mass.

15.3 Rotation of a rigid body

In order to describe the general motion of a rigid body, for example, a stick thrown up into the air, it is necessary to know both the motion of the stick's centre of mass and the way in which the stick is rotating. The earth's motion relative to the sun can be described by the motion of the earth's centre in its annual orbit round the sun, together with its rotation about its polar axis.

Before dealing with such general motion, rotation will be considered separately, and we commence with the consideration of the rotation of a rigid body about a fixed axis. Suppose that the rigid body illustrated in Fig. 15.3 is rotating with angular speed ω about the axis LL'. The path of any point P of the body is a circle in a plane perpendicular to the axis, the centre O of this circle being the point of this plane on the axis LL'. The speed of the point P using equation 10.1 is ωr_i, where $OP = r_i$. The direction of the velocity of P is along the tangent to this circle.

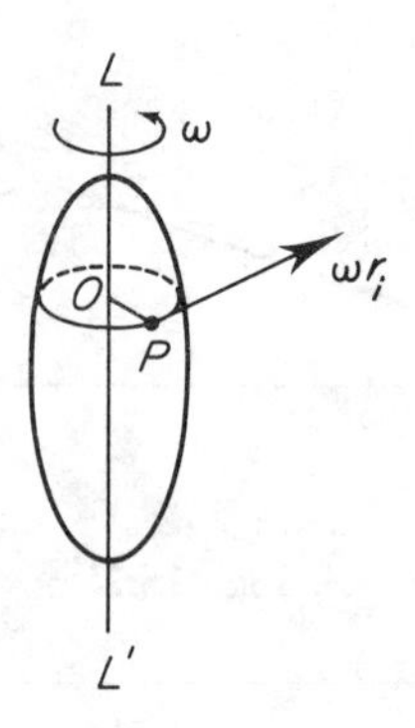

Fig. 15.3

Kinetic Energy

Regarding the rigid body as consisting of n separate particles (one of which with mass m_i is shown at P in Fig. 15.3), the kinetic energy of the body is given by

$$\begin{aligned}\text{K.E.} &= \sum_{i=1}^{n} \tfrac{1}{2}m_i(\omega r_i)^2 \\ &= \tfrac{1}{2}\omega^2(m_1r_1^2 + m_2r_2^2 + \ldots + m_nr_n^2) \\ &= \tfrac{1}{2}\omega^2 \sum_{i=1}^{n} m_ir_i^2.\end{aligned}$$

Definition The quantity $\sum_{i=1}^{n} m_ir_i^2$ is defined as the *moment of inertia, I*, of the rigid body about the axis of rotation. Note that the moment of inertia varies with the choice of the axis. The unit for moment of inertia is kg m^2.

The kinetic energy of a rigid body which is rotating with angular speed ω about a fixed axis is $\frac{1}{2}I\omega^2$, where I is the moment of inertia of the body about the axis of rotation. 15.3

When the body is free to rotate about the fixed axis under the action of external forces which are functions of position only (as in §6.2), it can be shown that the total work done by the internal bonding forces in any movement of the body is zero and, therefore, the principle of energy can be applied. For the solution of problems using the principle of energy, it is necessary to know the moment of inertia of the body about the axis of rotation. In the first set of examples the moments of inertia will be given. The methods of calculation of moments of inertia are given in §15.4.

EXAMPLE 1 *A flywheel of radius* 0·5 m *is free to rotate smoothly about an axis through its centre. The moment of inertia of the flywheel about this axis is* 2·4 kg m^2.

A light rope is attached at one end to a point P on the rim of the wheel and the rope is wrapped once round the flywheel so that the free end lies along the tangent to the flywheel at P. If the free end is pulled with a constant force of 8 N *so as to unwrap the string, find the angular speed of the wheel when it has rotated through one half of a revolution.*

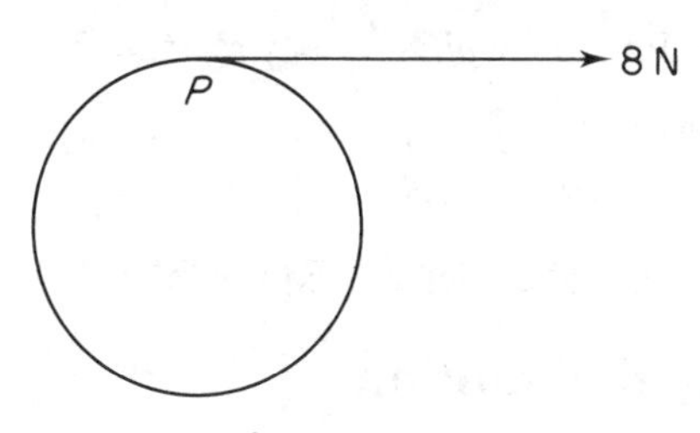

Fig. 15.4

When the wheel has rotated through half a turn, the point of application of the force has moved $\pi \times 0{\cdot}5$ m. Hence,

$$\text{work done} = 8 \times \pi 0{\cdot}5\,\text{J} = 4\pi\,\text{J}.$$

This work will be converted into kinetic energy of the wheel. Therefore,

$$4\pi = \tfrac{1}{2}2{\cdot}4\omega^2$$
$$\omega^2 = 4\pi/1{\cdot}2$$
$$\omega \approx \mathbf{3{\cdot}24\,rad\,s^{-1}}.$$

EXAMPLE 2 *A uniform rod is free to rotate about a horizontal axis which passes through one end of the rod and is perpendicular to the length of the rod. Given that the moment of inertia of the rod about the given axis is* $4Ml^2/3$, *where M is the mass of the rod and 2l its length, find the set of possible values of the angular speed when the rod is in its lowest position so that any point on the rod may describe complete circles in a vertical plane.*

Suppose that each point describes complete circles. Fig. 15.5 shows two positions of the rod – the lowest position when it has an angular speed of ω, and the highest position when it is vertically above the axis and moving with angular speed ω_1.

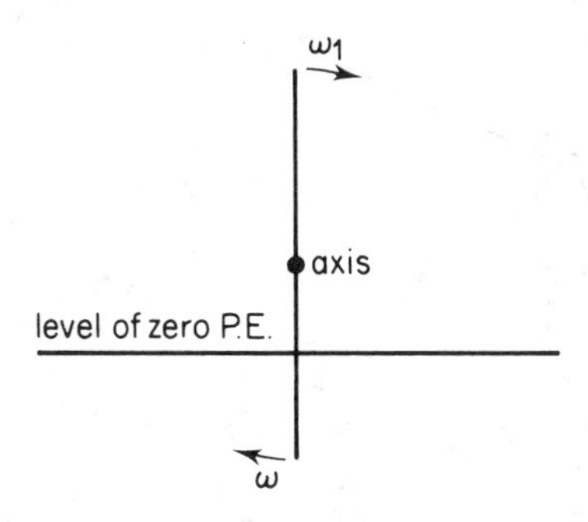

Fig. 15.5

We take the level of the centre of the rod when it is in its lowest position as the level of zero gravitational potential energy.
In the lowest position,

$$\text{P.E.} = 0, \ \text{K.E.} = \tfrac{1}{2}4Ml^2\omega^2/3.$$

In the highest position,

$$\text{P.E.} = 2Mgl, \ \text{K.E.} = \tfrac{1}{2}4Ml^2\omega_1^2/3.$$

By the principle of energy

$$2Ml^2\omega^2/3 = 2Ml^2\omega_1^2/3 + 2Mgl$$

To describe complete vertical circles we require $\omega_1 > 0 \Rightarrow \omega^2 > 3g/l$

Equation of angular motion

The term 'moment of inertia' emphasises an analogy with the mass of a body. If a given force acts in turn on each of several particles of differing mass, the greater the mass of the particle involved, the less the magnitude of the acceleration produced by the force. This is the *inertial* effect of mass.

In the study of the rotation of a body about a fixed axis and of the rate of change of the angular speed produced by a given force, clearly the line of action of the given force is of importance. If the line of action meets the axis, no turning effect is produced and no angular acceleration results. Try to open a door by exerting a force that passes through the line of the hinges! The greater the distance between the line of action of a force of given magnitude and the fixed axis, the greater the resulting angular acceleration. It is the *moment* of the force that is the effective unit in giving the body an angular acceleration.

Consider the motion at time t of one constituent particle P_i, of mass m_i, of a rigid body which is rotating about a fixed axis LL' with angular speed ω. Fig. 15.6(b) represents a plane section through P_i perpendicular to the axis. Particle P_i will describe a circle of radius r_i, centre O, where r_i is the distance of P_i from the axis.

By (10.5), the components of the acceleration of P_i are $\omega^2 r_i$ along $\overrightarrow{P_iO}$

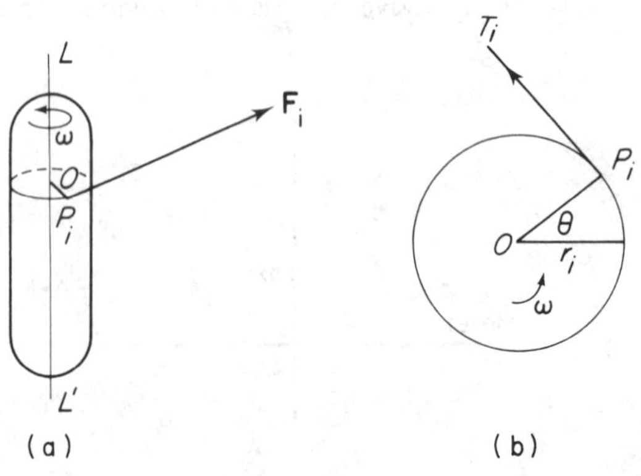

Fig. 15.6

and $r_i \dfrac{d\omega}{dt}$ along the tangent $\overrightarrow{P_iT_i}$ to the circle in the direction of θ increasing. Let $\mathbf{F}_i$ be the resultant external force acting on particle P_i, and let G_i be the resultant of the internal bonding forces acting on the particle. Then, by Newton's law ($\mathbf{F} = m\mathbf{a}$), these two forces $\mathbf{F}_i$ and $\mathbf{G}_i$ are together equivalent to

$$-m_i\omega^2 r_i\hat{\mathbf{r}}_i + m_i r_i \frac{d\omega}{dt}\hat{\mathbf{t}}_i,$$

where $\hat{\mathbf{r}}_i$ and $\hat{\mathbf{t}}_i$ are unit vectors along the radius P_iO and along the tangent $\overrightarrow{P_iT_i}$ respectively, that is,

$$\mathbf{F}_i + \mathbf{G}_i = -m_i\omega^2 r_i\hat{\mathbf{r}}_i + m_i r_i \frac{d\omega}{dt}\hat{\mathbf{t}}_i.$$

As the two sides of the equation are equivalent, we may take moments about the axis for each side separately and equate the results.

For the left-hand side, the internal bonding forces of all the separate constituent particles form a system in equilibrium. Referring to §3.7, we see that the sum of the moments of these bonding forces about the axis is zero. Thus, the sum of the moments of the left-hand side about the axis is the sum of the moments of the external forces about the axis.

For the right-hand side, the vector $-m_i r_i\omega^2\hat{\mathbf{r}}_i$ along $\overrightarrow{P_iO}$ intersects the axis and so has no moment about the axis. The vector $m_i r_i \dfrac{d\omega}{dt}\hat{\mathbf{t}}_i$ along $\overrightarrow{P_iT_i}$ has a moment about O of magnitude $m_i r_i^2 \dfrac{d\omega}{dt}$. Thus,

the sum of the moments of the external forces about the axis

$$= \sum_{i=1}^{n} m_i r_i^2 \frac{d\omega}{dt} = I\alpha. \qquad 15.4$$

where α is the angular acceleration about the axis.

Note the analogies between linear motion and angular motion. Let C denote the sum of the moments of the external forces about the axis.

linear motion	*angular motion*
$\mathbf{F} = m\mathbf{a}$	$C = I\alpha$
K.E. $= \frac{1}{2}mv^2$	K.E. $= \frac{1}{2}I\omega^2$
Uniform acceleration equations $v = u + at$ and so on	$\omega_1 = \omega_0 + \alpha t$

Later it will be shown that there are the following analogous results for work, momentum and impulse.

momentum $= mv$	angular momentum $= I\omega$
impulse $= Ft$ or $\int_{t_1}^{t_2} F\,\mathrm{d}t$	Angular impulse $= Ct$ or $\int_{t_1}^{t_2} C\,\mathrm{d}t$
work done $= Fs$ or $\int_{s_1}^{s_2} F\,\mathrm{d}s$	Work done $= C\theta$ or $\int_{\theta_1}^{\theta_2} C\,\mathrm{d}\theta$

If we refer to example 2 dealing with a rod hinged about one end, we can find the angular acceleration of the rod in any position. Suppose that the rod is inclined at an angle θ to the upward vertical through the hinge (see Fig. 15.7). The only external force acting on the rod that has a moment about the axis is its weight, which is of magnitude mg. Taking moments about the axis,

$$mgl\sin\theta = (4ml^2/3)\frac{\mathrm{d}\omega}{\mathrm{d}t}$$

$$\frac{\mathrm{d}\omega}{\mathrm{d}t} = 3g\sin\theta/(4l)$$

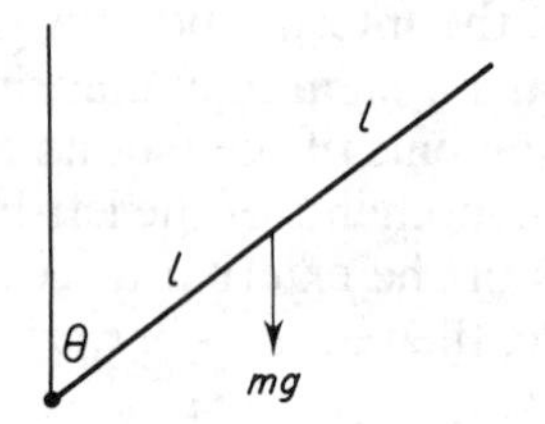

Fig. 15.7

EXAMPLE 3 *The moment of inertia of a flywheel about a smooth axis through its centre is* 1·4 kgm^2. *When the wheel is rotating with angular speed* 16 rad s^{-1} *a constant couple* G Nm *is applied to the wheel and as a result it is brought to rest in* 0·5 s. *Find the value of G. Find*
(a) the angle through which the wheel turns in this time and
(b) the angular speed when the wheel has turned through an angle $\pi/2$.

The deceleration = 32 rad s^{-2}. Therefore,

$$G = 1{\cdot}4 \times 32\,\text{Nm} = \mathbf{44{\cdot}8\,Nm}.$$

(a) Corresponding to the equation $s = ut + \frac{1}{2}at^2$, we have for angular motion

$$\theta = \omega_0 t + \tfrac{1}{2}\alpha t^2,$$

where θ is the angle through which the wheel turns in time t, ω_0 is the initial angular speed and α the constant angular acceleration. Hence,

$$\theta = 16 \times 0{\cdot}5 - 16 \times (0{\cdot}5)^2 = \mathbf{4\ radians}.$$

(b) Also corresponding to $v^2 - u^2 = 2as$, we have

$$\omega^2 - \omega_0^2 = 2\alpha\theta,$$

so when $\theta = \pi/2$,

$$\omega^2 - 16^2 = 2 \times (-32) \times \pi/2$$

$$\omega^2 = 256 - 32\pi.$$

$$\omega \approx \mathbf{12{\cdot}5\,rad\,s^{-1}}.$$

EXERCISE 15.3

Unless otherwise stated, take g as $10\,\text{ms}^{-2}$.

1 A gyroscope, whose moment of inertia about its axis is $6 \times 10^{-2}\,\text{kg}\,\text{m}^2$, is rotating at $10\,\text{rad}\,\text{s}^{-1}$ about its axis, which is fixed. Initially the gyroscope was at rest and it was set in motion by pulling one end of a string which was wrapped around the axis with a constant force of magnitude F newtons. The length of the string is 60 cm. Find the magnitude of the force if the string was pulled completely from the gyroscope and there was no resistance to the rotation of the gyroscope.

2 A bicycle wheel has moment of inertia of $0{\cdot}75\,\text{kg}\,\text{m}^2$ about its axis of rotation. A particle P of mass 0·04 kg is rigidly attached to a spoke at a distance of 30 cm from the axis. If the wheel is released when the spoke through P is horizontal and the wheel is free to rotate, find the angular speed when P reaches its lowest position.

3 A uniform rod AB, of mass 0·6 kg and length 1 m, is hinged at A so that the rod can rotate freely about a smooth horizontal axis. The rod is released when AB makes an angle of $60°$ with the upward vertical. Given that the moment of inertia of the rod about the axis of rotation is $0{\cdot}2\,\text{kg}\,\text{m}^2$, find the maximum angular speed in the subsequent motion.

4 A flywheel is rotating about its axis of symmetry at 240 revs/min. The motor driving the engine is shut off, and the flywheel slows down and comes to rest under the action of a frictional couple in $2\frac{1}{2}$ min. Given that the moment of inertia of the wheel about its axis of symmetry is $2{\cdot}5\,\text{kg}\,\text{m}^2$, find the magnitude of the frictional couple.

5 A uniform flywheel, which can rotate freely about its axis, is set in motion by a couple of constant moment C. The moment of inertia of the flywheel about its axis is I. Find the time taken for the angular speed of the flywheel to reach the value ω from rest, and find the angle through which it has turned in this time. (*L*)

6 A rope passing over a pulley, of radius 0·2 m and with moment of inertia $1\,\text{kg}\,\text{m}^2$ about its axis, connects two equal masses of 10 kg hanging freely at rest. If 5 kg is removed from one of the masses, determine

(a) the speed of the 10 kg mass when it has fallen a distance of 50 cm,

(b) the angular acceleration of the pulley.

7 A uniform disc of radius 20 cm is free to rotate about a horizontal axis, perpendicular to the plane of, and through the centre of, the disc. A particle of mass 0·05 kg is rigidly attached to a point on the circumference of the disc. Given that the moment of inertia of the disc about the axis is $0{\cdot}01\,\text{kg}\,\text{m}^2$, and that the system is released when the radius to the particle is horizontal, find the maximum angular speed of the system in the subsequent motion.

8 An electric motor drives a flywheel by exerting a torque of 8 kg m. As a result the flywheel reaches an angular speed of 600 revs/min in 20 seconds, starting from rest. Assuming that there is no resistance to motion, find the moment of inertia of the flywheel.

9 A pulley of radius 20 cm has one end of a string attached to a point on the rim. The string is wrapped round the pulley a number of times and it supports at its other end a particle of mass 1 kg, which hangs freely. Given that the moment of inertia of the pulley about its axis is $0{\cdot}2\,\text{kg}\,\text{m}^2$ and that there is no resistance to motion, find the angular acceleration of the pulley when the system is released with the string taut and the part supporting the particle.

10 You are given that the moment of inertia of a uniform square lamina, of mass M and of side a, about a diagonal is equal to $\frac{1}{3}Ma^2$. A uniform square lamina $ABCD$ of mass M and side a is free to rotate about a fixed vertical axis which coincides with the diagonal AC. The lamina is given an initial angular velocity ω_0 and, under the action of a constant driving torque G against a constant frictional torque L, completes 10 revolutions in the first second and 20 revolutions in the next second. Show that $\omega_0 = 10\pi\,\text{rad}\,\text{s}^{-1}$.

The constant driving torque G is then removed and the lamina is brought to rest by the frictional torque L which has been constant throughout the motion. The lamina is thus brought to rest in a further 15 revolutions. Find L and G. *(L)*

15.4 The calculation of moments of inertia

The simplest moment of inertia to calculate is that of an (ideal) hoop of uniform thin wire in the form of a circle, about an axis through its centre perpendicular to its plane. All the particles of this ideal hoop are at a distance r from the axis. Hence the moment of inertia is Mr^2, where M is the total mass of the hoop and r its radius. In almost all other cases, integration is required.

Uniform thin rod

To find the moment of inertia of a uniform thin rod, of mass M and length $2l$, about an axis perpendicular to its length through its centre O. The mass per unit length $= M/(2l)$.

Consider an element of length δx containing points of the rod with distance from the axis lying between x and $x + \delta x$ on one side of the axis and suppose that all the points in the element have distance from O within these limits (see Fig. 15.8).

$$\text{Mass of element} = M\delta x/(2l).$$

The contribution to $\Sigma m_i r_i^2$ lies between

$$\frac{Mx^2\delta x}{2l} \quad \text{and} \quad \frac{M(x+\delta x)^2}{2l}\,\delta x.$$

Hence,

$$\sum mr^2 \approx \sum_{x=-l}^{x=l} \frac{Mx^2\delta x}{2l}$$

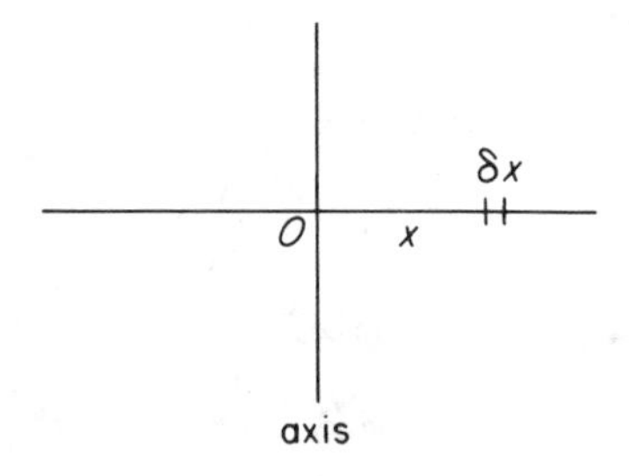

Fig. 15.8

and, in the limit, $$I = \int_{-l}^{l} \frac{Mx^2}{2l}\,\mathrm{d}x = \tfrac{1}{3}Ml^2$$

Uniform lamina in the form of a circular disc

To find the moment of inertia of a uniform lamina in the form of a circular disc, of mass M and radius r, about an axis through its centre O and perpendicular to its plane. The mass per unit area $= \dfrac{M}{\pi r^2}$.

Consider the element bounded by circles of radii x and $x + \delta x$ with centre O (see Fig. 15.9).

$$\text{Mass of element} = (2\pi x \delta x + (\delta x)^2)\frac{M}{\pi r^2} \approx \frac{2Mx}{r^2}\,x.$$

$$\text{Contribution to } \Sigma\, mr^2 \approx \frac{2Mx^3}{r^2}\,\delta x$$

and $$I = \int_0^r \frac{2Mx^3}{r^2}\,\mathrm{d}x = Mr^2/2$$

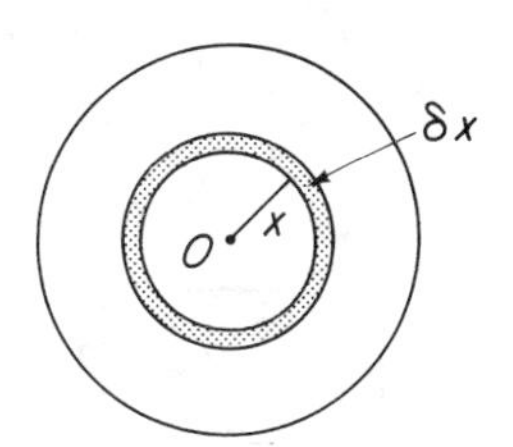

Fig. 15.9

Uniform solid sphere

To find the moment of inertia of a uniform solid sphere of mass M and radius r about a diameter.

The surface of the sphere is generated by revolving the circle $x^2 + y^2 = r^2$ about Ox. The elements used here are disc-like and are bounded by

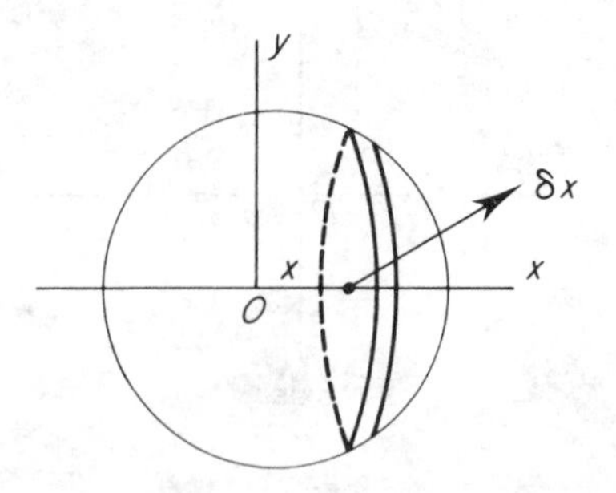

Fig. 15.10

the two circular faces with centres on Ox at distances of x and $x + \delta x$ from the centre of the sphere, together with the surface generated by the arc of the circle which is revolved (see Fig. 15.10).

Using the result just found for a circular disc, the contribution to the moment of inertia for such an element is approximately equal to $\rho\pi y^2 \delta x \cdot \frac{1}{2}y^2$, where ρ is the density. Hence,

$$I = 2\int_0^r \tfrac{1}{2}\pi\rho\,(r^2 - x^2)^2\,\mathrm{d}x = \pi\rho\left[r^4x - 2r^2\frac{x^3}{3} + \frac{x^5}{5}\right]_0^r = 8\pi\rho r^5/15$$

The total mass $M = 4\pi r^3\rho/3$, so

$$I = 2Mr^2/5.$$

Uniform circular cylinder

The result for a disc can be extended to find the moment of inertia of a uniform circular cylinder of mass M and radius r about its axis by summing a number of disc-like elements (see Fig. 15.11).

$$I = Mr^2/2$$

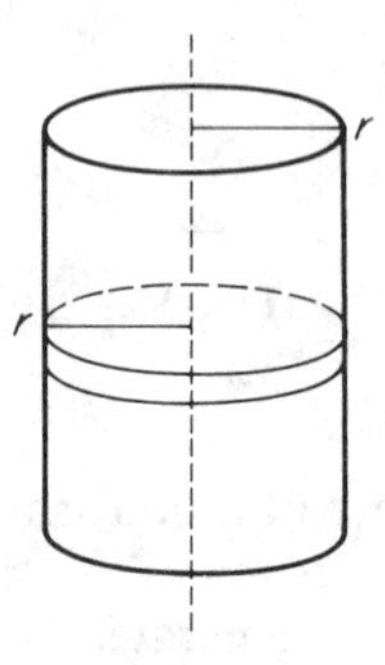

Fig. 15.11

Uniform lamina in the form of a rectangle

The result for a thin rod can be extended to find the moment of inertia of a uniform lamina in the form of a rectangle of mass M, length $2a$ and width $2b$, about an axis parallel to the sides of length $2b$ and through its centre (see Fig. 15.12).

$$I = Ma^2/3$$

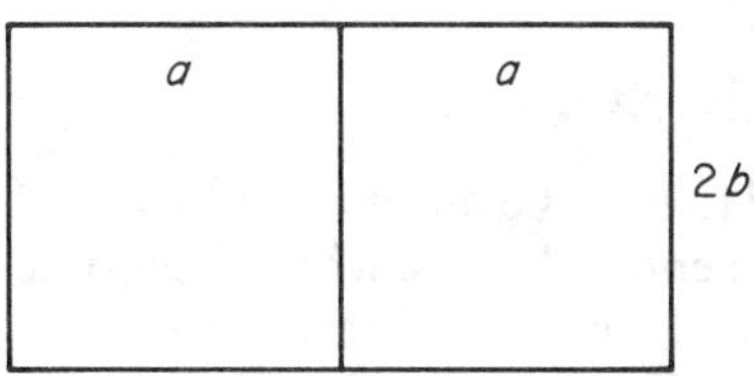

Fig. 15.12

Summary of results for uniform bodies.

body	axis through centre of mass	moment of inertia
hoop	perpendicular to plane	Mr^2
rod, length $2l$	perpendicular to its length	$Ml^2/3$
disc, lamina	perpendicular to plane	$Mr^2/2$
cylinder	axis of cylinder	$Mr^2/2$
sphere	diameter	$2Mr^2/5$
rectangular lamina sides $2a$, $2b$	parallel to side $2b$	$Ma^2/3$

The perpendicular axis or lamina theorem

If the moment of inertia of a lamina about two perpendicular axes in its plane are known, the moment of inertia about an axis through the point of intersection of the given axes and perpendicular to the plane of the lamina may be found.

Let Ox and Oy be taken as the given axes. Then I_z, the moment of inertia about Oz, is required. For a typical particle of the lamina at the point P_i with coordinates (x_i, y_i) referred to the given axes and with $OP_i = r_i$ (see Fig. 15.13)

the contribution to $I_x = m_i y_i^2$,

the contribution to $I_y = m_i x_i^2$,

the contribution to $I_z = m_i r_i^2 = m_i(x_i^2 + y_i^2)$

Summing for all particles of the lamina

$$I_z = I_y + I_x. \qquad 15.5$$

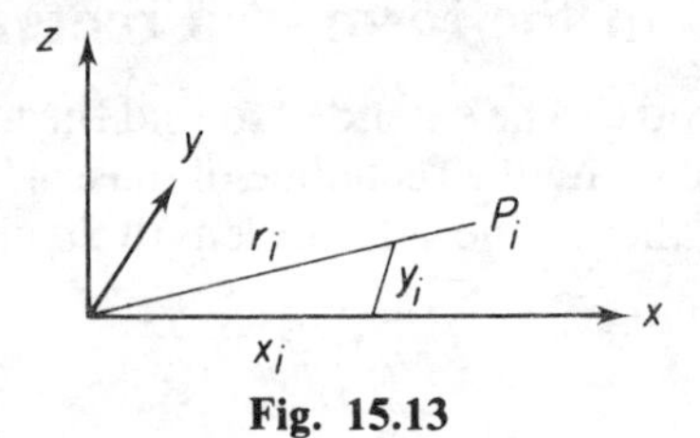

Fig. 15.13

The parallel axis theorem

This theorem is used to find the moment of inertia of a rigid body about an axis not through the centre of mass, when the moment of inertia about the parallel axis through the centre of mass is known.

In Fig. 15.14(a), MM' is the axis about which the moment of inertia is required and LL' is the parallel axis through the centre of mass G. Let the distance between the parallel axes be h. Consider one particle at P_i and a section of the body through P_i perpendicular to the given axis. Let this plane meet the axis at O_i and the parallel axis through the centre of mass at A_i. The mass of the particle is m_i (see Fig. 15.14(b)).

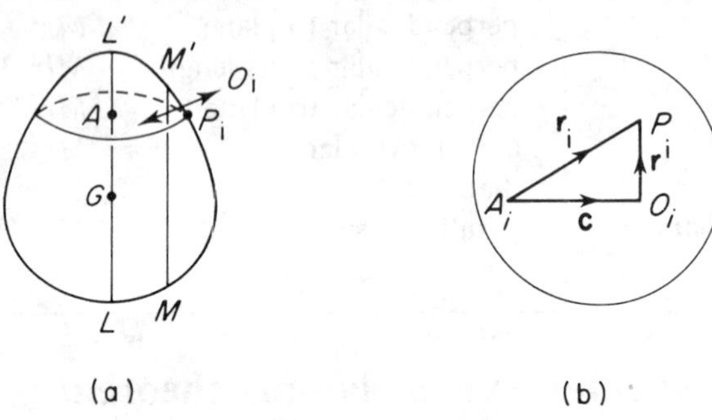

Fig. 15.14

Let the vectors $\overrightarrow{A_iP_i} = \mathbf{r}_i$, $\overrightarrow{O_iP_i} = \mathbf{r}$, $\overrightarrow{A_iO_i} = \mathbf{c}$. The contribution to the moment of inertia about the required axis $= m_i r_i^2$. Now

$$\mathbf{r} = \mathbf{r}_i - \mathbf{c}$$

$$I = \sum m_i(\mathbf{r}_i - \mathbf{c})^2$$

$$= \sum m_i \mathbf{r}_i^2 - 2\sum m_i \mathbf{r}_i \,.\, \mathbf{c} + \sum m_i \mathbf{c}^2.$$

Note that the vector $\mathbf{c}$ is the same for all the sections perpendicular to the axis. Now,

$$\sum m_i \mathbf{r}_i \,.\, \mathbf{c} = \mathbf{c} \,.\, (\sum m_i \mathbf{r}_i),$$

and, as $\mathbf{r}_i$ is the position vector of P_i with reference to a point on an axis through the centre of mass,

$$\sum m_i \mathbf{r}_i = \mathbf{0}.$$

(For, if the given axis is chosen as the z-axis, then for the ith particle $\mathbf{r}_i = x_i\mathbf{i} + y_i\mathbf{j}$ and $\sum m_i x_i = \sum m_i y_i = 0$ from equation 12.3.) Also

$$\sum m_i \mathbf{c}^2 = (\sum m_i)\mathbf{c}^2,$$

and as $\mathbf{c}$ is a constant and equal to h.

$$\sum m_i \mathbf{c}^2 = Mh^2$$
$$I = \sum m_i \mathbf{r}_i^2 + Mh^2$$
$$I = I_G + Mh^2, \qquad 15.6$$

where I_G is the moment of inertia about a parallel axis through the centre of mass.

EXAMPLE 1 *Find the moment of inertia of a rectangular plate of mass M and sides 2a, 2b about an axis perpendicular to its plane (a) through its centre, (b) through one of its corners.*

Let O be the centre of the plate and let the axes Ox and Oy be parallel to the sides of length $2a$ and $2b$ respectively (see Fig. 15.15).

$$I_x = \tfrac{1}{3}Mb^2 \quad \text{and} \quad I_y = \tfrac{1}{3}Ma^2.$$

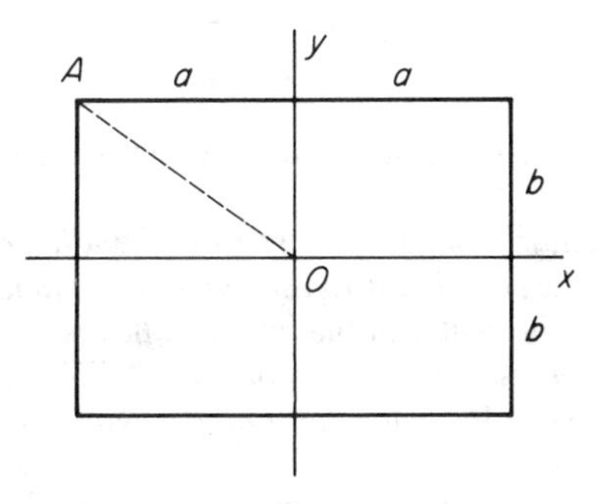

Fig. 15.15

By the perpendicular axis theorem,

$$I_z = \tfrac{1}{3}M(a^2 + b^2).$$

To find the moment of inertia I about the axis through one corner A and perpendicular to the plate, we use the parallel axis theorem.

$$I = I_z + M(a^2 + b^2)$$
$$I = 4M(a^2 + b^2)/3.$$

EXAMPLE 2 *Find the moment of inertia of a uniform circular lamina of mass M and radius r about a diameter.*

First, it should be noted that, due to the symmetry of a circle, the moment of inertia of the lamina about any diameter is the same. Thus, if Ox and Oy are two

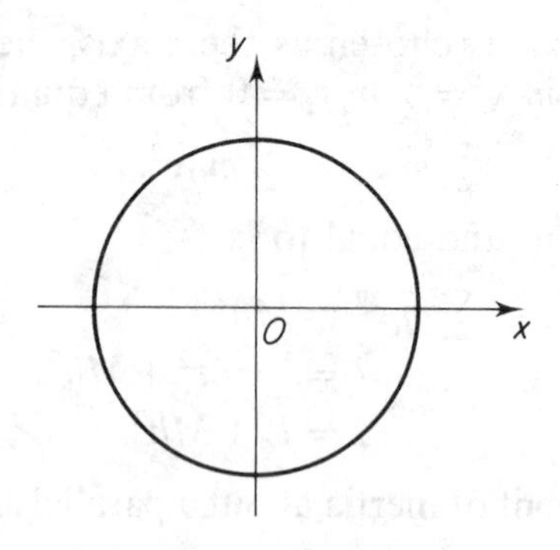

Fig. 15.16

perpendicular diameters, the moments of inertia about Ox and Oy, I_x and I_y respectively, satisfy

$$I_x = I_y$$

and, by the lamina theorem,

$$I_z = 2I_x.$$

From the result found for a circular disc (§15.4)

$$I_z = \tfrac{1}{2}Mr^2$$
$$I_x = \tfrac{1}{4}Mr^2$$

The moment of inertia of the disc about any diameter is $\frac{1}{4}$ $\boldsymbol{Mr^2}$.

EXAMPLE 3 *A circular lamina, of mass M and radius r, is pinned to a vertical board by a smooth pin, perpendicular to the plane of the lamina and through a point A at a distance r/2 from the centre of the lamina. The disc is free to rotate about the pin and it is released from rest when OA is horizontal. Find the angular speed and the angular acceleration when AO makes an angle θ with the downward vertical.*

By the parallel axis theorem, the moment of inertia of the disc about the axis through A is

$$\tfrac{1}{2}Mr^2 + M(r/2)^2 = \tfrac{3}{4}Mr^2.$$

Taking the level of the centre O when it is vertically below A as the level of zero gravitational potential energy (Fig. 15.17), at the start,

$$\text{P.E.} = Mgr/2, \text{ K.E.} = 0.$$

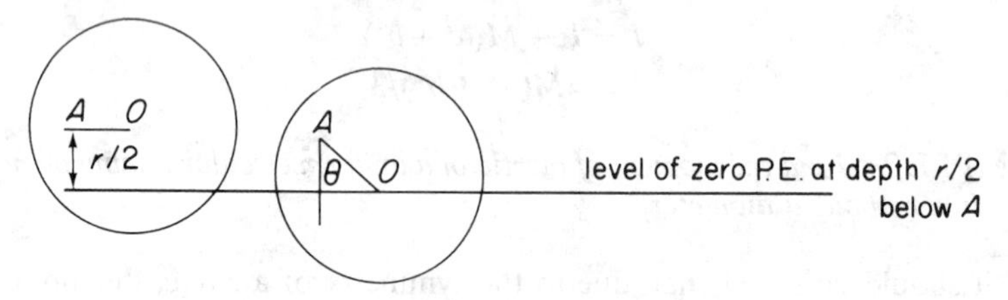

Fig. 15.17

In the second position,

$$\text{P.E.} = Mgr(1-\cos\theta)/2, \text{ K.E.} = \tfrac{1}{2}\cdot\tfrac{3}{4}Mr^2\omega^2.$$

By the principle of energy,

$$\tfrac{3}{8}Mr^2\omega^2 + Mgr(1-\cos\theta)/2 = Mgr/2,$$

$$\tfrac{3}{8}Mr^2\omega^2 = Mgr\cos\theta/2$$

$$\omega^2 = \frac{4g\cos\theta}{3r}$$

$$\omega = \sqrt{\frac{4g\cos\theta}{3r}}.$$

Using equation 15.4

$$-\frac{Mgr\sin\theta}{2} = \frac{3}{4}Mr^2\frac{d\omega}{dt}$$

$$\frac{d\omega}{dt} = -\frac{2}{3}\frac{g}{r}\sin\theta.$$

EXERCISE 15.4

1 Prove, by integration, that the moment of inertia of a thin uniform rod AB, of mass m and length $2a$, about an axis through A and perpendicular to its length is $4ma^2/3$. *(O&C)*

2 Prove, by integration, that the moment of inertia of a uniform circular disc, of mass m and radius a, about an axis through its centre and perpendicular to its plane is $\frac{1}{2}ma^2$. *(O&C)*

3 A uniform rectangular plate has sides of lengths $2a$, $2b$ and mass m. Prove that its moment of inertia about an axis perpendicular to it and through its centre is $\frac{1}{3}m(a^2+b^2)$. *(O&C)*

4 Show, by integration, that the moment of inertia of a uniform thin square plate $ABCD$, of mass m and side $2a$, about the edge AB is $4ma^2/3$. Deduce the moment of inertia of the plate about an axis through its centre and perpendicular to its plane. *(L)*

5 Find the moment of inertia of a uniform disc of mass m and radius r about a tangent line in the plane of the disc.

6 Show, by integration, that the moment of inertia of a uniform triangular lamina PQR about QR is $Mh^2/6$, where M is the mass of the lamina and h is the length of the altitude through P. *(L)*

7 Prove that the moment of inertia of a uniform solid sphere of radius a and mass M about a diameter is $\frac{2}{5}Ma^2$. *(L)*

8 A uniform plane lamina OAB of mass M is in the form of a quadrant of a circle of centre O and radius a. By use of integral calculus, find the position of its centre of gravity and prove that its moment of inertia about an axis perpendicular to its plane through O is $\frac{1}{2}Ma^2$.

The lamina is smoothly hinged at O about a horizontal axis so that it can swing freely with its plane vertical. It is held with its lower edge horizontal and then released. Find the greatest angular velocity of the lamina. *(O&C)*

9 AB is a uniform rod of mass m. The distances of A and B from an axis coplanar with AB are p and q. Prove that the moment of inertia of the rod about this axis is

$$\tfrac{1}{3}m(p^2+pq+q^2)$$

if the axis does not intersect the rod. Find a similar expression for the moment of inertia if the axis does intersect the rod. *(O&C)*

10 Find the moments of inertia of a uniform cuboid, of mass M and with sides of lengths $2a$, $2b$ and $2c$; about the following axes:
(i) an axis through its centre of mass parallel to the sides of length $2c$,
(ii) an edge of length $2c$.

11 A uniform lamina has the shape of the region bounded by the curve $ay = x^2/9$, the line $x = 3a$ and the x-axis. Show that, if k is the mass/unit area of the lamina,
(a) the mass of the lamina M is equal to ka^2,
(b) the coordinates of the centre of mass of the lamina are $(2\tfrac{1}{4}a, 0{\cdot}3a)$,
(c) the moment of inertia of the lamina about the y-axis is $5{\cdot}4Ma^2$, and about an axis parallel to the y-axis through the centre of mass is $0{\cdot}3375Ma^2$.

12 Prove that the moment of inertia of a uniform lamina of mass m, in the shape of an equilateral triangle ABC of height h, about an axis through one vertex and perpendicular to the plane ABC, is $\tfrac{5}{9}mh^2$.

The lamina is free to move in a vertical plane about a smooth horizontal axis through A. While it is in equilibrium, with BC below A, the mid-point of BC is given a velocity u in the direction BC. Prove that the lamina will make complete revolutions if $u^2 > \tfrac{24}{5}gh$. *(O&C)*

13 A uniform circular disc has mass $4M$, radius $4a$ and centre O, and AOB is a diameter. A circular hole of radius $2a$ is made in the disc, the centre of the hole being at a point C on AB, where $BC = 5a$. The resulting lamina (of mass $3M$) is free to rotate about a fixed smooth horizontal axis through B perpendicular to the plane of the lamina. Show that the moment of inertia of the lamina about this axis is $69Ma^2$.

The lamina is set in motion with angular velocity $(11g/69a)^{\frac{1}{2}}$ from the position in which A is vertically below B. Find the angle through which it has turned when it first comes to rest instantaneously. *(O&C)*

14 A uniform circular disc of mass $4m$ and radius $2a$ is rotating with angular velocity ω about a vertical axis through its centre perpendicular to its plane, which is horizontal. Write down expressions for its kinetic energy and its angular momentum about the axis.

The disc is brought to rest by a tangential force of magnitude $4mg$ applied at its rim. Find the time taken and the angle the disc turns through in that time. *(L)*

15 A uniform solid of revolution is obtained by revolving the region bounded by the curve $cy^2 = x^3$ and the line $x = c$ about the x-axis through four right angles. Find the moment of inertia of this solid about the axis of revolution in terms of its mass M, and c.

16 Find the moment of inertia of a uniform solid right circular cone of mass M, height h and base radius r, about (a) the axis, (b) a line through the vertex parallel to the base.

If these results are equal, show that $r = 2h$ and that the minimum moment of inertia about a line parallel to the base is $51Mh^2/80$. *(O&C)*

15.5 Use of the energy equation to determine the angular acceleration

$$\frac{d\omega}{dt} = \frac{d\omega}{d\theta}\cdot\frac{d\theta}{dt} = \omega\frac{d\omega}{d\theta}. \qquad 15.7$$

This result corresponds to $\frac{dv}{dt} = v\frac{dv}{dx}$ in rectilinear motion.

If the energy equation is expressed in terms of the angle θ through which the body has rotated in time t, then it will be of the form

$$\tfrac{1}{2}I\omega^2 + V(\theta) = \text{constant},$$

where $V(\theta)$ is the potential energy of the system. By differentiating this equation with respect to θ, we have

$$I\omega\frac{d\omega}{d\theta} + \frac{dV}{d\theta} = 0$$

and, as $\frac{d\omega}{dt} = \omega\frac{d\omega}{d\theta}$, the angular acceleration is obtained, as illustrated in the following example.

EXAMPLE 1 *A uniform rod, of mass M and length 2l, is rotating smoothly about an axis through one end and perpendicular to its length. The rod is released when horizontal. Write down the energy equation when the rod is inclined at an angle θ to the horizontal and, hence, determine the angular acceleration in terms of θ.*

The moment of inertia of the rod about the axis of rotation is $4ml^2/3$. Taking the level of the centre of the rod when it is at its lowest position to be the level of zero gravitational potential energy (Fig. 15.18), the energy equation is

$$\tfrac{1}{2}(4ml^2/3)\omega^2 + mgl(1 - \sin\theta) = \text{constant}$$

If the equation is only to be used to find the angular acceleration, there is no need to record the value of the constant (which is, in fact, mgl). Differentiating

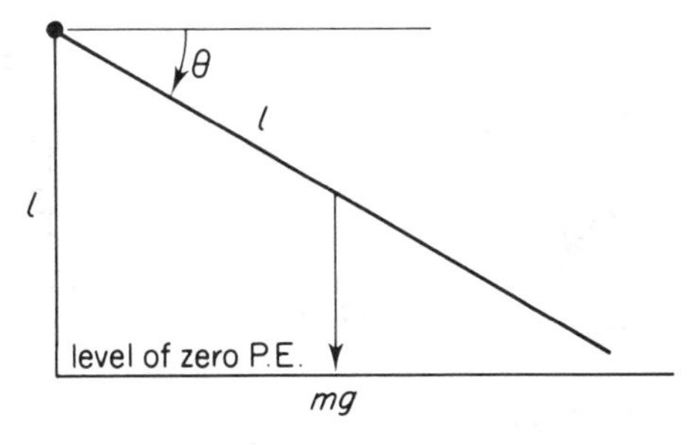

Fig. 15.18

with respect to θ,

$$\tfrac{2}{3}ml^2 \cdot 2\,\omega \frac{d\omega}{d\theta} - mgl\cos\theta = 0$$

$$\Rightarrow \omega \frac{d\omega}{d\theta} = \frac{d\omega}{dt} = \frac{3}{4}\frac{g}{l}\cos\theta.$$

The same result can be obtained by use of the equation of angular motion,

$$mgl\cos\theta = \frac{4}{3}ml^2\frac{d\omega}{dt} \qquad \frac{d\omega}{dt} = \frac{3}{4}\frac{g}{l}\cos\theta.$$

The reason that the equation of angular motion may be obtained by differentiation of the energy equation is that the energy equation is derived in the first place by integration of the equation of motion, that is

$$\text{the equation of motion} = \frac{d}{d\theta}\,(\text{the equation of energy}).$$

Some problems involve composite bodies in which the calculations of the moments of inertia require the addition of the moments of inertia of the separate parts, and the moments of the forces acting on the parts have to be added. This is illustrated in the following example.

EXAMPLE 2 *A pendulum consists of a uniform rigid rod, of mass 3M and length 4r, attached to a circular lamina of mass 2M and radius r. The lamina and the line of the rod lie in the same plane, the centre of the lamina lying on the line of the rod produced. The lamina is attached to the end of the rod at a point on the rim. The pendulum is free to rotate about a smooth horizontal axis through the other end O of the rod. If the rod is released when the rod is inclined at an angle β to the downward vertical, find (a) the angular speed when the pendulum is in its lowest position, (b) the angular acceleration when the rod is inclined at an angle θ to the downward vertical.*

The moment of inertia of the pendulum about the axis through O is found by considering the parts – the rod and the lamina – separately.

For the rod, the moment of inertia about a parallel axis through its centre of mass is

$$3M(2r)^2/3 = 4Mr^2.$$

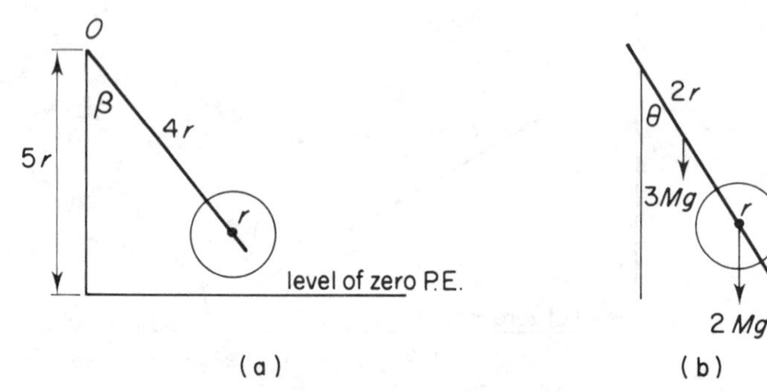

Fig. 15.19

The moment of inertia about the required axis is

$$4Mr^2 + 3M(2r)^2 = 16Mr^2.$$

For the lamina, the moment of inertia about a parallel axis through its centre is

$$\tfrac{1}{2}2Mr^2 = Mr^2,$$

and so the moment of inertia about the required axis is

$$Mr^2 + 2M(5r)^2 = 51Mr^2.$$

Hence, the total moment of inertia I about the axis through O is $67Mr^2$.
(a) Taking the level of the lowest point of the path of the centre of the disc as the level of zero gravitational potential energy (Fig. 15.19), the energy equation gives initially,

$$\text{P.E.} = 2Mg\,.\,5r(1-\cos\beta) + 3Mg(5r - 2r\cos\beta),\ \text{K.E.} = 0,$$

and at the lowest position,

$$\text{P.E.} = 3Mg\,.\,3r,\ \text{K.E.} = \tfrac{1}{2}67Mr^2\omega^2$$

and
$$Mgr(10 - 10\cos\beta + 15 - 6\cos\beta) = 9Mgr + \tfrac{1}{2}67Mr^2\omega^2$$

$$16Mgr(1-\cos\beta) = \tfrac{1}{2}67Mr^2\omega^2$$

$$\omega^2 = \frac{32g}{67}(1-\cos\beta)$$

$$\omega = \sqrt{\left[\frac{32g}{67}(1-\cos\beta)\right]}$$

(b) See Fig. 5.19(b). The equation of angular motion is

$$3Mg\,.\,2r\sin\theta + 2Mg\,.\,5r\sin\theta = -I\frac{d^2\theta}{dt^2},$$

where $I = 67Mr^2$. The negative sign is necessary as the moment of the forces is in the direction of θ decreasing. Hence,

$$16Mgr\sin\theta = -67Mr^2\frac{d^2\theta}{dt^2}$$

The angular acceleration is given by

$$\frac{d^2\theta}{dt^2} = -\frac{16g}{67r}\sin\theta.$$

15.6 The compound pendulum.

A rigid body which is free to rotate about a fixed axis through a point O of the body, which is not the centre of mass, can rest in stable equilibrium with its centre of gravity vertically below the axis of suspension.

In Fig. 15.20(a), the two forces acting on the body – the force at the axis and the weight – are equal in magnitude, opposite in direction and act in

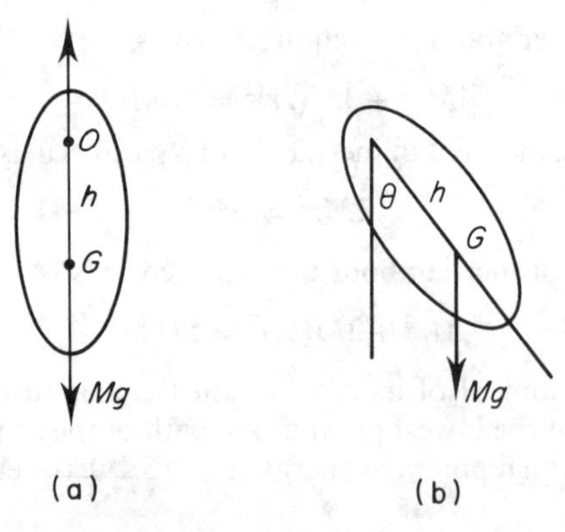

Fig. 15.20

the same vertical line. If the body is slightly displaced from this position, as in Fig. 15.20(b), the weight of the body provides a restoring moment tending to give the body an angular acceleration to return it to the equilibrium position.

If the angular displacement from the equilibrium position is θ, the distance OG is h and the moment of inertia about the axis is I, the equation of angular motion is

$$Mgh \sin\theta = -I\frac{d^2\theta}{dt^2}.$$

The angular acceleration is negative, as the restoring moment acts in the opposite sense to that in which θ is measured. Hence,

$$\frac{d^2\theta}{dt^2} = -\frac{Mgh}{I}\sin\theta. \qquad 15.8$$

When θ is small, $\sin\theta \approx \theta$, and in this case

$$\frac{d^2\theta}{dt^2} \approx -\frac{Mgh}{I}\theta.$$

Referring to §14.1, we see that this motion is approximately simple harmonic with period $2\pi\sqrt{[I/Mgh]}$.

It is also possible to derive equation 15.8 from the energy equation. When OG is inclined at θ to the downward vertical, the energy equation is

$$\tfrac{1}{2}I\omega^2 - Mgh\cos\theta = \text{constant}.$$

Differentiating with respect to θ,

$$\omega\frac{d\omega}{d\theta} + Mgh\sin\theta = 0.$$

Using equation 15.7

$$\frac{d^2\theta}{dt^2} = -\frac{Mgh}{I}\sin\theta.$$

EXAMPLE 1 *A uniform rod, of mass M and length 2l, is free to rotate about a smooth axis perpendicular to its length and passing through one end. Find*
(a) the period of the motion for small oscillations,
(b) if, in a second case, the point of the rod through which the axis passes is at a distance x (< 1) *from its centre, the period of small oscillations,*
(c) the value of x for which this period is least.

(a) When the rod is inclined at θ to the vertical, the restoring moment of the weight is $Mgl\sin\theta$. So, the equation of angular motion is

$$Mgl\sin\theta = -\frac{4}{3}Ml^2\frac{d^2\theta}{dt^2}$$

$$\Rightarrow \frac{d^2\theta}{dt^2} \approx -\frac{3g}{4l}\theta,$$

for small values of θ. That is, the motion is approximately simple harmonic with period $\mathbf{2\pi\sqrt{[4l/(3g)]}}$.
(b) When the axis of rotation is at a distance x from the centre, by the parallel axis theorem the moment about the axis is $Ml^2/3 + Mx^2$ (see Fig. 15.21(b)). The restoring moment is $Mgx\sin\theta$. Hence,

$$Mgx\sin\theta = -M(l^2/3 + x^2)\frac{d^2\theta}{dt^2}$$

$$\Rightarrow \frac{d^2\theta}{dt^2} \approx -\frac{3gx}{l^2 + 3x^2}\theta$$

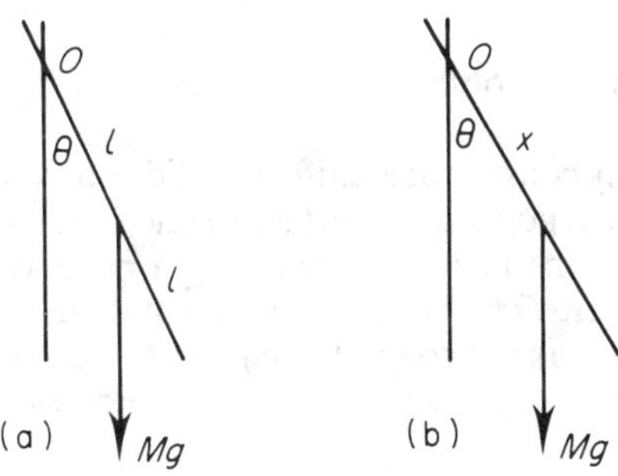

Fig. 15.21

for small θ. So, motion is approximately simple harmonic for small values of θ, and with period $\mathbf{2\pi\sqrt{[(l^2+3x^2)/(3gx)]}}$.
(Note that this result can be checked by putting $x = l$).
(c) The period will be least when $(l^2 + 3x^2)/x$ is least, where $0 < x < l$. Put

$$y = l^2/x + 3x.$$

$$\frac{dy}{dx} = -l^2/x^2 + 3, \quad \text{when } \frac{dy}{dx} = 0,\ x = l/\sqrt{3}.$$

$$\frac{d^2y}{dx^2} = 2l^2/x^3,$$

and $x = l/\sqrt{\mathbf{3}}$ gives a minimum value of the period.

15.7 The radius of gyration

In some texts, the moment of inertia of a body is defined by stating its radius of gyration. If M is the mass of a given body and k is the length of a line such that Mk^2 is the moment of inertia about a given axis, then k is called the *radius of gyration* of the body about that axis.

Thus, for a rod of mass M and length $2l$, the moment of inertia about an axis through its centre and perpendicular to its length is $Ml^2/3$. Hence

$$k^2 = l^2/3$$

and the radius of gyration is $l/\sqrt{3}$.

For a disc about an axis through the centre perpendicular to its plane, $k = l/\sqrt{2}$.

In general, $I = Mk^2$.

EXERCISE 15.7

1 A uniform rod AB, of length $2a$ and mass m, is free to rotate in a vertical plane about a smooth horizontal axis through the end A and perpendicular to the rod. The rod is released from rest when it is horizontal. Show that when the rod has turned through an angle θ, the angular speed ω acquired is given by

$$\omega^2 = 3g \sin \theta/(2a).$$

Hence, or otherwise, obtain the angular acceleration of the rod when $\theta = \pi/3$.

2 A composite body consists of a uniform solid sphere, of mass m and radius r, rigidly attached at a point on its surface to one end B of a light thin rod AB, of length $3r$, so that the line of the rod, when produced, passes through the centre of the sphere. The body is free to rotate in a vertical plane about a smooth horizontal axis through the end A of the rod and perpendicular to the rod. Show that the moment of inertia of the rod about this axis is equal to $16{\cdot}4mr^2$.

If the body is released from rest when AB makes an angle α with the downward vertical through A, show that the angular speed ω acquired when AB makes an angle θ with the downward vertical is given by

$$\omega^2 = 2g(\cos \theta - \cos \alpha)/(4{\cdot}1r).$$

Hence, or otherwise, prove that the angular acceleration of the body about the axis in this position is $-(g \sin \theta)/(4{\cdot}1r)$.

3 A composite body consists of a uniform rod AB, of length $2a$ and mass $3m$, rigidly attached at one end B to the centre of a uniform plane circular lamina, of mass m and radius a, such that the rod is perpendicular to the plane of the lamina. The body is free to rotate in a vertical plane about a fixed smooth horizontal axis through the other end A of the rod and perpendicular to the rod. Show that

(i) the moment of inertia of the body about the axis is $33ma^2/4$,

(ii) if the body is released from rest when AB makes an angle α with the downward vertical, the angular speed acquired when the angle made by AB with the downward vertical is θ is given by

$$33\omega^2 = 40ag(\cos\theta - \cos\alpha).$$

Hence, or otherwise, show that

$$\frac{d^2\theta}{dt^2} = -20g\sin\theta/(33a).$$

Hence determine the period of the simple harmonic motion to which this motion approximates when α is small.

4. A flywheel with moment of inertia about its axle $\frac{2}{3}ma^2$ and radius a is free to rotate about its axle which is smooth, horizontal and fixed. A light inextensible string is wrapped around the wheel and has a particle of mass m attached to the free end. The system is set in motion with the hanging part of the string taut and vertical. Show that when the flywheel has turned through an angle θ, the angular speed acquired, ω, is given by

$$\omega^2 = 6g\theta/(5r)$$

Hence find the acceleration of the particle.

5 Prove by use of integral calculus that the moment of inertia of a plane circular lamina of radius a and mass M about a diameter is $Ma^2/4$.

Such a circular lamina is fixed at its centre to one end of a light rod of length $b(b > a)$ in its plane, the other end of the rod being freely pivoted on a horizontal axis perpendicular to the rod and parallel to a diameter of the lamina. The rod is moved through an angle α from the vertical and then released from rest. If α is small, find the period of oscillation of the approximate simple harmonic motion. *(O&C)*

6 A uniform lamina of mass m has the shape of an isosceles triangle of height h. Prove that its moment of inertia about the line parallel to its base through its vertex is $\frac{1}{2}mh^2$.

If the lamina oscillates about this axis write down the equation of motion, and hence find the length of the equivalent simple pendulum. *(O&C)*

7 A uniform solid sphere, of radius a and density ρ, is of mass M. Show, by integration, that $M = 4\pi\rho a^3/3$. Assuming that the moment of inertia of a thin uniform disc, of mass m and radius r, about its central axis is $\frac{1}{2}mr^2$, show also that the moment of inertia of the sphere about a diameter is $2Ma^2/5$. Deduce the moment of inertia of the sphere about a tangent line.

The sphere is freely pivoted about a smooth horizontal axis which coincides with a tangent line.

(i) Prove that the periodic time of small oscillations of the sphere about its equilibrium position is

$$2\pi\sqrt{\{(7a)/(5g)\}}.$$

(ii) The sphere is held so that its centre O is at the same level as the axis and is then released. Find the velocity of O when it passes vertically below the axis. *(O&C)*

8 A uniform wire AOB, of total length $3a$ and mass m, is bent to form a right angle at O with straight arms AO and OB, of length a and $2a$ respectively.

(a) Show that the moment of inertia of the wire about an axis through O perpendicular to the plane of the wire is ma^2.
(b) The wire is freely suspended from O and makes small oscillations in its own vertical plane. Show that the length of the equivalent simple pendulum is $6a/\sqrt{17}$.
(c) The system is freely suspended from O with OAB in a vertical plane and is released from rest with A vertically below O. Find the speed of B when it is vertically below O. (O&C)

9 ABC is a uniform isosceles triangular lamina in which $AB = AC = 5a$ and $BC = 6a$. The mass of the lamina is m. Find its moment of inertia
(i) about an axis through A parallel to BC,
(ii) about an axis through A perpendicular to BC and lying in the plane of ABC.
The lamina is free to rotate about a horizontal axis through A perpendicular to the plane of the triangle. Find the period of small oscillations about the position of equilibrium in which BC is below A. (L)

10 A piece of heavy uniform wire of mass M and length $20a$ is bent in the form of an isosceles trapezium $ABCD$ in which the equal sides BC and AD are each of length $5a$ and CD is of length $8a$. Prove that the moment of inertia of the wire about an axis through N, the midpoint of CD, perpendicular to the plane of $ABCD$ is $149\ Ma^2/15$.
The wire is freely suspended so that it can oscillate in a vertical plane about the same axis through N. Calculate the period of small oscillations about the position of stable equilibrium. (L)

11 Show that the radius of gyration of a uniform rectangular lamina with sides of length a and b about an axis through the mid-points of the sides of length a is $a/\sqrt{12}$.
This lamina is smoothly hinged along one of its sides of length b and can perform small oscillations about this hinge which is fixed and horizontal. The period of these oscillations is T_1. When the lamina is smoothly hinged to a fixed pivot at one of its corners it can perform small oscillations in its own vertical plane with period T_2. Find T_1 and T_2 and show that, if $T_2 = 2T_1$, then $b = a\sqrt{15}$. (L)

MISCELLANEOUS EXERCISE 15

1 A uniform solid circular cylinder of radius a and mass m can turn freely about its axis, which is fixed in a vertical position. A horizontal force of magnitude mg acts on the cylinder, its line of action being at a distance a from the axis. Find the time taken to make n complete revolutions from rest. (L)

2 Show that the radii of gyration of a uniform square lamina $ABCD$ about a diagonal and about PQ, where P and Q are the mid-points of AB and CD, are equal. (L)

3 A uniform solid circular cylinder of mass M can rotate freely about a horizontal axis which coincides with a diameter of an end face of the cylinder. The length and the radius of the cylinder are both equal to a. Show that its moment of inertia about the axis is $7Ma^2/12$.

The cylinder is slightly disturbed from rest when its centre of mass is vertically above the axis. Show that, when the axis of the cylinder makes an angle θ with the upward vertical,

$$7a\dot{\theta}^2 = 12g(1-\cos\theta),$$

and obtain an expression for the angular acceleration $\ddot{\theta}$. (*L*)

4 Show that the moment of inertia about one edge of a uniform square lamina of mass m and side $2a$ is $4ma^2/3$.

A hollow cube of edge $2a$ is constructed from six squares of uniform sheet metal of negligible thickness. The cube is revolving freely about a fixed horizontal axis which bisects two opposite edges of one face. Find the maximum value of the angular acceleration of the cube. (*L*)

5 Two heavy particles of equal mass are fixed to the corners A and C of a light square framework $ABCD$ of side $2a$, AC being a diagonal. The framework can rotate freely in a vertical plane about a fixed horizontal axis which bisects AD at right angles. Find the length of the equivalent simple pendulum for small oscillations about the position of stable equilibrium.

If the corner C is held vertically above the mid-point of AD and is then released from rest, show that the maximum angular velocity ω of the framework is given by

$$15a\omega^2 = (10+4\sqrt{5})g. \qquad (L)$$

6 A uniform plane rectangular lamina $ABCD$ has $AB = 2a$ and $BC = 2b$, where $a > b$. Prove that the radius of gyration of the lamina about BC is $2a/\sqrt{3}$.

Find the period of small oscillations when the lamina is suspended so that it can swing freely in its own plane, which is vertical, about a smooth pivot at the mid-point of AB.

This period is unchanged when the pivot is moved to the mid-point of BC. Show that $a/b = (3+\sqrt{5})/2$. (*L*)

7 A flywheel has moment of inertia I about its axis of rotation, which is vertical. When a constant frictional torque is applied to the flywheel it is brought to rest from angular velocity Ω_1 in time t_1. When a particle of mass m is attached to the flywheel at a distance a from the axis, the flywheel comes to rest from angular velocity Ω_2 in time t_2 against the same frictional torque. Show that

$$I = \frac{ma^2\Omega_2 t_1}{\Omega_1 t_2 - \Omega_2 t_1}.$$

If the numbers of revolutions of the flywheel in the respective cases are n_1, n_2, show that

$$I = \frac{ma^2 n_2 t_1^2}{n_1 t_2^2 - n_2 t_1^2}. \qquad (L)$$

8 When a uniform solid cylinder of radius a and mass M is rotating about its fixed horizontal axis, there is a constant resisting torque of magnitude G. A light inextensible string has one end attached to the curved surface of the cylinder and is wound several times round this surface. Hanging from the free end of this string is a particle of mass km. When the system is released from rest the particle has an acceleration $\frac{1}{4}g$ if $k = 1$ and an acceleration $\frac{1}{2}g$ if

$k = 2$. Find M and G in terms of a, m and g. Find the acceleration of the particle when $k = 4$. *(L)*

9 Two particles of mass $2m$ and m are connected by a light inextensible string passing over a pulley of mass $2m$ which may be regarded as a uniform circular disc, free to rotate about a fixed axis through its centre. The system is released from rest when the particles are at an equal height. If the string does not slip on the pulley, prove that the speed of the particles when each has moved a vertical distance h is $\sqrt{(\frac{1}{2}gh)}$.

Find the tension in each part of the string, and the acceleration of the heavier particle. *(L)*

10 A uniform rod AB of mass 0·2 kg and length 1 m is freely hinged at end A to a vertical wall. The other end B of the rod is attached by means of an elastic string, of modulus k newtons and natural length 1 m, to a point C on the wall vertically above A with $AC = 1$ m. The system is released from rest with angle $CAB = \pi/3$. Show that when the angle $CAB = 2\theta$ and ω is the angular speed of the rod, the energy equation gives

$$\tfrac{1}{2} = \omega^2/30 + \cos 2\theta + k(2\sin\theta - 1)^2/2.$$

Hence show that if $k = 3 + 2\sqrt{2}$, the rod will come to instantaneous rest in the horizontal position.

11 A uniform disc has mass m and radius a. Show that its moment of inertia about an axis perpendicular to its plane and through its centre is $\frac{1}{2}ma^2$.

The disc is free to rotate without friction about the axis, which is horizontal. One end of a light string is attached to a point on the circumference of the disc and part of the string is wound on the circumference. The other end of the string carries a particle A of mass $2m$ hanging freely. The system is released from rest and a restoring couple of moment $mga\theta$ acts on the disc, when θ is the angular displacement of the disc. Write down the equation of motion of the disc about its centre and the equation of motion of A. Hence show that the tension in the vertical part of the string is $\frac{2}{5}(1 + 2\theta)mg$. Show also that

$$5a\frac{d^2\theta}{dt^2} = 4g - 2g\theta.$$

Deduce that the motion of A is simple harmonic and find the period of this motion.

(You may assume that part of the string always remains wound on the disc during the motion.) *(O&C)*

12 A flywheel of radius a m turns about its axle, which is horizontal. A light inextensible string is wrapped round the wheel and has a mass attached to its free end. A constant frictional couple resists the motion of the flywheel. When the mass is m kg, it falls a distance $2h$ m from rest in one second; when the mass is $2m$ kg, it falls $3h$ m from rest in one second. Show that the moment of inertia of the flywheel is $\frac{1}{2}ma^2(g/h - 8)$ kg m^2 and the frictional couple resisting motion is $ma(12h - g)$ N m. *(O&C)*

13 A bell is rigidly attached to a wheel and axle, the bell-rope being attached to the rim of the wheel. The wheel and axle are symmetric about the axis of rotation, which is horizontal, but the bell is attached so that the centre of mass of the bell is 0·75 m from the axis. The mass of the bell is 480 kg; the radius of the wheel is 1·25 m; the moment of inertia of the wheel, bell and axle about the axis of rotation is 750 kg m^2. The bell is stationary with its centre of mass vertically above the axis of rotation. When the bell is slightly disturbed, it begins to rotate so that the bell-rope is wound up on the wheel. The rope, whose mass can be ignored, lifts a load of 80 kg attached to its free end.

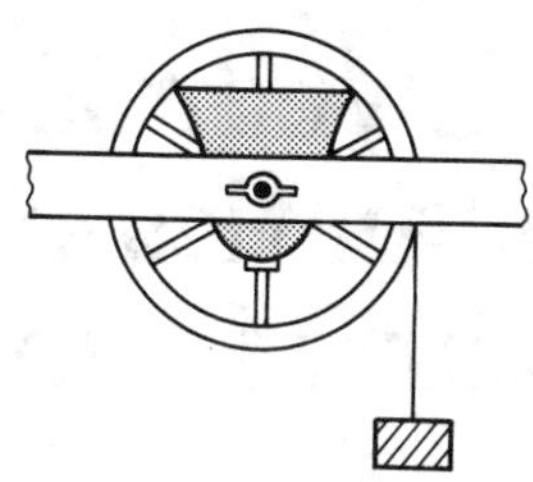

When the bell has rotated through an angle θ radian, its angular velocity is ω radian per second. Write down expressions for the total loss of potential energy and the total gain in kinetic energy of the system up to this instant. Assuming that there has been no loss of energy due to friction or any other cause, find the speed of the load on the end of the bell-rope at the instant when the centre of mass of the bell is vertically below the axis of rotation for the first time. Find also the acceleration of the load at this instant. (Take g as 9·8 ms^{-2}.) *(MEI)*

14 Two similar gear wheels are mounted on parallel axes and run permanently in mesh. Each wheel is of radius a and has moment of inertia I about its axis. When in motion, a frictional couple of moment P acts on each wheel in its plane opposing motion. A couple of constant moment G, $(G > 2P)$, is applied to one of the wheels in its plane; find

(a) the tangential force between the wheels,
(b) the angular acceleration of each wheel,
(c) the number of revolutions made by each wheel in acquiring from rest a speed of n revolutions per second.

Show that, if when the speed is n rev/s the couple G ceases to act, the wheels will come to rest after a further time T, where

$$T = (2n\pi I)/P. \qquad (O\&C)$$

16 Further Dynamics and Kinematics of Rectilinear Motion

In Chapter 4, methods were given for finding the displacement from the acceleration when the latter was either a constant or an explicit function of time. This condition severely restricts the types of problems that can be considered and, in particular, excludes important classes of problems that occur in practice. We now extend the problems that can be considered to those where the acceleration can depend explicitly on either the speed or the displacement. Problems of the first type occur in the study of resisted motion, whilst a simple example of the second type of problem is simple harmonic motion. The basic methods are given in §16.1, and are applied in §16.2 to solve simple resisted motion problems and in §16.3, to obtain, from first principles, the general solution of the simple harmonic equation.

16.1 Acceleration dependent on velocity or displacement

It is possible, in principle, when the acceleration is given explicitly as a function of velocity or displacement, to obtain a formal relation between the displacement and time. The methods that have to be used are fairly simple but do, unfortunately, often lead to integrals which can not be evaluated simply. In both cases, the general approach to be adopted will be developed from particular examples.

We start with the case when the acceleration is $-v\,\mathrm{ms}^{-2}$, where $v\,\mathrm{ms}^{-1}$ is the speed at time t s, and at time $t = 0$, v is given to be unity. We now have

$$\frac{\mathrm{d}v}{\mathrm{d}t} = -v.$$

It is impossible to integrate the right-hand side directly with respect to t as v is, as yet, unknown. The equation can, however, be re-written in the separable form (cf *Pure Mathematics for Advanced Level* §25.3).

$$\frac{1}{v}\frac{\mathrm{d}v}{\mathrm{d}t} = -1.$$

Direct integration of this latter equation gives

$$\int \frac{dv}{v} = -\int dt$$

$$\Rightarrow \ln v = -t+b,$$

where b is a constant. Substituting the values at $t = 0$, gives $b = 0$.

In this case, we could have avoided introducing the constant by integrating the left-hand side from 1 to v and the right-hand side from 0 to t. This gives

$$\int_1^v \frac{dw}{w} = -\int_0^t dt'.$$

substituting w for v as the variable of integration. (Usually it is better not to use the same letter for a limit of integration and the variable of integration.) Therefore

$$\ln v = -t$$

and

$$v = \dot{x} = e^{-t}.$$

The right-hand side is now a known function of t and

$$\frac{dx}{dt} = e^{-t}.$$

Integrating with respect to t from 0 to t gives, taking $x = 0$ at $t = 0$,

$$x-1 = -e^{-t}.$$

We can now look at this example and see that the steps necessary in a more general case are as follows.

(i) Separate the variables, the v-dependence on one side and the t-dependence on the other.

(ii) Integrate both sides and remember to introduce an arbitrary constant or integrate between definite limits if v is given for a particular value of t.

(iii) Re-arrange the result obtained so that v is written as being equal to a function of t.

(iv) Write v as dx/dt and integrate, either introducing a second arbitrary constant or integrating between definite limits if possible.

The above procedure will give v and x in terms of t. However, if v is required in terms of x, then equation 4.1 should be used to replace $\ddot{x}$ by $\frac{v\,dv}{dx}$, and then the above steps can still be followed with t replaced by x.

EXAMPLE 1 *At time t s a particle has an acceleration $(1/v^2)\,\mathrm{ms}^{-2}$, where $v\,\mathrm{ms}^{-1}$ is its velocity and, at $t = 0$, $v = 1$. The displacement of the particle from a fixed point O is 2 m at $t = 0$. Find an expression for the displacement x m from O at time t s and a relation between velocity and displacement.*

We have
$$\frac{dv}{dt} = \frac{1}{v^2},$$
therefore, separating the variables gives
$$\int v^2\,dv = \int 1\,dt.$$
Integrating the left-hand side from 1 to v and the right-hand side from 0 to t gives
$$\frac{1}{3}(v^3 - 1) = t.$$
Rearranging to obtain v in terms of t we have
$$v = \frac{dx}{dt} = (3t+1)^{1/3}$$
Integrating with respect to t from 0 to t gives, since $x = 2$ at $t = 0$,
$$x - 2 = \int_0^t (3w+1)^{1/3}\,dw = \frac{1}{4}[(3t+1)^{4/3} - 1]$$
so that
$$\mathbf{x = \frac{1}{4}[(3t+1)^{4/3} + 7].}$$

The velocity can be obtained in terms of x by eliminating t between x and v, but it is simpler to obtain it directly by using equation 4.1. This gives
$$v\frac{dv}{dx} = \frac{1}{v^2},$$
and
$$\int v^3\,dv = \int dx.$$
Inserting the appropriate limits gives
$$\frac{1}{4}(v^4 - 1) = x - 2$$
or
$$\mathbf{v = (4x - 7)^{1/4}.}$$

We now look at the corresponding problem of finding x from $\ddot{x}$ when the latter depends only on x, and we start with the case when the acceleration $a\,\text{ms}^{-2}$ at time t s is e^{2x}, where the displacement is x m. At $t = 0$, we take $\dot{x} = 1$ and $x = 0$. The basic equation is
$$\ddot{x} = e^{2x}.$$
It is impossible to integrate this equation as it stands. The situation is, however, simplified by using equation 4.1 so that
$$v\frac{dv}{dx} = e^{2x},$$
where the velocity is denoted by $v\,\text{ms}^{-1}$. This can be integrated with

respect to x immediately to give

$$v^2 = e^{2x} + b,$$

where b is a constant. The condition $v = 1$ at $x = 0$ gives $b = 0$ and, therefore,

$$v = \frac{dx}{dt} = e^x,$$

($\dot{x}$ is given to be 1 at $x = 0$, so the positive root has to be taken). The variables can be separated so that x only occurs on the left-hand side and t only on the right-hand side,

$$\int e^{-x}\,dx = \int dt.$$

Inserting the relevant limits gives

$$\begin{cases} \displaystyle\int_0^x e^{-w}\,dw = t \\ \Rightarrow 1 - e^{-x} = t \\ \Rightarrow x = -\ln(1-t). \end{cases}$$

The solution is only valid for $t < 1$ and $x \to \infty$ as $t \to 1$. From the point of view of mechanics this is not a particularly suitable example, as the acceleration is continuously increasing, but the mathematics is sufficiently simple for each step to be fairly clear.

The above example can be used to see the steps necessary in the general case, and these are as follows.

(i) Replace $\ddot{x}$ by $\dfrac{v\,dv}{dx}$ and integrate with respect to x, introducing either an arbitrary constant or integrating between definite limits.

(ii) Obtain $\dot{x}$ by taking a square root. The sign of this has to be carefully considered. Now re-arrange this so that the x-dependence is on one side and the t-dependence on the other.

(iii) Integrate this equation, again introducing an arbitrary constant or integrating between definite limits.

EXAMPLE 2 *The acceleration of a particle free to move on the x-axis is* $-x^{-3}\,\text{ms}^{-2}$ *and it is released from rest at the point where* $x = 1$. *Find the time taken to reach the point where* $x = \frac{1}{2}$.

The basic equation is $$\frac{v\,dv}{dx} = -x^{-3}.$$

Integrating with respect to x from 1 to x gives

$$v^2 = (1 - x^2)/x^2.$$

This equation shows that $|x| \leqslant 1$ and, therefore, since the particle is released from rest at $x = 1$, it can only move to the left. So, $\dot{x}$ is negative for $x < 1$. This also follows from the acceleration being negative near $x = 1$. Hence,

$$\dot{x} = -(1-x^2)^{\frac{1}{2}}/x$$

and

$$-\frac{x}{(1-x^2)^{\frac{1}{2}}}\frac{dx}{dt} = 1.$$

This equation is in separable form, and integrating with respect to x from $x = 1$ to $x = \frac{1}{2}$ gives

$$\int_1^{\frac{1}{2}} \frac{x\,dx}{(1-x^2)^{\frac{1}{2}}} = -t.$$

Therefore, the time taken is $(\sqrt{3}/2)$ s.

EXERCISE 16.1

Questions 1 to 6 refer to a particle P moving along Ox with acceleration a ms^{-2} in the direction of x increasing, x m denotes the displacement of P from O at time t s and at this time the velocity of P, in the direction of increasing x, is v ms^{-2}.

1 $a = -2v$, $x = 2$ and $v = 4$ when $t = 0$; find x as a function of t.
2 $a = 1/v$; find v and x as functions of t given that $x = 3$ and $v = 2$ when $t = 0$.
3 $a = 4v^3$; find v in terms of x given that $v = -0{\cdot}25$ when $x = 0$.
4 $a = -v^2/4$, and at $t = 0$, $x = 0$ and $v = 2$; prove that when $x = 8$, $v = 2e^{-2}$ and $t = 2e^2 - 2$.
5 $a = 6x^2 + 4x$; given that $v = (10)^{\frac{1}{2}}$ when $x = 1$, find v^2 when $x = 2$.
6 $a = k/x$, where k is constant; given that $v = 1$ when $x = 2$, and $v = 3$ when $x = 3$, find v, correct to two significant figures, when $x = 5$.
7 The acceleration due to gravity is given to be $10(6{\cdot}4 \times 10^6/x)^2$, where x m is the distance measured from the centre of the earth. The earth is taken to be a sphere of radius $6{\cdot}4 \times 10^6$ m. Find, to the nearest km, the maximum height reached above the earth by a rocket projected with speed 1 000 ms^{-1} from the earth's surface.

16.2 Resisted motion

In many circumstances, the resistance to bodies moving in air (and other media) can be approximated by a fairly simple function of velocity. Therefore, if the only other forces acting are constant, the acceleration will, by Newton's law, be a function of velocity. Hence, such problems are soluble by the first method described in the previous section. Before applying any method of solution one should make certain that the equation that is being solved is the correct one, and, for problems involving resistance, it is necessary to consider very carefully the directions of resistive forces. The following examples will illustrate some of the points that can occur in solving problems of resisted motion. Conventionally, a resistive force is one which is in the opposite direction to the velocity.

EXAMPLE 1 *A particle of mass 3 kg is acted upon by a resistive force, which is proportional to its speed and such that it experiences a force of 300 N when moving with a speed $10\,\mathrm{ms^{-1}}$. Determine the time taken for the particle speed to drop from $15\,\mathrm{ms^{-1}}$ to $5\,\mathrm{ms^{-1}}$.*

It is important to have in mind a particular reference direction, and we take this to be in the direction of a unit vector $\mathbf{i}$. The velocity at time t s, is therefore $\dot{x}\mathbf{i}\,\mathrm{ms^{-1}}$ where the displacement is x m. The resistive force is in the opposite direction to the velocity and proportional to it, that is, it is $-k\dot{x}\mathbf{i}\,\mathrm{N}$, where k is a positive constant. Therefore,

$$3\ddot{x} = -k\dot{x}$$

or

$$3\dot{v} = -kv,$$

where $v = \dot{x}$. We are given that

$$10k = 300$$

so that

$$\dot{v} = -10v.$$

Separating the variables gives

$$\int \frac{\mathrm{d}v}{v} = -\int 10\,\mathrm{d}t$$

that is,

$$\ln v = -10t + c.$$

Assuming that $t = 0$ at the instant the particle has a speed of $15\,\mathrm{ms^{-1}}$, $c = \ln 15$. Therefore,

$$v = 15\mathrm{e}^{-10t}.$$

Substituting $v = 5$ gives $t = \dfrac{1}{10}\ln 3 \approx 0{\cdot}11$. The time taken for the speed to drop from $15\,\mathrm{ms^{-1}}$ to $5\,\mathrm{m^{-1}}$ is, therefore, approximately **11 s.**

EXAMPLE 2 *A stone is thrown vertically upwards with initial speed u. The resistance of the air is $(v/u)\,g$ per unit mass, where v is the speed. Find the time taken to reach the highest point and the greatest height reached.*

During the period that the stone is rising, the forces acting on it are as shown in Fig. 16.1. If we take the reference direction to be upwards, then Newton's law gives

$$m\frac{\mathrm{d}v}{\mathrm{d}t} = -mg\left(1 + \frac{v}{u}\right).$$

The next step, as described in §16.1, is to separate the variables which gives

$$\int \frac{\mathrm{d}v}{1 + \dfrac{v}{u}} = -\int g\,\mathrm{d}t.$$

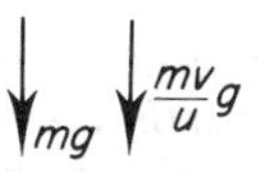

mg $\quad \dfrac{mv}{u}g$

Fig. 16.1

Inserting the limits $v = u$ and $v = 0$ on the left-hand side and the corresponding values $t = 0$ and $t = T$ (the total time taken) on the right-hand side, gives

$$\left[u \ln\left(1 + \frac{v}{u}\right)\right]_u^0 = -gT$$

$$\Rightarrow T = (u/g) \ln 2.$$

To find the time T_1 taken to any value v_1 of v, the limits for v would be u and v_1 whilst the corresponding ones for t would be 0 and T_1. Integrating between limits can, as mentioned in §16.1, avoid the algebra necessary to find a constant of integration.

One method of finding the greatest height would be to find, as above, v for a general value of t, and then replace v by $\dot{x}$ and carry out a further integration. This is unnecessarily complicated, and it is far simpler to use equation 4.1 which gives

$$m\frac{v\mathrm{d}v}{\mathrm{d}x} = -mg - m\frac{v}{u}g.$$

Separating the variables gives

$$\int \frac{v\mathrm{d}v}{1 + \frac{v}{u}} = -\int g\mathrm{d}x.$$

In order to carry out the integration, this equation has to be re-arranged by dividing once on the left-hand side

$$\int \left(u - \frac{u}{1 + \frac{v}{u}}\right)\mathrm{d}v = -\int g\,\mathrm{d}x.$$

Integrating between $v = u$ and $v = 0$, and $x = 0$ and $x = H$ (the maximum height), gives

$$\int_u^0 \left(u - \frac{u}{1 + \frac{v}{u}}\right)\mathrm{d}v = -gH$$

which gives

$$H = \frac{u^2}{g}(1 - \ln 2).$$

EXAMPLE 3 *Determine the maximum height reached by a stone projected vertically upwards with speed u, when the air resistance at speed v is gv^2/u^2 per unit mass. Determine also the speed with which the stone returns to the point of projection.*

If the reference direction is taken upwards, then the forces acting will be as shown in Fig. 16.3. The equation of motion is

$$m\frac{dv}{dt} = -mg - mg\frac{v^2}{u^2}.$$

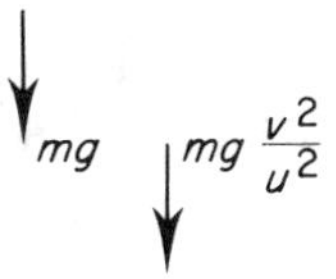

mg

$mg\frac{v^2}{u^2}$

Fig. 16.3

On using equation 4.1, this becomes

$$v\frac{dv}{dx} = -g - g\frac{v^2}{u^2},$$

which separates to give

$$\int\frac{v\,dv}{1+v^2/u^2} = \int -g\,dx.$$

Integrating with respect to v from $v = u$ to $v = O$, and from $x = O$ to $x = H$ (the maximum height)

$$H = \frac{u^2}{2g}\ln 2.$$

If we now consider the downward path, using the same reference direction, the forces acting will be as shown in Fig. 16.4. The equation of motion is

$$mv\frac{dv}{dx} = -mg + mg\frac{v^2}{u^2},$$

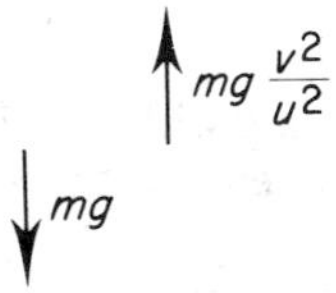

$mg\frac{v^2}{u^2}$

mg

Fig. 16.4

(the terminal speed is seen, from this equation, to be u) which gives

$$\int_0^{-w}\frac{v\,dv}{1-v^2/u^2} = -\int g\,dx.$$

Therefore, the speed of return (w) is given by

$$\int_0^{-w}\frac{v\,dv}{(1-v^2/u^2)} = -g\int_H^0 dx = gH.$$

This gives, on substituting for H,

$$\log(1 - w^2/u^2) = -\ln 2$$

that is,

$$1 - \frac{w^2}{u^2} = \tfrac{1}{2}$$

so that

$$w = \frac{u}{\sqrt{2}}.$$

EXAMPLE 4 *A car of mass m moves on a level road against a constant resistance of magnitude R. The engine of the car works at the constant rate RU, where U is a constant. Find the time taken, starting from rest, to reach a speed of $\frac{1}{2}U$.*

The driving force when the car is moving with speed v is RU/v, so the equation of motion is

$$m\frac{dv}{dt} = \frac{RU}{v} - R.$$

This equation can be written in the separable form

$$m\frac{v}{(U-v)}\frac{dv}{dt} = R.$$

Therefore, the total time T is given by

$$m\int_0^{U/2} \frac{v\,dv}{(u-v)} = RT$$

that is,

$$T = \frac{mU}{R}(\log 2 - \tfrac{1}{2}).$$

EXERCISE 16.2

In numerical exercises, take $g = 10\,\text{ms}^{-2}$ and give answers correct to three significant figures.

1 A particle of mass 2 kg moving with speed $v\,\text{ms}^{-2}$ is resisted by a force of $8v$ newtons. Find the time taken for the speed to reduce from $20\,\text{ms}^{-1}$ to $2\,\text{ms}^{-1}$.

2 A car of mass 1·5 tonne moves on a level road with the engine working at an effective constant rate of 15 kW. Find the distance travelled while the speed increases from $4\,\text{ms}^{-1}$ to $30\,\text{ms}^{-1}$.

3 A particle of mass 2 kg is projected with speed $10\,\text{ms}^{-1}$ along a rough table, the coefficient of friction being 0·5. The particle is also subject to an air resistance which, when it is moving with speed $v\,\text{ms}^{-1}$, is v newtons. Find the time taken for the particle to come to rest.

4 A particle subject to a resistive force proportional to the square of its speed has its speed reduced from $10\,\text{ms}^{-1}$ to $5\,\text{ms}^{-1}$ in 5 s. Find the distance travelled during this period.

5 An aeroplane of mass 25 tonnes, takes off from a runway under a constant thrust of 280 kN. The drag is $12v^2$ N when the speed is $v\,\text{ms}^{-1}$. Find the

minimum length of runway for take off given that the take-off speed is $100\,\mathrm{m\,s^{-1}}$.

6 A car of mass m moves under the action of a driving force F, and a resistive force which at a speed v is mkv, where k is a constant. Find the maximum speed U that the car could attain and find the distance travelled from rest before three-quarters of this speed is attained. Find also the time taken in this case.

7 A parachutist jumps from a ballon. The air resistance with the parachute open is assumed to be proportional to the speed of the parachutist and is such that if he were to fall indefinitely he would attain a speed of $4\,\mathrm{m\,s^{-1}}$ (that is, his terminal speed is $4\,\mathrm{m\,s^{-1}}$). Find the distance he drops in 3 seconds, assuming the parachute opens immediately.

8 If the resistance in question 7 is assumed to be proportional to the square of the speed, with the terminal speed being unchanged, find the speed after the parachutist has been dropping for t s.

9 A car of mass m, whose engine works at an effective rate k, moves along a horizontal road under the action of a resistance ku^2, where u is its speed. Find the distance travelled as the speed changes from $0{\cdot}25\,(R/k)^{1/3}$ to $0{\cdot}5\,(R/k)^{1/3}$.

10 A raindrop of mass m falls from rest and is subjected, when moving at speed v, to a resistance kv^2. Assuming its mass remains constant, show that when it has dropped a distance h

$$kv^2 = g(1 - \mathrm{e}^{-2kh}).$$

11 A car exerts a constant driving force such that it can climb a slope of 1/10 at a steady speed of $15\,\mathrm{m\,s^{-1}}$. The resistance to motion is proportional to the car speed. The car freewheels down a slope of inclination $\sin^{-1}(1/20)$ at a steady speed of $25\,\mathrm{m\,s^{-1}}$. Find, assuming the same tractive force is exerted going down a slope of $\sin^{-1}(1/10)$, the time taken for the speed to increase from $10\,\mathrm{m\,s^{-1}}$ to $20\,\mathrm{m\,s^{-1}}$.

16.3 Simple harmonic motion

Simple harmonic motion was introduced in Chapter 14 by considering a particular form of displacement and showing that it satisfied a particular differential equation. The form assumed involved two independent arbitrary constants, and was such that x and $\dot{x}$ could be prescribed arbitrarily for a particular value of t and, therefore, could be regarded as a solution of the differential equation. We now start directly from the equation and show how the form of displacement could have been deduced directly from the equation without any assumptions.

The equation governing simple harmonic motion is

$$\ddot{x} = -\omega^2 x,$$

which is a particular example of the type of equation considered in the latter part of §16.1. Following the procedure given there, the equation is re-written as

$$\frac{v\,dv}{dx} = -\omega^2 x.$$

Integration gives

$$\dot{x}^2 = \omega^2 (a^2 - x^2),$$

where a is a constant. Therefore,

$$\dot{x} = \omega (a^2 - x^2)^{1/2},$$

where we take the positive root, and separating the variables gives

$$\int \frac{dx}{(a^2 - x^2)^{\frac{1}{2}}} = \int \omega\, dt.$$

The integral of $(a^2 - x^2)^{-\frac{1}{2}}\,dx$ can be found by setting $x = a \sin \theta$, and it is equal to $\sin^{-1} x/a$. Therefore,

$$\sin^{-1}\left(\frac{x}{a}\right) = \omega t + \varepsilon,$$

where ε is a constant. Hence,

$$x = a \sin (\omega t + \varepsilon),$$

which is the form of the general solution quoted.

MISCELLANEOUS EXERCISE 16

1 A particle P moves along Ox with variable velocity $v\,\text{m s}^{-1}$. When $OP = x$ m, the acceleration of P in the direction of x-increasing is $-v\,\text{m s}^{-2}$. Given that $v = 10$ when $x = 0$, find v in terms of x. (*L*)

2 A particle moves in a straight line along a horizontal table, the retardation being $kv^{\frac{3}{2}}$ when the speed is v. If u is the initial speed, prove that the speed after a time t is

$$u/(1 + \tfrac{1}{2}kt\sqrt{u})^2.$$

Find the distance which the particle has travelled in this time. (*O*)

3 (a) A particle moves on the positive x-axis under the action of a force directed away from the origin. The magnitude of the force is proportional to the square root of the time. The particle starts from rest at time $t = 0$. Prove that the time taken to accelerate from speed u to speed $2u$ is a little less than three-fifths of the time taken to accelerate from rest to speed u.

Find, to two significant figures, the ratio of the corresponding distances travelled by the particle.

(b) A second particle also moves on the positive x-axis under the action of a single force directed away from the origin, but in this case the magnitude of the force is proportional to the square root of the distance, x, of the particle from the origin. This particle has speed u when $x = a$, and speed $2u$ when $x = 4a$. Find its speed when $x = 9a$. (*C*)

4 (i) A particle moves in a straight line Ox in the direction of increasing x so that its distance x from O and its corresponding speed v at any instant satisfy

$$e^{x/a} = 1 + v/V,$$

where a and V are positive constants. Show that the acceleration of the particle is given by $v(v+V)/a$.
(ii) A particle P moves in a straight line Ox so that at time t its acceleration in the direction of increasing x is $U\omega^2 te^{\omega t}$, where ω and U are positive constants. Given that P starts from rest at O when $t = 0$, find its distance from O at time t. Find also the speed of P and its distance from O when $t = 1/\omega$. *(L)*

5 A particle moves along the x-axis. For all values of x its retardation is $1/(2v^2)\,\text{ms}^{-2}$ where $v\,\text{ms}^{-1}$ is its speed. At time $t = 0$ seconds the particle is projected from the origin with speed $u\,\text{ms}^{-1}$ in the direction x increasing. Show that the speed is halved when $t = 7u^3/12$ seconds and find the value of x in terms of u at this instant. *(L)*

6 A particle leaves a point A at time $t = 0$ with speed u and moves towards a point B with a retardation λv, where v is the speed of the particle at time t. The particle is at a distance s from A at time t. Show that (i) $v = u - \lambda s$, (ii) $\log_e(u - \lambda s) = \log_e u - \lambda t$.

At $t = 0$ a second particle starts from rest at B and moves towards A with acceleration $2 + 6t$. The particles collide at the mid-point of AB when $t = 1$. Find the distance AB and the speeds of the particles on impact. *(AEB 1971)*

7 (i) A particle starts with speed $20\,\text{ms}^{-1}$ and moves in a straight line. The particle is subjected to a resistance which produces a retardation which is initially $5\,\text{ms}^{-2}$ and which increases uniformly with the distance moved, having a value of $11\,\text{ms}^{-2}$ when the particle has moved a distance of 12 m. Given that the particle has speed $v\,\text{ms}^{-1}$ when it has moved a distance of x m, show that, while the particle is in motion,

$$v\frac{dv}{dx} = -(5 + \tfrac{1}{2}x).$$

Hence, or otherwise, calculate the distance moved by the particle in coming to rest.

(ii) At time t seconds, where $t > 0$, the position vector of particle A with respect to a fixed origin O is $(\mathbf{i}t - \mathbf{j}\cos t + \mathbf{k}\sin t)$ m and the position vector of particle B with respect to O is $(\mathbf{i} + \mathbf{j}e^{-t} - \mathbf{k}e^{-t})$ m. Find the least value of t for which the velocities of A and B are perpendicular and show that, at this instant, the accelerations of the particles are parallel. *(L)*

8 Given that s, defined by

$$s = ut + \tfrac{1}{2}at^2,$$

where u and a are constants, represents the displacement of a particle at time t show, by differentiation, that u is the velocity at time $t = 0$ and that the acceleration is equal to a.

A train starting from rest is uniformly accelerated during the first minute of its journey when it covers 600 m. It then runs at a constant speed until it is brought to rest in a distance of 1 km by applying a constant retardation.
(i) Find the maximum speed attained by the train.
(ii) Determine the magnitude of the retardation.
(iii) Given that the total journey time is 5 minutes, determine the distance covered at constant speed.

(iv) Given that the magnitude of the retardation, instead of being constant, is directly proportional to the speed and the train speed is halved from the same constant speed in a distance of 500 m, find the magnitude of the retardation, in m/s^2, when the train's speed is 10 m/s. *(AEB 1982)*

9 A bead moves on a very long straight horizontal wire which is lightly greased. When a current is passed through the wire to melt the grease the retardation of the bead is $ba^2 \exp(-at/2)$, where a and b are positive constants and t is the time that has elapsed since the current was switched on. At time $t = 0$, when the current is switched on, the bead is projected with a speed $u(> 2ab)$ from a point O on the wire. Determine
(i) the subsequent speed of the bead at time t, and show that this is always greater than $u - 2ab$,
(ii) the displacement of the bead from O at time t.
If, instead, the retardation of the bead is equal to av, where v is its speed, and the bead is again projected with a speed u from O at $t = 0$, determine
(iii) the speed of the bead at time t,
(iv) the time taken for the speed to reduce to $\frac{1}{2}u$. *(AEB 1982)*

10 The stopping distance, x, is the distance travelled by a vehicle in the interval between the driver seeing an obstacle and the vehicle coming to rest. It is calculated on the assumptions that
(i) a time interval T elapses between the obstacle being seen and the brakes being applied, during which time there is zero acceleration;
(ii) the brakes, when applied, immediately produce a retardation equal to that produced when the vehicle slides along the road. The coefficient of friction between the road and tyres is a constant value μ.
Given that the initial speed is $u = 20\,\text{ms}^{-1}$, $T = 2\,\text{s}$, $\mu = 0{\cdot}3$ and $g = 10\,\text{ms}^{-2}$, find x in metres.

A more realistic model of the braking effect is that the retardation is dependent on the speed of the vehicle. With the above values of u and T, and given that the retardation at speed $v\,\text{ms}^{-1}$ is $\dfrac{660}{(v+200)}\,\text{ms}^{-2}$, calculate the new value of x in metres. *(AEB 1983)*

11 A particle moving in a straight line is subject to a resisting force which produces a retardation kv^3 where v is the speed and k is a constant. If u is the initial speed, s is the distance moved in time t, and v is the speed at time t, find equations
(i) connecting the variables v and t,
(ii) connecting the variables v and s, and deduce that

$$ks^2 = 2t - 2s/u.$$

A bullet is fired horizontally from a rifle at a target 3 000 m away. The bullet is observed to take 1 second to travel the first 1 000 m and $1\frac{1}{4}$ seconds to travel the next 1 000 m. Assuming that the air resistance is proportional to v^3, and neglecting gravity, find the time taken to travel the last 1 000 m and show that the bullet reaches the target with speed $8\,000/13\,\text{ms}^{-1}$. *(JMB)*

12 A particle of mass m moves in a straight line under no forces except a resistance mkv^3, where v is the velocity and k is a positive constant. If the initial velocity is $u > 0$ and x is the distance covered in time t, prove that

$$kx = \frac{1}{v} - \frac{1}{u}, \quad t = \frac{x}{u} + \frac{1}{2}kx^2. \qquad (L)$$

13 A particle of mass m moves along a horizontal straight line under the action only of a resisting force of magnitude mv^2/a, where v is its speed and a is a positive constant. Given that the particle is projected from a point O at time $t = 0$ with speed u, show that, when it is at a distance x from O, its speed is $ue^{-x/a}$. (*L*)

14 A particle of mass m moves on the x-axis subject to a resistance of magnitude mkv^4, where k is a positive constant and v is the speed of the particle at time t. Given that the particle passes through the origin at time $t = 0$ with speed u show that it will travel a distance $3/(2ku^2)$ before its speed is reduced to $u/2$. Find the time taken in slowing down to this speed. (*L*)

15 A shell is fired vertically upwards with velocity u from a point A on the surface of the earth. When its distance from the centre of the earth is x, it is subject to a gravitational force k/x^2 per unit mass directed towards the centre of the earth. Show that when $u^2 < 2k/a$, where a is the radius of the earth, the shell rises to a maximum height $a^2u^2/(2k - au^2)$ above A. What happens when $u^2 > 2k/a$?

Find the distance travelled by the shell in time t when $u^2 = 2k/a$. (*W*)

16 A particle of mass m is set in motion with speed u. Subsequently, the only force acting upon the particle directly opposes its motion and is of magnitude $k(1 + v^2)$, where v is its speed at time t and k is constant. Show that the particle is brought to rest after a time $(m/k) \tan^{-1} u$ and find an expression in terms of m, k and u for the distance travelled by the particle in this time. (*JMB*)

17 A man pushes a loaded cart of total mass m along a straight horizontal path against a constant resisting force λm. The cart starts from rest, and the man exerts a force, in the direction of the motion, of magnitude $K\,e^{-\alpha x}$, where x is the distance moved, α is a positive constant and K is a constant greater than λm. Write down a differential equation, involving x and the speed v, for the motion of the cart (while it continues to move), and hence obtain v in terms of x and the given constants. (*JMB*)

18 A car of mass 1 000 kg starts from rest and moves on level ground against a constant resistance of 1 000 newtons. The pull of the engine increases uniformly with the distance travelled, starting at 1 500 newtons and increasing to 4 500 newtons when the car has travelled 150 metres. Sketch the acceleration-distance graph. Hence, or otherwise, calculate the speed of the car when it has travelled 150 metres. (*L*)

19 A load P of mass m is being raised vertically by an engine working at a constant rate kmg. Denoting by v the speed of the load when it has been raised a distance x, show that

$$v^2 \frac{dv}{dx} = (k - v)g.$$

Initially P is at rest. Show that

$$gx = k^2 \log_e \frac{k}{k - v} - kv - \tfrac{1}{2}v^2.$$

Write down the work done by the engine in time t and hence, or otherwise, show that the time taken for the load to reach the speed v is

$$\frac{k}{g} \log_e \frac{k}{k - v} - \frac{v}{g}.$$

(*JMB*)

20 A car, of mass M kg, starts from rest and moves in a straight line on a horizontal plane on which the resistance to motion is kv N when the speed of the car is $v\,\text{ms}^{-1}$, k being a constant. Throughout the motion, the power of the engine is kept at a constant P W. Show that the speed $v\,\text{ms}^{-1}$ when the car has been in motion for a time t s is given by the differential equation

$$P - kv^2 = Mv\frac{dv}{dt}.$$

Hence show that

$$v = \left\{\frac{P}{k}\left[1 - \exp\left(-\frac{2kt}{M}\right)\right]\right\}^{\frac{1}{2}}.$$

Deduce that the car takes T s to reach half its maximum speed under these conditions, where

$$T = \frac{M}{2k}\ln\tfrac{4}{3}.$$ (C)

21 A car of mass M is driven with constant power P against a constant frictional resistance K. Show that its velocity v when it has travelled a distance x satisfies the equation

$$Mv\frac{dv}{dx} = P\left(\frac{1}{v} - \frac{1}{V}\right)$$

where $V = P/K$. Show also that in order to raise its speed from $\frac{1}{2}V$ to $\frac{3}{4}V$ it must travel a distance

$$\left(\log_e 2 - \frac{13}{32}\right)\frac{MV^3}{P}.$$ (O&C)

22 A car of mass M is driven by an engine working at constant power and the resistance to motion is proportional to the square of the speed. If R is the resistance at full speed V on a horizontal road, prove that the distance covered as the speed increases from v_1 to v_2 is

$$\frac{MV^2}{3R}\log_e\left(\frac{V^3 - v_1^3}{V^3 - v_2^3}\right).$$ (O)

23 In an automated railway marshalling yard a moving wagon is continuously retarded as it travels along a siding. One design scheme for such a yard suggests that a retarding force proportional to the square of the speed of the wagon should be used. Show that this implies that the wagon will not come to rest in a finite time.

An alternative scheme is proposed in which the retarding force is of magnitude $mk(v^2 + a^2)$, where v is the speed of the wagon, m is the mass of the wagon and k and a are positive constants. If the speed of the wagon on entering a siding is U, find an expression for the time taken for the wagon to come to rest. Find also an expression for the distance travelled in this time. (L)

24 The resistance to the motion of a lorry of mass m is kv at speed v, where k is a constant.

(i) Find the acceleration of the lorry when it is climbing a hill of inclination α to the horizontal at speed v under full power S. Given that it can climb the same hill under full power at a steady speed u, show that

$$ku^2 = S - mgu \sin \alpha.$$

(ii) The lorry can travel at steady speed w on a level road under full power and the same law of resistance. Show that the time taken for it to accelerate under full power from speed v_1 to speed v_2 (where $v_1 < v_2 < w$) on the level road is

$$\frac{mw^2}{2S} \log_e \frac{w^2 - v_1^2}{w^2 - v_2^2}. \qquad (JMB)$$

25 A particle of unit mass moves on a straight line subject to a force $k^2/2x^2$ towards a point O on the line, where x is the distance of the particle from O and k is a positive constant. The particle is projected from the point $x = \frac{1}{4}a (> 0)$ away from O with a speed $(3k^2/a)^{\frac{1}{2}}$. Show that during the motion

$$\left(\frac{dx}{dt}\right)^2 = k^2\left(\frac{1}{x} - \frac{1}{a}\right)$$

and that the particle comes to rest at the point $x = a$.

Express the time T for the particle to reach the point $x = a$ as a definite integral. By evaluating this integral, using the substitution $x = a \sin^2 \theta$ or otherwise, show that

$$T = \left(\frac{\pi}{3} + \frac{\sqrt{3}}{4}\right)\frac{a^{3/2}}{k}. \qquad (JMB)$$

26 An aeroplane of mass 5 000 kg lands on an aircraft carrier at a relative velocity of $32\,\text{m s}^{-1}$ parallel to the deck. It is immediately caught by an arrester gear which exerts a decelerating force directly proportional to the distance the aeroplane travels along the deck, the constant of proportionality being 3 000 N/m. When it has been slowed to $8\,\text{m s}^{-1}$ the arrester gear releases the aeroplane. The pilot then brings it to rest with a constant braking force of 10 000 N.

Find the total landing distance measured from the point of contact with the arrester gear.

Show also that the time during which the arrester gear is operating is

$$\left(\frac{5}{3}\right)^{\frac{1}{2}} \sin^{-1}\left(\frac{\sqrt{15}}{4}\right) \text{ seconds.}$$

(The mass of the carrier is assumed to be very large compared with that of the aeroplane.) (*JMB*)

27 A parachutist of mass m steps out of an aeroplane in level flight at a height of V^2/g above horizontal ground, V being a constant. He does not open his parachute until the vertical component of his velocity is V. Neglecting air resistance, show that he will have fallen through a height $V^2/(2g)$ when he opens his parachute and that a time V/g has elapsed.

When opened, the parachute produces a drag to vertical motion of magnitude $2mgv/V$, where v is the vertical component of his velocity. Find the height of the parachutist above the ground when the vertical component of his velocity is $3V/4$. (*L*)

Answers

Exercise 1.2 (p. 2)

1 $\{(1, 4), (2, 3), (3, 2), (4, 1)\}$, $\{(1, 3), (3, 1), (2, 4), (4, 2), (3, 5), (5, 3), (4, 6), (6, 4)\}$.
2 (a) (i) $\{1, 2, 3, 4\}$ (ii) $\{(C, D), (C, H), (C, S), (D, H), (D, S), (H, S), (C, C), (D, D), (H, H), (S, S)\}$.
(iii) $\{(A, A), (A, K), \ldots, (A, 2)$
$(K, K), (K, Q), \ldots, (K, 2)$
.
.
.
$(3, 3), (3, 2)$
$(2, 2)\}$.
(b) (i) $\{B, W, R\}$ (ii) $\{(B, B), (B, W), (B, R), (W, W), (W, R), (R, R)\}$.
(iii) $\{(B, B), (B, W), (W, B), (B, R), (R, B), (W, W), (W, R), (R, W), (R, R)\}$.
(c) $\{n: 3 \leqslant n \leqslant 18; n \in Z^+\}$. (d) (i) {Yes, No}, (ii) $\{n: 0 \leqslant n \leqslant 12, n \in Z^+\}$.

Exercise 1.4A (p. 5)

1 22100. **2** $\dfrac{131.39!}{6!(7!)^2.32!}$ **3** 27. **4** 4096. **5** 24. **6** 21.

Exercise 1.4B (p. 6)

1

	1	2	3	4	5	6
1	2	6	4	10	6	14
2	6	4	10	6	14	8
3	4	10	6	14	8	18
4	10	6	14	8	18	10
5	6	14	8	18	10	22
6	14	8	18	10	22	12

(a) 7/36 (b) 7/36.
2 $\{(H, H), (H, T), (T, H), (T, T)\}$ (a) $\frac{1}{4}$ (b) $\frac{1}{2}$. **3** $95/663 = 0{\cdot}143$.
4 $72/5525 = 0{\cdot}013$. **5** (a) $5/108 = 0{\cdot}0463$ (b) $25/216 = 0{\cdot}116$.
6 25, (a) $1/25 = 0{\cdot}04$ (b) $8/25 = 0{\cdot}32$ **7** $1/125 = 0{\cdot}008$ (b) $12/125 = 0{\cdot}096$
(c) $64/125 = 0{\cdot}512$. **8** (a) 0·79 (b) 0·44.
9 (a) 0·00137 (b) 0·787 (a) 0·00109 (b) 0·783.
10 (a) $1{\cdot}54 \times 10^{-6}$ (b) $5{\cdot}00 \times 10^{-6}$ (c) $1{\cdot}44 \times 10^{-3}$ (d) $1{\cdot}965 \times 10^{-3}$
(e) $1{\cdot}54 \times 10^{-5}$. **11** (a) $5/36 = 0{\cdot}139$ (b) $1/12 = 0{\cdot}083$
(c) $5/36 = 0{\cdot}139$, $6{\cdot}17 \times 10^{-3} = 1/162$. **12** (a) $\frac{1}{4}$ (b) $\frac{1}{4}$.

13

X	5	6	7	8	9	10	11
$P(\bar{X})$	0·1	0·1	0·2	0·2	0·2	0·1	0·1

(a) 0·6 (b) 0·3, 0·1.

14

X	0	1	2	3	4	5	6
$P(X)$	1/6	5/18	2/9	1/6	1/9	1/18	0

15 (a) 0·0219 (b) 0·501. **16** $\frac{1}{3}$. **17** 360 (a) $\frac{1}{3}$ (b) 1/15 (c) 1/15. **18** (a) $1/72 = 0{\cdot}0139$,
(b) $5/108 = 0{\cdot}0463$.
19 $\{(B, B), (B, W), (W, B), (W, W)\}$, $13/18 = 0{\cdot}722$.

20

total score (X)	3	4	5	6	7	8	9	10	11	12
$P(X)$	1/64	3/64	6/64	10/64	12/64	12/64	10/64	6/64	3/64	1/64

(a) 1/16 (b) 11/16 (c) 1/16 (d) 9/32.

Exercise 1.5 (p. 13)

1 $P(A \cup B) = P(A) + P(B) - P(A \cap B)$, (a) 1/6, $\frac{2}{3}$ (b) 0, 5/6.
2 (a) $\frac{1}{3}$ (b) $\frac{1}{2}$ (c) 1/10 (d) 3/10 (e) 7/30.
3 (a) $1/221 = 0{\cdot}0045$ (b) $33/221 = 0{\cdot}149$ (c) $1/17 = 0{\cdot}0588$.
4 (i) 0·134 (ii) 0·0191 (iii) 0·332 (iv) 0·0399.
5 (a) 1/16 (b) $9/64 = 0{\cdot}141$ (c) $15/4^5 = 0{\cdot}0146$ (d) 0·274. **6** $1/17 = 0{\cdot}0588$.
7 $5/24 = 0{\cdot}208$. **8** 0, 3/10, 1/15, 1/20, not independent.
9 $\frac{1}{3}$. **10** (a) 1/20 (b) 2/5. **11** (a) independent (b) not independent.
12 (a) $\frac{1}{3}$ (b) 5/12 (c) $1{\cdot}41 \times 10^{-4}$ (d) 0·237. **13** (a) 0·3 (b) 0·28 (c) 0·82.
14 (a) $\frac{1}{3}$ (b) 5/12 (c) 0. **15** 7/8.

Exercise 1.6 (p. 16)

1 (a) 0·34 (b) 0·063 (c) 0·19 (d) 0·97, no black marbles.
2 (a) $64/243 = 0{\cdot}263$ (b) $8/243 = 0{\cdot}033$ (c) $160/229 \approx 0{\cdot}219$.
3 (a) 6/7 (b) $6/35 = 0{\cdot}171$. **4** 1·45%.
5 (a) 0·297 (b) 0·703 (a) 0·196 (b) 0·804.
6

score	1	2	3
probability	4/7	2/7	1/7

(a) $12/49 = 0{\cdot}245$ (b) $20/49 = 0{\cdot}408$.
7 (a) 5/18 (b) 4/9 (c) 0 (d) 1/8.
8 (a) $1/21 = 0{\cdot}0476$ (b) 0·0794 (c) 0·0714 (d) 0·0446.

Exercise 1.7 (p. 19)

1 0·379. **2** 0·64. **3** 0·459. **4** 0·652. **5** 0·619. **6** (a) $\frac{1}{3}$ (b) $\frac{1}{3}$.
7 (a) 1/256 (b) 225/256, 0·09.

Exercise 1.8 (p. 23)

1 (a) 125/216 (b) 75/216 (c) 15/216 (d) 1/216.
2 (a) $(1/6)^5 = 1{\cdot}29 \times 10^{-4}$ (b) 0·965. **3** (a) 0·259 (b) 0·663.
4 (a) 27/64 (b) 27/64 (c) 9/64 (d) 1/64. **5** 0·212.
6 (a) $1{\cdot}15 \times 10^{-3}$ (b) 0·797. **7** (a) 8/729 (b) 160/729 (c) 80/729.
8 (a) 0·265 (b) 0·760. **9** 0·516.

Miscellaneous Exercise 1 (p. 23)

1 7/16. **2** 0·524. **3** (a) 4 368 (b) 858 (c) 0·294.
4 (a) 24 (b) 28 (c) 15/32. **5** (i) 5 040 (ii) 50 400.
6 (a) $9! = 362\,880$ (b) 7/36 (c) 1 260 (d) 5/9.
7 (a) 81/256 (b) 1/256 (c) 3/128 (d) 3/32. **8** (a) 60, 24 (b) 0·1, 1/7.
9 (i) 0·0414 (ii) 0·957. **10** (a) $1{\cdot}98 \times 10^{-3}$ (b) $1{\cdot}44 \times 10^{-3}$, 0·0102.
11 (i) (a) $4{\cdot}52 \times 10^{-3}$ (b) 0·287 (ii) 0·102. **12** (a) 0·276 (b) 0·456, 0·106.
13 (a) 0·812 (b) $4{\cdot}33 \times 10^{-3}$.
14 (a) 0·0659 (b) 0·896, $\dfrac{(n-1)(n-2)}{2}(\frac{1}{4})^3(\frac{3}{4})^{n-3}$, 8 or 9.
15 (a) $\frac{1}{3}$ (b) $\frac{1}{3}$ (c) $\frac{1}{2}$ (d) 5/12. **16** (a) 5/8 (b) 1/6 (c) 5/24.
17 (a) $\binom{25}{3}(0{\cdot}05)^3(0{\cdot}95)^{22}$ (b) $(0{\cdot}95)^{25}$ (c) $1 - 2{\cdot}2(0{\cdot}95)^{24}$, 1, 5.
19 0·496, 0·6976. **20** (a) 1/6 (b) 0·177 (c) 0·490 (d) 0·0625.
21 $\binom{800}{10}(0{\cdot}01)^{10}(0{\cdot}99)^{790}$, 0·13, $9{\cdot}14 \times 10^{-3}$. **22** (a) 0·0106 (b) 0·105.
23 $2p^3 + p^2 + p$. **24** $c^9(10 - 9c)$, £3·51, 0·428.

Exercise 2.4 (p. 37)

1 $(5\mathbf{i} + 2\mathbf{j})$ N, 5·39 N, 22°. **2** $(12{\cdot}5\mathbf{i} + 4{\cdot}33\mathbf{j})$ N, 13·2 N, 19°.
3 $(2{\cdot}83\mathbf{i} - 1{\cdot}5\mathbf{j})$ N, 3·21 N, −28°. **4** $(-3{\cdot}46\mathbf{i} - 6\mathbf{j})$ N, 6·93 N, −120°.
5 $(1{\cdot}84\mathbf{i} + 3{\cdot}23\mathbf{j})$ N, 3·72 N, 60°. **6** 3 N. **7** $\alpha = 0$, $P = 5$ N.
8 $P = 17{\cdot}1$ N, $Q = 12{\cdot}8$ N. **9** $P = 4{\cdot}64$ N, $Q = 3{\cdot}42$ N.
10 $P = 6{\cdot}14$ N, $\theta = -9°$. **11** $\beta = 90°$, $\alpha = 53°$.

Exercise 2.5 (p. 43)

1 0·4. **2** 1 600. **3** 1. **4** 20 N. **5** 10 N, 17·3 N. **6** 50 N, 40 N.
7 17·7 N, 2·17 N. **8** 6 N, 8 N. **9** $\sqrt{5}$. **10** $\cos^{-1}(1/4)$. **11** $mg\tan\frac{1}{2}\theta$.
12 $2mg\cos\theta$. **13** $8a/5$.

Exercise 2.6 (p. 47)

1 (i) 20 N (ii) 40 N. **2** 0·4 **3** (a) 8 N (b) 16 N. **4** 0·103.
6 $(2\sqrt{3}-1)/(2-\sqrt{3})$. **7** $a/b(1+a^2)^{\frac{1}{2}}$.

Miscellaneous Exercise 2 (p. 48)

1 53° (a) 25 (b) 25/7. **2** $\sqrt{22}$, $\tan^{-1} 11$.
3 $3\mathbf{i}+4\mathbf{j}$, $3(y-1)=4(x-1)$, $\mathbf{r}=\mathbf{i}+\mathbf{j}+\lambda(3\mathbf{i}+4\mathbf{j})$, $1/\sqrt{2}$.
4 (i) $\frac{1}{13}(5\mathbf{i}+12\mathbf{j})$, $\frac{1}{5}(4\mathbf{i}+3\mathbf{j})$ (ii) $35\mathbf{i}+84\mathbf{j}$, $64\mathbf{i}+48\mathbf{j}$, 165 N. **5** $2\sqrt{5}$ N, $4\sqrt{2}$ N.
6 (i) $\sqrt{3}P$ (ii) $5\pi/6$ (iii) $2\pi/3$, $\pi/3$. **7** (i) 815 N, 455 N
(ii) 335 N in same direction as force of 600 N. **9** Line through C and mid-point AB, line through N and point of intersection of medians of triangle KLM.
14 $1{\cdot}6a$, $1{\cdot}8mg$, $0{\cdot}85mg$. **15** $3mg$, $2mg$, $\sqrt{3}mg$.
16 $2\sqrt{2}mg/3$. **17** $mg\sin\lambda$ **18** slips when $X=\dfrac{143W}{150}$.

Exercise 3.1 (p. 56)

1 −6 Nm, −1·8 Nm. **2** 3·8 Nm, −7·6 Nm. **3** 1 Nm, 8 Nm. **4** −1·9 Nm, −2·8 Nm.
5 20 Nm. **6** 93 Nm. **7** 51 Nm. **8** $2Ql\cos\theta-Wl\cos\theta$, $Ql\cos\theta-Pl\sin\theta$.
9 $2Ql\cos\theta-2Fl\sin\theta-Wl\cos\theta$, $Ql\cos\theta-Pl\sin\theta-Fl\sin\theta$.
10 $4Sl\cos\theta-10Wl\cos\theta$, $6Wl\cos\theta-4Rl\cos\theta$. **11** $2aF\sin\theta-aW\sin 2\theta$, $2aR-aW\sin 2\theta$.

Exercise 3.4 (p. 61)

1 10 N, 6 N. **2** 13 N, 4 N. **3** −11 N, 4 N. **4** 6 N, 3 N.
5 28 N, 20 N, 48 N. **6** $m\tan 15°$.

Exercise 3.5 (p. 63)

1 −12 N, 3 m. **2** 6 N, 6 m. **3** 10 N, 6 m. **4** −5 N, 6·8 m. **5** −5 N, 4·2 m.
6 System is in equilibrium.

Miscellaneous Exercise 3 (p. 65)

1 (a) 80 N (b) 2·25 N (c) 150 N, 70 N. **2** 200 N, 20 N.
3 $3W$, W (a) $4W$ (b) $22W/3$. **4** 40 N, 60 N. **5** 5 kg, 2:3. **7** 73·5 Nm.
8 $\dfrac{\pi}{4}$, $\dfrac{5\pi}{4}$, $8m$, $\dfrac{10mg}{3}$. **9** $m=1$, $n=2$. **10** $(n+1)a/n$, $W(n-2)/3$.

Exercise 4.3 (p. 75)

1 2. **2** 3, −13·5. **3** 14. **4** 27·5. **5** 2·5. **6** 2. **7** 2. **8** 31/3. **9** 4.
10 1·6, 2. **11** $u=12$, $a=0{\cdot}5$; $u=13$, $a=0$. **12** 27 m. **13** $t=6$ s.
14 $t=3$ s, 6 s. **15** 32 s, 12 s, 16 s. **16** 5 s. **17** 15 ms^{-1}, 0·15 ms^{-2}.

Exercise 4.4 (p. 77)

1 20 m, 4 s. **2** 30 ms^{-1}, 2 s. **3** (a) 1 s, 8 s. (b) 3 s, 6 s. **4** 40. **5** 6 s. **6** 15 ms^{-1}, 1 s.
7 441 m. **8** $(u+gT)/g$.

Exercise 4.5 (p. 79)

1 $9t^2+8t$. **2** $84t^2+12t$. **3** $2t^4+3t^3+t^2+2t$. **4** t^5+t^4-3t+1.
5 $1+5t+e^{-t}$. **6** 2, 9. **7** $5t^3+t+2$. **8** 8π. **9** $10\pi-2$. **10** $4x$.

Exercise 4.6 (p. 83)

1 82, 249, 496, 831; 42 m, 207 m, 580 m, 1 243 m. **2** 13·62, 13·51. **3** 0·57 s. **4** 0·92 s.

Miscellaneous Exercise 4 (p. 84)

1 $\frac{192}{(u+6)}$ s, $\frac{60}{u}$ s (i) $u = 10$ (ii) $\frac{1}{3}$ ms^{-2}, $\frac{5}{3}$ ms^{-2}.
3 $\frac{10}{u}, \frac{14}{u}, \frac{6}{u}$; $u = 5$, $x = \frac{5}{2}$, $y = \frac{25}{6}$. **4** (i) $2/k$, $5/k$ (ii) 0·7.
5 $\frac{af}{V}\left[1-\left(1-\frac{2V^2}{af}\right)^{\frac{1}{2}}\right]$ **6** (i) $\frac{1}{2}ft^2+b-ut$ (ii) $\frac{V^2}{2f}+(V-u)t+b$, $b+\frac{u^2}{2f}$.
7 $\frac{15}{4}\left(\frac{d}{3\alpha}\right)^{\frac{1}{2}}$. **9** 47/3 **10** 6, 2/3, $X > 64$. **12** 80 s, 90 km h^{-1}, 37 s.
13 $351u^2/800g$, 50/4. **14** 8 cm. **15** (32/27) m, 2, 4 ms^{-1}.
16 $\left(u+\frac{k}{\omega}\right)t-\frac{k}{\omega^2}\sin\omega t$ **17** 10(ln 2 − 1). **18** (ii) $2 < t < 5$. (iv) 8·4.
19 81. **20** (a) 1·16 ms^{-1}, 2·64 ms^{-1} (b) 9·71 m. **21** 3430 m, 2457 kW. **22** 1·45 min.
23 24·4 ms^{-1}.

Exercise 5.3 (p. 93)

1 2000 N. **2** (a) 2·4 s (b) 36 m. **3** 136 N.
4 (a) 2·5 ms^{-2} (b) 1600 kg (c) 125 m. **6** 0·2 ms^{-2}. **7** (a) 360 N (b) 860 N.
8 102·5 N, 100 N, 95 N. **9** (i) (a) 2 400 N (b) 1 600 N (c) 800 N (ii) (a) 17 400 N (b) 11 600 N (c) 5 800 N. **10** 0·056.

Exercise 5.4 (p. 96)

1 5 ms^{-2}. **2** 5 ms^{-2}, 10 m. **3** (i) 12 m (ii) $\frac{3}{4}$. **4** 4 ms^{-2}. **5** 2 ms^{-2}. **6** 5 ms^{-2}.

Exercise 5.5 (p. 98)

1 4 ms^{-1}. **2** (a) 18 N s (b) 0 (i) 9 ms^{-1}, 0. **3** 0·38 ms^{-1}, 0.364 m.
4 (i) 6 ms^{-1} (ii) $6t-3t^2/2$, (iii) 8 m.

Exercise 5.6 (p. 100)

1 6 N s, 120 N. **2** $2\frac{1}{2}$ ms^{-1}, 12·5 m. **3** 4 ms^{-1}. **4** (a) 8400 N s (b) 42 000 N.
5 1·19 N s, 19·8 N. **6** 28·3 N s. **7** $u+I/m$. **8** 7 ms^{-1}.

Miscellaneous Exercise 5 (p. 101)

1 3 300 N, 9 300 N. **2** 1 000 N, 50 s. **3** (a) 2·4 s (b) 36 m.
4 $\frac{2}{3}$ ms^{-2}, 384 N, 360 N, 312 N. **5** 440 N. **6** 0·5 ms^{-2}, 0·75 ms^{-2}, 6 m.
7 1·2 kg, 5 ms^{-2}. **8** $T = m(f+g)$. **9** (i) 4 m (ii) 23·6 m. **11** 5/16. **12** $u^2/16$, $u/\sqrt{2}$.
15 10 N s, 30 N s. **16** 8 ms^{-1}, 850 000 N. **17** 27·7 m, $-2t$ ms^{-2}.
18 (a) 5 ms^{-1}, 20 ms^{-1} (b) 2 m, 9 m, 11 m.
19 (a) $2{\cdot}4\times10^{-4}$ N s, (b) 40, (c) $9{\cdot}6\times10^{-3}$ N. **20** $(70t-t^2)/240$ ms^{-1}.

Exercise 6.4 (p. 111)

1 (a) 3·25 m (b) 8·06 ms^{-1}. **2** (i) 4 J (ii) 0·2. **3** 10·78 m. **4** (i) 3·16 ms^{-1} (ii) 2·90 ms^{-1}. **5** $2E/(mg)$, $(mg)^2l/(2E)$. **6** $\sqrt{(gl/8)}$. **8** $2a\sqrt{(g/l)}$.

Exercise 6.5 (p. 114)

1 22·9 kW. **2** $4{\cdot}5\times10^4$ N, 450 kW, 24 ms^{-1}. **3** **25** ms^{-1}, 9/49.
4 $7{\cdot}643\times10^6$ J, 156 kW. **5** 45 ms^{-1}, 2·695 ms^{-2}.
6 (a) 102 kW (b) 0·03 ms^{-2} (c) 0·028. **8** 18 kW, 1/12 ms^{-2}.
9 (a) 20 ms^{-1} (b) 2 000 J (c) 2 000 J (d) 4 000 W.
10 62 500 J, 10 000 J, 72·5 kW. **11** (a) 29 400 J, 45 000 J, 74·4 kW.
12 2 100 J, 7 kW. **13** $\frac{3}{4}$ ms^{-2}. **14** 19·69 ms^{-1}, 0·079 ms^{-2}.

Miscellaneous Exercise 6 (p. 116)

1 36 kW, 20 ms^{-1}, 900 N, 0·075 ms^{-2}. **2** 400 N, 10 kW.
3 20 kW, arcsin (1/20), $\frac{1}{4}$ ms^{-2} **4** 400 kg, 21·9 ms^{-1}.
5 $8mg$. **6** $41\frac{2}{3}$, 10 000 N, 20 000 N, 4 800 m.
7 (a) $(1000H - MgV\sin\alpha)/(MV)$ (b) $(1000H + MgV\sin\alpha)/(MV)$, $\sin\alpha = 1000H/(3MV)$, $3V/4$.
8 40 567 J. 27 m, 23·1 ms^{-1}. **9** 81 kW. **10** (i) $7a/9$ below O (ii) $\frac{2}{3}(ag)^{1/2}$.
11 $(9+\sqrt{17})a/8$. **12** $2a$. **13** $(2ga)^{\frac{1}{2}}$.

Exercise 7.1 (p. 122)

1 $6u$. **2** $2u$. **3** $4u/3$. **4** 5 ms^{-1}. **5** $2{\cdot}5u$. **6** $4u$, $4{\cdot}5u$. **7** 4 ms^{-1}. **8** $10u$.
9 13·5 kNs. **10** $2u$, $2mu$.

Exercise 7.2 (p. 126)

1 $h_1 = 0{\cdot}45$ m, 0·6 s. **2** 0·5. **3** 0·2. **4** 4 ms^{-1}, 1 ms^{-1}. **5** 125 m. **6** 2·5 m.
7 1·7 s. **8** 3 Ns, 0·75 Ns. **9** 3·1, 4·6. **10** 0·25, 4.
11 0·25, 3. **12** 2·5, 5. **13** 9, 5. **14** 2, 0. **15** $e > 0{\cdot}25$. **16** 0·5.

Exercise 7.3 (p. 127)

1 200 N. **2** 750 kN. **3** 7 N. **4** 1·2 m.

Exercise 7.4 (p. 130)

1 1 ms^{-2}, 500 N. **2** 1·25 ms^{-2}, 400 N. **3** 0·9 ms^{-2}, 385 N.
4 10 ms^{-1}, 350 N. **5** 2·5 ms^{-2}, 37·5 N. **6** 4 ms^{-2}, 42 N. **7** $0{\cdot}2g$, $4{\cdot}8\,Mg$.
8 6 ms^{-2}, 12 N. **9** $15\sqrt{2}$ N. **10** 31·5 N. **11** 6·25 ms^{-2}. **12** 33·75 N.
13 4·6 ms^{-2}. **14** (5/7) ms^{-2}. **15** (110/59) ms^{-2}. **16** (35/6) ms^{-2}.

Exercise 7.5 (p. 133)

1 0·3125 m. **2** 1 s, 1·875 ms^{-1}. **3** 6 s. **4** 0·098 m. **5** $4M^2h/(2M+m)^2$.

Miscellaneous Exercise 7 (p. 134)

1 $\frac{1}{2}mgh$, $m(gh)^{\frac{1}{2}}[1+\sqrt{2}]$. **2** 3. **3** (i) 1·4 ms^{-1}, 39·2 J, 19·6 J (ii) 67·2 J.
4 10 ms^{-1}, 750 N, 1 ms^{-1} and 2 ms^{-1}, 40 kJ. **5** (i) $\frac{1}{2}$ (ii) $\frac{1}{4}mu^2$ (iii) $\frac{2}{3}$ (iv) u.
7 4·5 ms^{-1}, 6 ms^{-1}, 21·5 J. **8** $2mg$, $2(1+e)(ag)^{\frac{1}{2}}/5$. $(ag)^{\frac{1}{2}}(2-3e)/5$.
13 82·4 W, 5·67 N. **14** 0·99 kg
15 (i) $\frac{1}{8}$ ms^{-2} (ii) 312·5 N (iii) 50 kW (iv) 950 N.
16 $41\frac{2}{3}$ (a) 10 kN (b) 20 kN, 4 800 m. **17** $u/3$ ms^{-1}, 0·5 ms^{-2}. **18** $g/2$, $3mg/2$.
19 12 m. **20** (a) 2 ms^{-2}, 40 N, 36 N (b) 1 ms^{-2}, 45 N, 33 N.
21 (i) $\left(\frac{4gh}{13}\right)^{\frac{1}{2}}$ (ii) $\frac{2h}{9}$ (iii) $\left(\frac{13}{15}\right)^{\frac{1}{2}}$, $m(gh)^{\frac{1}{2}}\left[2^{\frac{1}{2}}+\left(\frac{30}{13}\right)^{\frac{1}{2}}\right]$.

Exercise 8.2 (p. 144)

1 $7\mathbf{i}$. **2** (a) The lines do not meet (b) meet at $\mathbf{r} = 7\mathbf{i}+3\mathbf{j}+11\mathbf{k}$.
3 $21\mathbf{i}+30\mathbf{j}$. **4** $33\mathbf{i}+19\mathbf{j}+37\mathbf{k}$. **5** (a) 11·05 N (b) 11.
6 $3\mathbf{i}+4\mathbf{j}$, $\mathbf{r} = \mathbf{i}+\mathbf{j}+\lambda(3\mathbf{i}+4\mathbf{j})$, $3y+1 = 4x$, $1/\sqrt{2}$. **7** $-9\mathbf{i}/4$. **8** 25 N, 164°.
9 (a) 14·4 N (b) $\frac{2}{3}$ (c) $2{\cdot}4\mathbf{i}+1{\cdot}6\mathbf{j}$. **10** (a) $\mathbf{r} = \mathbf{i}+2\mathbf{k}+(t/3)(2\mathbf{i}+2\mathbf{j}+\mathbf{k})$
(b) $\mathbf{r} = -\mathbf{j}+\mathbf{k}+(t/3)(2\mathbf{i}+2\mathbf{j}+\mathbf{k})$, $\cos\theta = (6t-2)/\sqrt{[2(9t^2-6t+18)]}$.
11 4·58 N, $\mathbf{r} = 4\mathbf{i}-5\mathbf{j}+9\mathbf{k}+\lambda(\mathbf{i}-4\mathbf{j}+2\mathbf{k})$. **12** 6·93, 10·39 N.

Exercise 8.3 (p. 148)

1 $\overrightarrow{RO}$, $\overrightarrow{RQ}$, 69·28 kmh^{-1} on a bearing 030°, 40 kmh^{-1} on a bearing 120°.
2 (a) $-(2\mathbf{i}+4\mathbf{j})$ ms^{-1}, (b) 5·1 m, 8, 4. **3** 73·7° (i) 6·72 km, (ii) 2·94 minutes, 332·2°.
4 $(-10+30t, 0)$, 073·7° $(0, 10-40t)$, $(2500t^2-1400t+200)^{\frac{1}{2}}$, 16·8 min. past noon, 50 ms^{-1} in direction of the vector $-(3\mathbf{i}+4\mathbf{j})$.
5 $(60\mathbf{i}-300\mathbf{j}-60\mathbf{k})$ kmh^{-1}, $(30-60t)\mathbf{i}+(30-300t)\mathbf{j}+(2-60t)\mathbf{k}$, 4 min. 30 s.
6 12·12. **7** 0·6 s, 45·46 m. **8** $\frac{3}{4}u$, $u/20$, $(u/20)[-17\mathbf{i}+16\mathbf{j}]$ ms^{-1}.
9 $Z°$ where $\tan Z° = (\tan\theta+\tan\phi)/2$.

Exercise 8.6 (p. 154)

1 (a) 14 m (b) 5 ms^{-1} in direction of $3\mathbf{i}-4\mathbf{j}$ (c) $2\mathbf{i}\,ms^{-2}$,
(d) 4·12 ms^{-2} in direction of $-\mathbf{i}-4\mathbf{j}$. **2** (a) $9\mathbf{i}\,ms^{-1}$, $(2\mathbf{i}+\mathbf{j})\,ms^{-1}$
(b) 7·07 ms^{-1} in direction $7\mathbf{i}-\mathbf{j}$ (c) $3\mathbf{i}\,ms^{-2}$, $(3\mathbf{i}-2\mathbf{j})\,ms^{-2}$.
3 $3\pi/(4w)$, $3\pi/(4w)$. **4** $\omega(-\mathbf{a}\sin t+\mathbf{b}\cos t)$, $-\omega^2(\mathbf{a}\cos t+\mathbf{b}\sin t)$, $-m\omega^2\mathbf{r}$.
5 $\pi/4$, $3\pi/4$. **8** $m(2\mathbf{i}-\mathbf{j})$ N. **9** $m\omega^2|\mathbf{a}|$, $m\omega^2|\mathbf{b}|$.
10 (a) $2|\omega-1|$ (b) $2\sqrt{(\omega^4+1)}$ (c) π/ω, $2\pi/\omega$. **11** $(\mathbf{i}+2\mathbf{j})t\,ms^{-1}$, $(4\mathbf{i}+9\mathbf{j})$ m.
12 $2\mathbf{i}(T^4-1)+3\mathbf{j}(T^2+1)$ m, $(94\mathbf{i}+30\mathbf{j})$ m.

Miscellaneous Exercise 8 (p. 155)

1 3·46 N along DA, where AD bisects angle BAC. **2** $(2\mathbf{i}+2\mathbf{j})\,ms^{-2}$.
3 (a) $7\sqrt{3}$ N (b) $(2\mathbf{i}-\mathbf{j}+3\mathbf{k})$ (c) $\mathbf{r}=2\mathbf{i}-\mathbf{j}+3\mathbf{k}+\lambda(\mathbf{i}+\mathbf{j}+\mathbf{k})$.
4 (i) 10 N m (ii) twice the area of triangle ABC,
(iii) $2\overrightarrow{CA}$, through the mid-point of the altitude AD. **5** $(6\mathbf{i}+26\mathbf{j})$ m.
6 $(\mathbf{i}+3\mathbf{j})$m, 22·9 N, $3\mathbf{j}$ m, 11·2 Nm. **7** $(128\mathbf{i}+16\mathbf{j}-112\mathbf{k})$ m.
8 (ii) N.W. (iii) 5·31 min. **9** 041·4°, 45 min., 048·2°, 7·45 km.
10 3, 8 min. 20 s, 3·162 km. **11** $(7\mathbf{i}+2\mathbf{j})$ knots, 12·30, $(4{\cdot}5\mathbf{i}+\mathbf{j})$ nautical miles.
12 From 16·6°, from 343·4°, 8·48 knots east, 8 knots. **13** 54 min., 084·3°.
14 $-3\mathbf{j}\,ms^{-1}$, $4\mathbf{i}\,ms^{-1}$, $(4\mathbf{i}+3\mathbf{j})\,ms^{-1}$, 5 ms^{-1}, $(\mathbf{i}-4\mathbf{j})\,ms^{-1}$, 0·6.
15 $(-3\mathbf{i}+4\mathbf{j})\,km\,h^{-1}$, $(6t-1)\mathbf{i}+(15-2t)\mathbf{j}$ km, 13·9 km, 8·24 km.
16 $(\mathbf{i}+\mathbf{j})\,ms^{-1}$, $-4\sqrt{2}$, 8 m. **17** $(2\mathbf{i}-\mathbf{j}+2\mathbf{k})\,ms^{-1}$, $(6\mathbf{i}+2\mathbf{j}+4\mathbf{k})$ m, 3·32 ms^{-1}, $(\mathbf{i}+\mathbf{j}+3\mathbf{k})$.
18 $(8\mathbf{i}-7\mathbf{j}+6\mathbf{k})\,ms^{-1}$, $(6\mathbf{i}-6\mathbf{j}+2\mathbf{k})\,ms^{-1}$, $(6\mathbf{i}+12\mathbf{j}+8\mathbf{k})$ m.
19 $(-2\mathbf{i}-4\mathbf{j})\,ms^{-1}$, 5·1 m. **20** $v=-[\omega(a\sin\omega t\,\mathbf{i}+a\cos\omega\,\mathbf{j})]\,ms^{-1}$,
$\dot{\mathbf{r}}=-\omega^2[(a\cos\omega t)\mathbf{i}+(a\sin\omega t)\mathbf{j}\,ms^{-2}\tfrac{1}{2}[(a\cos\omega t)\mathbf{i}+(3a\sin\omega t)\mathbf{j}]$ m, $n\pi/(2\omega)$.

Exercise 9.1 (p. 164)

1 $2\mathbf{i}+3\mathbf{j}$. **2** $4\mathbf{i}+(11-10t)\mathbf{j}$. **3** $6\mathbf{i}+5\mathbf{j}$. **4** $2{\cdot}2\mathbf{i}+4{\cdot}8\mathbf{j}$.
5 21, 41. **6** $2\sqrt{3}$, 38.

Exercise 9.2 (p. 169)

1 8·66 m, 1·25 m, 1 s. **2** 2·91 m, 0·548 m, 0·662 s. **3** 4·8 m, 1·8 m, 1·2 s.
4 7·12 m, 0·450 m, 0·597 s. **5** 16° above horizontal. **6** 61° below horizontal.
7 48° below horizontal. **8** 1 s, 1·4 s. **9** 4 s, 5·6 s. **10** 19°, 71°.
11 27°, 63°. **12** 2, 8. **13** 0·75, 8 s. **14** 16·9 m, 16·2 m.

Exercise 9.3 (p. 172)

1 182 m. **2** 1·5 km from P. **3** 60 m, 120 m. **4** 34·6 m, 80·0 m. **5** $4u^2/(7g)$.

Exercise 9.4 (p. 175)

1 $(2+5t)\mathbf{i}+(4-7t)\mathbf{j}\,ms^{-1}$. **2** $t(1+t)\mathbf{i}+3t(2t-1)\mathbf{j}$ m.
3 $(3t^2+t+2)\mathbf{i}+(-2t^2+t+5)\mathbf{j}$ m. **4** $t=2$ s. **5** $t=3$ s. **6** $(a/\sqrt{2}, 2a)$.

Miscellaneous Exercise 9 (p. 175)

1 26·3 ms^{-1} at 48° 49′ elevation, 15·3 m. **4** $26°34' < \alpha < 63°26'$, 98 m, 22·05 m.
5 $2u_Hu_V/g$, $u_V^2/2g$ (i) 18·75 ms^{-1}, 20 ms^{-1}, (ii) 8/15 (iii) 32 m (iv) 15/16.
6 45°, 30 ms^{-1}. **7** 2·2 s. **9** (ii) 10 s, 2 (iii) 240 m, 40 m.
10 (a) 1·8 m (b) 4·1 ms^{-1}. **11** 1·75, 31·5. **13** $\frac{47a}{96}$, $\frac{47a}{54}$.
14 $\tan^{-1}[(1-e)\tan\alpha/e(1+e)]$. **15** (iii) $d\tan\alpha/(1+e)$ (iv) $\sqrt{3}-1$.
18 $40\sqrt{5}$ m.
20 (i) $\frac{20}{29}$ (ii) 126 m (iii) 2 s (iv) 66·1 ms^{-1}, $\tan^{-1}20/63$ below horizontal.
22 $(1220/3)^{\frac{1}{2}}\,ms^{-1}$, $\tan^{-1}6/5$. **25** $2\tan\beta$, $2V\sec\beta/g$. **26** $\frac{200}{7}$ s, 1200 m, 3 ms^{-1}.
27 $(9t-t^2)\mathbf{i}+(\tfrac{1}{2}t^2-4t)\mathbf{j}$. **28** $-33/65$, $8(\mathbf{i}-\mathbf{j})$ N. **29** (a) $12\mathbf{i}+8\mathbf{j}$, $4\mathbf{i}+5\mathbf{j}$ (b) $-\mathbf{i}-3\mathbf{j}$.

Exercise 10.2 (p. 188)

1 54 N. **2** 3·5. **3** 0·0462 m. **4** 0·932 s. **5** $3{\cdot}62 \times 10^4$ km. **6** 0·324.
7 $2(5g/l)^{\frac{1}{2}}$. **8** $(gh)^{\frac{1}{2}}$. **9** 590 ms^{-1}. **10** $216\pi^2$. **11** $(10/\pi^2)$ m. **12** $768\pi^2$N.
13 1·65 s. **14** (11/7). **15** 14·1 rad s^{-1}. **16** $2\pi(a^2-b^2)^{\frac{1}{4}}g$.

Exercise 10.3 (p. 192)

1 0·324, 0·37. **2** $1{\cdot}25\,mg \sin\alpha$.

Exercise 10.4 (p. 197)

1 $(68\,gl)^{\frac{1}{2}}$. **2** 8 ms^{-1}. **3** 4·94 N. **4** $2mg\ (n-2+3\cos\theta)$, $n > 5$.
5 $n = 3{\cdot}5$. **6** $3mg\cos\theta$. **7** $3a/4$. **8** $a/3$.

Miscellaneous Exercise 10 (p. 197)

1 (i) $(8g/a)^{\frac{1}{2}}$, $(8ag)^{\frac{1}{2}}$, $2\pi(a/8g)^{\frac{1}{2}}$ (ii) $(3g/4a)^{\frac{1}{2}} \leqslant \omega < (33g/4a)^{\frac{1}{2}}$ (iii) $\omega \leqslant (3g/8a)^{\frac{1}{2}}$.
2 (a) $10mg/3$ (b) mg, $(10g/3h)^{\frac{1}{2}}$ **3** $\dfrac{ma}{2}(2\omega^2+\Omega^2)$, $\dfrac{ma}{2}(\Omega^2-2\omega^2)$.
4 $T = mg\cos\alpha + ma^2\omega^2\sin^2\alpha$,
$R = mg\sin\alpha - ma\omega^2\sin\alpha\cos\alpha$, $l = \dfrac{2a\cos\alpha}{(1+\cos\alpha)^2}$, $T = R = \dfrac{5mg}{7}$.
5 $\omega^2 = g/l\cos\theta$. **6** $2\pi(4R/5g)^{\frac{1}{2}}$, $2mg/3$, $7mg/12$. **8** (i) $5mg/3$, (ii) $5mg/12$.
10 $mg(\sqrt{3}-1)$. **12** (i) $5mg/3$, $10mg/3$ (iii) $2\pi(l/5g)^{\frac{1}{2}}$.
13 $m\left(g\sin\alpha + \dfrac{V^2}{l}\operatorname{cosec}\alpha\right)$. **14** $(gh)^{\frac{1}{2}}$. **17** 11·7°. **18** $(ag/2)^{\frac{1}{2}}$
19 $m\left[\dfrac{u^2}{a^2}+g(3\cos\theta-2)\right]$, $[e^2u^2+2ag(1-e^2)]^{\frac{1}{2}}$. **20** $(\frac{1}{2}ag)^{\frac{1}{2}}$.
24 $(a/\sqrt{2},\ a/\sqrt{2})$. **25** $\dfrac{m}{2}\left(\dfrac{V^2}{a}-4g-12\cos\theta\right)$. **26** $mg\sin\theta\dfrac{(3-\lambda)}{1+\lambda}$, $mg\cos\theta\dfrac{2\lambda}{1+\lambda}$.

Exercise 11.1 (p. 205)

1 $8\mathbf{i}+9\mathbf{j}$. **2** $-(0{\cdot}6\mathbf{i}+0{\cdot}8\mathbf{j})$ Ns. **3** $15\mathbf{i}-39\mathbf{j}$. **4** 0·25.
5 $24\mathbf{i}-6\mathbf{j}$. **6** $(8\mathbf{i}+14\mathbf{j})$ m.

Exercise 11.2 (p. 207)

2 1/3. **3** 54°, 34°. **4** $(2\mathbf{i}+7\mathbf{j})$ ms^{-1}, $20\mathbf{i}$ Ns. **5** $(12{\cdot}2\mathbf{i}-4{\cdot}6\mathbf{j})$ ms^{-1}.

Exercise 11.3 (p. 209)

1 (a) 58 J (b) 1 460 J (c) 10 N. **2** (a) 13 J (b) 2 J (c) 17 J.
3 (a) 6 J (b) 72 J (c) 87 J. **4** 37·008 kJ.

Exercise 11.4 (p. 214)

1 3 ms^{-1}. **2** 8·89 ms^{-1}. **3** 3·61 ms^{-1}. **4** 2·83 ms^{-1}. **5** 2 ms^{-1}. **6** $(15gl)^{\frac{1}{2}}/4$.

Exercise 11.5 (p. 216)

1 $(7\mathbf{i}+15\mathbf{j})$ Ns. **2** (a) 11 J (b) -6 J (c) 8 J.
3 (a) 5 J (b) $-38/3$ J (c) 0·5 J. **4** (a) 6 J (b) 1 328 J (c) 3 915 J.
5 $(9\mathbf{i}-2\mathbf{j}+2\mathbf{k})$ ms^{-1}.

Miscellaneous Exercise 11 (p. 217)

1 $2\mathbf{i}+2\mathbf{j}$ or $2\mathbf{i}-6\mathbf{j}$. **2** $a_1 = 1$, $b_1 = -3$, $a_2 = 3$, $b_2 = 2$, $-8m\mathbf{k}$.
3 (i) $27y^2 = x^3$ (ii) $12\mathbf{i}+12\mathbf{j}$, $\mathbf{r} = 12(1+\lambda)\mathbf{i}+(8+12\lambda)\mathbf{j}$, $20\sqrt{2}$, 40.
4 $2u(e^t\mathbf{i}+e^{-t}\mathbf{j})+u(-\mathbf{i}+2\mathbf{j})t-2u(\mathbf{i}+\mathbf{j})$, $3/\sqrt{(10)}$, $3mu^2(2-3\ln 2)$.
5 $\mathbf{r} = u[\mathbf{i}(2+\ln 2)]+\mathbf{j}(-1+2\ln 2)]+(t-\ln 2)u(3\mathbf{i}+\mathbf{j})$.
6 $\dfrac{\pi}{4}$, $2mue^{\frac{1}{2}}(1+e)/3$. **8** $(-30\mathbf{i}+30\mathbf{j})$ N, 7 J.

9 $(2t-4)\mathbf{i}-2t^2\mathbf{j}+(t^2+3)\mathbf{k}\ \mathrm{ms^{-1}}$, $2(5t^3+5t-4)\,\mathrm{W}$. **10** $4(\mathbf{j}+t\mathbf{k})$, $4t(2+t^2)$.

11 $-4a\sin 2t\mathbf{i}+2a\cos 2t\mathbf{j}-4ma(2\cos 2t\mathbf{i}+\sin 2t\mathbf{j})$, $12ma^2\sin 4t$, $6ma^2$, $\dfrac{\pi}{4}$.

13 (i) (a) $-m(\mathbf{i}+\pi\mathbf{j})$ (b) $-2m(\mathbf{i}-\pi\mathbf{j})$ (c) $\frac{1}{2}m(1+\pi^2)$ (ii) $-4/25$.
14 u, $2mu$, mu^2, $u^2/8g$. **15** $2ml^2\dot\theta^2(2+\cos^2\theta)+2mgl(\sin\theta-\sin 2\theta)=0$.

16 40 600 J, 15 m, 33 ms^{-1}. **18** $\dfrac{mg}{3}(4+\cos\theta)$, $mg(5\sin\theta-4\theta)/3$.

19 $\dfrac{3}{2}(ga)^{\frac{1}{2}}$, $\dfrac{5a}{4}$. **20** 2·75, 0·02, 0·06 m. **21** $\frac{1}{2}mg$. **22** $(80/7)^{\frac{1}{2}}$.

23 $\mathbf{i}+3\mathbf{j}+7\mathbf{k}$, $20\sqrt{59}$ N (i) (a) 1 800 J (b) 6 640 J (ii) 11 ms^{-1}, $\frac{1}{2}\mathbf{i}-\dfrac{21}{2}\mathbf{j}-\dfrac{63}{4}\mathbf{k}$.

24 $(-\mathbf{i}+3\mathbf{j}-4\mathbf{k})$ m, $(4\mathbf{i}+3\mathbf{j}+2\mathbf{k})$ ms^{-1}, 138 J, 60 J, 198 J.
25 $20\mathbf{i}+27\mathbf{j}-34\mathbf{k}$, $\sqrt{590}$ ms^{-1}, -144 J.
26 $(-6t^2+3)\mathbf{i}+(2t+2)\mathbf{j}+3t^2\mathbf{k}$, $(-2t^3+3t)\mathbf{i}+(t^2+2t+1)\mathbf{j}+(t^3+1)\mathbf{k}$, 304
(i) $-7\mathbf{i}+2\mathbf{j}+4\mathbf{k}$ (ii) $-17\mathbf{i}+\dfrac{8}{3}\mathbf{j}+9\mathbf{k}$.

Exercise 12.1 (p. 225)

1 4·1, 4. **2** (a) 2·5 (b) 2. **3** (a) 3·5 (b) $1\frac{2}{3}$. **4** 2·2, 2·6. **5** $\frac{1}{2}$, $\frac{1}{2}$. **6** 4·5 cm.
7 $\frac{2}{3}a$. **8** $2\frac{1}{2}a$, $2\frac{1}{2}a$. **9** $\dfrac{(4a^2\pm 2ax+x^2)}{3(2a+x)}$ (this checks for $x=0$ and $x=2a$).

Exercise 12.2 (p. 231)

1 (0, 9·6). **2** $(5{\cdot}4a, 0)$, $(5{\cdot}4a, 2\frac{1}{4}a)$. **3** $(1\frac{2}{3}a, 0)$. **5** $(\pi/2,\ 4{\cdot}19)$.
6 $(0, 1\frac{2}{3}a)$. **8** (a) $(1{\cdot}8a, 1{\cdot}8a)$ (b) $(\pi/2-1, \pi/8)$ (c) (2, 3·6).
9 (a) $(21a/88, 0)$ (b) $(45h/56, 0)$, $(0, \frac{2}{3}a)$. **10** $(0, 1{\cdot}8a)$.

Exercise 12.5 (p. 236)

1 $3r$. **2** $3(h^2+4lh+2l^2)/[4(h+3l)]$ from vertex. Check for $h=0$, $l=0$.
3 (a) $2/3a$ (b) 36·87°. **4** 3·21 cm. **5** 7 cm. **7** $4\frac{1}{2}$ cm, 6 cm, 14·9°. **8** (b) $4\frac{2}{3}$ cm.
10 18 cm, 26 cm. **11** 70·4°, 0·64 W. **12** $a\sqrt{3}$ from the vertex. **13** $11a/24$.

Miscellaneous Exercise 12 (p. 239)

1 (9, 3) (a) 18 units^2 (b) (5·4, 1·125), 3·2.
2 (a) 26·6° (b) magnitude of $G=2aW$. **3** $a/(2\pi-3\sqrt{3})$. **4** a/π.
5 $\frac{2}{3}$, 2. **8** $\frac{2}{3}a$, $W/2$, $W\sqrt{5}/2$ at θ to the horizontal, where $\tan\theta=2$, $\theta=63{\cdot}4°$.
11 $\frac{2}{3}l$ from the top, where l = length of ladder.

Exercise 13.1 (p. 245)

1 0·5. **2** $5W/4$. **3** W. **4** $Wl/2h$, $2h/l$, $h/2a$. **5** Wb/a, $W(a^2-b^2)^{\frac{1}{2}}/a$. **6** $5b/a$.
7 $2W$, $2\sqrt{5}W$. **8** $W/3$, $\sqrt{3/5}$. **9** $W\cos\alpha\,\mathrm{cosec}\,2\theta$, 45°. **10** $3/\mu$. **11** 5/3.
12 $3W/2$. **13** 0·5. **14** 2/3. **15** 9·6 m. **16** $7a/2$ from O.

Exercise 13.2 (p. 249)

1 $3W/4$. **2** $(9W/4)\tan\theta$, $(9W/4\tan\theta)$, $5W/8$. **3** 30°, 16°. **4** BC, 0·37.
5 $W\sec\alpha/2$, $W\sec\alpha$, $W\sin\alpha\tan\alpha$. **6** $\mu\geqslant 1$. **7** $\mu\geqslant 1/7$. **8** $W/(n^2-2n)^{\frac{1}{2}}$.

Exercise 13.3 (p. 254)

2 Wr/h, μW. **3** $2W/3$, μW, $\mu>2/3$. **4** $W/2\sqrt{2}$, $\mu\sqrt{2}W/(1+\mu)$.
5 $W/\sqrt{2}$, $\mu\sqrt{2}W/(1-\mu)$, $\mu>1/3$.

Exercise 13.4 (p. 258)

1 Both reactions 100 N, 125 N, 75·4 N, 75·4 N.
2 Both reactions 125 N, 144·3 N, 72·3 N, 0, 144·3 N.

3 300 N, 500 N, 346·6 N, 173·2 N, 115·5 N, 577·4 N.
4 100 N, 400 N, 266·7 N, 100 N, 166·7 N, 500 N.
5 Reaction at D, 224 N; at A, 200 N; AB, 200 N; BD, 140 N; BC, 140 N; DC, 100 N.
6 BD, 173·2 N; DC, 81·6 N; AB, 157·7 N; BC, 111·6 N.

Exercise 13.5 (p. 262)

1 $4\sqrt{2}$ N along AC, 5 Nm, 1·25 m.
2 $2\sqrt{2}$ N, parallel to CA, 2·5 m from A on BA produced.
3 $10P$ at 53° 8′ to BA, cutting it at 1 m from A. **4** $4x+3y+4=0$.
5 $8\sqrt{10}$ N, $x+3y=8$. **6** $2\sqrt{7}$ N, 0·25 m, 0·5 m. **7** $8x+4y=17$.
8 2 N. **9** $3\sqrt{7}$ N through centre at $\tan^{-1}(\sqrt{3}/2)$ to FC.

Miscellaneous Exercise 13 (p. 264)

1 $a/6$. **2** $(5W/2)\operatorname{cosec}\theta$, $(5W/2)\cot\theta$, $W/2$.
4 $F\cos\theta$, $F\sin\theta$, $M_A = F(h\sin\theta + b\cos\theta)$, $M_B = Fb\cos\theta$, $\tan^{-1}h/b$, 0.
5 0·3, 0·3W, $75a/28$. **6** $W(\sin\theta+\mu\cos\theta)$, $(1-\mu)/(1+\mu)$, $a(1-2\mu)$.
9 $W/\sqrt{3}, \dfrac{8a}{5}$. **10** $\frac{1}{2}W\tan\theta$.
12 $W/\sqrt{3}$, $W(17-8\sqrt{3})^{\frac{1}{2}}/2\sqrt{3}$, $\tan^{-1}2(\sqrt{3}-1)$ to AB, $4W$. **13** 35·26°.
14 4/11, 7°2. **16** $W/54\sqrt{13}$. **23** 70° 26′, $16W/45$.
24 $3a^3-x^3/3(2a^2-x^2)$, $p\dfrac{x}{p}<1-\tan\alpha$. **17** $\dfrac{4a(h+2ab)}{|h^2-3a^2b|}$, $b>20$.
19 (i) $w+\frac{1}{2}kw(1\pm p)$ (ii) (a) C (b) $\dfrac{(2\mu-\tan\theta)}{\tan\theta-\mu}$, $\mu\leqslant\tan\theta\leqslant 2\mu$.
21 $\dfrac{2W}{3}\sqrt{3}$, both equal to $\frac{1}{3}W\sqrt{3}$.
24 $\tan\alpha\leqslant 2/3$, $\tan\alpha\geqslant\frac{1}{2}$. **25** (i) (a) $2P\sqrt{3}$, 30° to BC (b) $3a$ (ii) $2K/a\sqrt{3}$.
26 2, $6a$, $3a/2$. **27** (iii) $3\overrightarrow{OG}$ (iv) forces in equilibrium.
28 $m=1$, $n=4$, 5 Pa in sense $ADCB$. **29** $2\sqrt{2}P$, $3Pa$, $3a/2\sqrt{2}$.
31 $AB=(P+Q)/\sqrt{2}$, $BD=Q/\sqrt{2}$, $CB=\sqrt{2}Q$; $P/\sqrt{2}$, $AD=Q$, $CD=Q+\frac{1}{2}P$.
32 $AB=FA=3W$, $BC=EF=4W$, $BD=DF=W$, $CD=DE=2W/\sqrt{3}$, $AD=W$.
33 $AD=W/3\sqrt{3}$, $BD=2W/3\sqrt{3}$, $CD=4W/3\sqrt{3}$, $AB=2W/3\sqrt{3}$, $BC=2W/3\sqrt{3}$.

Exercise 14.2 (p. 279)

1 π s. **2** 0·3 m. **3** $(1/\pi)$ ms^{-1}. **4** $(5/3\pi)$ m. **5** 5 m, π s.
6 (1/3) s, (2/3) s. **7** 13 m. **8** 1·5 m.

Exercise 14.3 (p. 283)

1 24·6 N, 5·4 N. **2** $2\pi^2$. **3** 1·6 m. **4** 2·5 N. **5** $(3gl/8)^{\frac{1}{2}}$, $\pi(l/6g)^{\frac{1}{2}}$.
6 Extensions $\pm 0{\cdot}5\,l$, $0{\cdot}25\,l\cos(2g/l)^{\frac{1}{2}}t + l(3/16)^{\frac{1}{2}}\sin(2g/l)^{\frac{1}{2}}t$.
7 $(\pi/10^{\frac{1}{2}})$ s, 1·85 m, 1.65 m. **8** $(gb)^{\frac{1}{2}}/2$, $g/2$. **9** $\pi(a/g)^{\frac{1}{2}}/4$, $a/16$.
10 $(\pi/5)$ s, 0·9 m. **11** 0·575 m. **12** 0·7 m, 0·17 m. **13** $\pi(a/16g)^{\frac{1}{2}}$.

Miscellaneous Exercise 14 (p. 285)

1 π s, 2π s, $3\pi/2$ s. **2** $4\sqrt{6}$ cm/s^{-1}, $12\sqrt{10}$ cms^{-2}, 0·13 s, $2{\cdot}25\times10^{-3}$ J.
3 12 s, $(24/\pi)$ m, 5 s, 17·0 s. 1·41 ms^{-1}. **4** (i) $2\sqrt{3}/\pi$, $2/\sqrt{3}$. (ii) $\frac{1}{2}a$.
5 $-\pi$, $3\pi^2/16$, 2·36 s. **6** (a) (i) $\dfrac{2\pi}{n}$ (ii) na (b) (i) 1·05 mh^{-1}
(ii) $1513\frac{1}{3}$. **7** $\sqrt{13}$, 2π. **9** $5\pi^2/18$, 2 s. **10** $x=1$, π, 2·5.
11 $2\pi/3$ s, 2 m, 6 ms^{-1}, 18 ms^{-2}.
12 (i) 12, -20 (ii) 0·29. **13** 3, $l/2$. **15** $2\pi(l/2g)^{\frac{1}{2}}$, $(\frac{1}{2}gl)^{\frac{1}{2}}$, $\left(\dfrac{3gl}{8}\right)^{\frac{1}{2}}$, $\frac{1}{2}mg$.
18 (i) $\dfrac{5c}{4}$ (ii) $2c$. **19** $2(ga^2/l)^{\frac{1}{2}}$.

20 $\dfrac{2\pi}{3}\left(\dfrac{Ma}{kmg}\right)^{\frac{1}{2}}, \left(\dfrac{3kmg}{Ma}b^2\right)^{\frac{1}{2}}$ **21** $2(2gl)^{\frac{1}{2}}/3$.

22 (i) $\dfrac{1}{3}(8gl)^{\frac{1}{2}}$ (ii) $\dfrac{4}{15}(8gl)^{\frac{1}{2}}$ (iii) $\left(\dfrac{l}{8g}\right)^{\frac{1}{2}}\cos^{-1}\dfrac{3}{5}$.

23 $(a\lambda/m)^{\frac{1}{2}}$. **24** (i) a (ii) $\frac{1}{2}\pi\left(\dfrac{a}{2g}\right)^{\frac{1}{2}}$ (iii) $(2ag)^{\frac{1}{2}}/9$ (iv) $2\sqrt{2a/3}$.

Exercise 15.3 (p. 299)

1 5 N. **2** 0·57 rad s^{-1}. **3** 6·7 rad s^{-1}. **4** 0·42 N m. **5** $I\omega/C$, $I\omega^2/(2C)$.
6 1·12 ms^{-1}, 6·25 rad s^{-2}. **7** 4·08 rad s^{-1}. **8** 2·55 kg m^2. **9** $8\frac{1}{3}$ rad s^{-2}.
10 $125Ma^2\pi/9$, $185Ma^2\pi/9$.

Exercise 15.4 (p. 307)

4 $2ma^2/3$ **5** $5mr^2/4$ **8** $4\sqrt{2}a/(3\pi)$ from centre, $4[g(1+\sqrt{2})/(3\pi a)]^{\frac{1}{2}}$.
9 $\frac{1}{3}m(p^2-pq+q^2)$. **10** $M(a^2+b^2)/3$, $4M(a^2+b^2)/3$. **13** $\pi/3$.
14 $4ma^2\omega^2$, $8ma^2\omega$, $a\omega/g$, $a\omega^2/(2g)$. **15** $2Mc^2/7$. **16** $3Mr^2/10$, $3M(r^2+4h^2)/20$.

Exercise 15.7 (p. 314)

1 $3g/(8a)$. **3** $2\pi\sqrt{[33a/(20g)]}$. **4** $3g/5$ **5** $\pi\sqrt{[(a^2+4b^2)/(bg)]}$.
6 $2\pi\sqrt{(3h/4g)}$. **7** (ii) $\sqrt{(10ag/7)}$. **8** (c) $2\sqrt{(ag)}$.
9 (i) $8ma^2$, $3ma^2/2$, $2\pi\sqrt{[57a/(16g)]}$. **10** $2\pi\sqrt{[149a/(21g)]}$.
11 $T_1 = 2\pi\sqrt{[2a/(3g)]}$, $T_2 = 2\pi\sqrt{\frac{2}{3}}[\sqrt{(a^2+b^2)}/g]$.

Miscellaneous Exercise 15 (p. 316)

1 $\sqrt{(2an\pi/g)}$. **3** $6g\sin\theta/(7a)$. **4** $9g/(19a)$. **5** $3a$. **6** $2\pi\sqrt{[(a^2+4b^2)/(3gb)]}$.
8 $M = 2m$, $G = mag/2$, $0{\cdot}7g$. **9** $3mg/2$, $5mg/4$, $g/4$. **11** $2\pi\sqrt{[5a/(2g)]}$.
13 P.E. change $= -0{\cdot}75(1-\cos\theta)$. $480g-80g$. $1{\cdot}25\theta$, K.E. $= 437{\cdot}5\,\omega^2$, 6·02 ms^{-1}, 1·4 rad s^{-2}.
14 $G/(2a)$, $(G-2P)/(2I)$, $2n^2\pi I/(G-2P)$.

Exercise 16.1 (p. 324)

1 $4-2e^{-2t}$. **2** $2^{\frac{1}{2}}(t+2)^{\frac{1}{2}}$, $[2^{3/2}(t+2)^{3/2}+1]/3$. **3** $-0{\cdot}25/(x+1)$. **5** 50. **6** 6·7.
7 50 km.

Exercise 16.2 (p. 328)

1 0·58 s. **2** 898 m. **3** 0·805 s. **4** 250 m. **5** 583 m.
6 (F/mk), $(\ln 4 - 0{\cdot}75)U/k$, $\ln 4/k$. **7** 10·4 m. **8** $4(e^{5t}-1)/(e^{5t}+1)$ ms^{-1}.
9 $\ln(1{\cdot}25)/3k$. **11** 5 s.

Miscellaneous Exercise 16 (p. 330)

1 $10-x$. **2** $tu/(1+\frac{1}{2}kt\sqrt{u})$. **3** 2·2, $u(85/7)^{\frac{1}{2}}$.
4 $U[(\omega t-1)e^{\omega t}+1]/\omega$, U, $1/\omega$. **5** $15u^4/32$. **6** 4, 5, $a-2\lambda$.
7 (i) 20 m (ii) $\pi/4$. **8** (i) 20 ms^{-1} (ii) 0·2 ms^{-2} (iii) 2·8 km (iv) 0·2 ms^{-2}
9 $u-2ab+2ab\exp(-at/2)$, $(u-2ab)t+4b-4b\exp(-at/2)$, $u\exp-at$, $a^{-1}\ln 2$.
10 106·7 m, 104·7 m. **11** $\dfrac{1}{v^2}-\dfrac{1}{u^2}=2kt$, 1·5 s. **14** $8u^3/3k$.
15 $[a^{3/2}+3t(\frac{1}{2}k)^{\frac{1}{2}}]^{2/3}-a$. **16** $\dfrac{m}{2k}\ln(1+u^2)$.
17 $m\dfrac{v\,dv}{dx} = Ke^{-\alpha x}-\lambda m$, $v = 2K(1-e^{-\alpha x})/(\alpha m)-2\lambda x$. **18** $10\sqrt{6}$ ms^{-1}. **19** $kmgt$.
23 $\dfrac{1}{ak}\tan^{-1}\dfrac{U}{a}, \dfrac{1}{2k}\ln\left(1+\dfrac{U^2}{a^2}\right)$. **25** $\dfrac{1}{k}\displaystyle\int_{a/4}^{a}\left(\dfrac{ax}{a-x}\right)^{\frac{1}{2}}dx$. **26** 56 m.

Index